Quantitative *In Silico* Chromatography
Computational Modelling of Molecular Interactions

RSC Chromatography Monographs

Series Editor:
R.M. Smith, *Loughborough University of Technology, UK*

How to obtain future titles on publication:
A standing order plan is available for this series. A standing order will bring delivery of each new volume immediately on publication.

For further information please contact:
Book Sales Department, Royal Society of Chemistry, Thomas Graham House, Science Park, Milton Road, Cambridge, CB4 0WF, UK
Telephone: +44 (0)1223 420066, Fax: +44 (0)1223 420247
Email: booksales@rsc.org
Visit our website at www.rsc.org/books

Quantitative In Silico Chromatography
Computational Modelling of Molecular Interactions

Toshihiko Hanai
Health Research Foundation, Kyoto, Japan
Email: thanai@attglobal.net

THE QUEEN'S AWARDS
FOR ENTERPRISE:
INTERNATIONAL TRADE
2013

RSC Chromatography Monographs No. 19

ISBN: 978-1-84973-991-7
ISSN: 1757-7055

A catalogue record for this book is available from the British Library

Published by The Royal Society of Chemistry,
Thomas Graham House, Science Park, Milton Road,
Cambridge CB4 0WF, UK

Registered Charity Number 207890

For further information see our web site at www.rsc.org

Preface

Modern chromatography began when Martin and Synge developed "partition chromatography" in 1941. In 1958, Kovats proposed an index to predict gas chromatography retention times, leading to discussion of the mechanism of retention and studies to validate retention time prediction. With the separation of saccharides using a combination of ion-exchange resin and aqueous ethanol without ion-exchange by Samuelson *et al.* in 1965, a variety of combination systems were investigated for liquid chromatographic separation. In particular, the relationship between the structure and the chromatographic behavior of a variety of mono- and di-substituted benzenes on an anion-exchange resin was studied with anhydrous ethanol as the eluent. It was found that retention depends on the type and position of the substituents, with the degree of adsorption being, at least partially, related to electron-withdrawing ability. The results suggested that molecular adsorption on the solid phase occurred by the formation of either charge-transfer or hydrogen-bonding interactions between the analyte and the anionic groups on the resin. These findings were confirmed in studies using a variety of packing materials, ion exchangers, bonded-phase silica gels, and organic polymers for the chromatographic separation of organic acids and saccharides in organic solvent mixtures. These results also supported the importance of hydrogen bonding for analyte retention on the solid phase. Similar experiments were carried out to analyze the chromatographic behavior of phthalate esters in aqueous solvent systems, and showed that elution order was related to solubility (hydrophobicity), leading, in 1974, to a liquid chromatography classification scheme based on solubility factors, as proposed by Hanai. Furthermore, in 1979, Hanai *et al.* demonstrated an optimized method for reversed-phase liquid chromatography using the octanol–water partition coefficient. The quantitative analysis of retention mechanisms proved to be difficult, however, in the absence of fast personal

RSC Chromatography Monographs No. 19
Quantitative *In Silico* Chromatography: Computational Modelling of Molecular Interactions
By Toshihiko Hanai
© Toshihiko Hanai, 2014
Published by the Royal Society of Chemistry, www.rsc.org

computers and computational software designed for general use by non-specialists.

In recent years, analytical chemists have increasingly turned their attention to drug discovery and drug analysis and to solve fundamental biologically significant questions in physiology and genetics. New technologies have been developed, and a variety of instruments have been redesigned for biomedical applications. For example, the development of capillary column gas chromatography, in which separation power is based on a very high theoretical plate number, answered questions about sample purity, and the development of high-performance liquid chromatography opened a new era in bio-related fields by allowing faster separations of unstable macromolecules. The improved thin layer liquid chromatography technique, high-performance planar liquid chromatography, is capable of the simultaneous analysis of many samples, providing an overview of the components present in complex mixtures, and permitting two-dimensional separation under different conditions with the possibility of multiplexed detection.

Capillary column gas chromatography (GC)/mass spectrometry (MS) has also been used to achieve more difficult separations and to perform the structural analysis of molecules, and laboratory automation technologies, including robotics, have become a powerful trend in both analytical chemistry and small molecule synthesis. On the other hand, liquid chromatography (LC)/MS is more suitable for biomedical applications than GC/MS because of the heat sensitivity exhibited by almost all biomolecules. More recent advances in protein studies have resulted from combining various mass spectrometers with a variety of LC methods, and improvements in the sensitivity of nuclear magnetic resonance spectroscopy (NMR) now allow direct connection of this powerful methodology with LC. Finally, the online purification of biomolecules by LC has been achieved with the development of chip electrophoresis (microfluidics).

As a complementary approach to these technological advances, computational chemical analysis is a promising technique with the potential to analyze the mechanisms of molecular interaction between analytes and solid phases, especially given the feasibility of modeling the three-dimensional structures of biological macromolecules, such as proteins. Importantly, this technology can be easily used to study the retention mechanisms in chromatography for a variety of model phases. Furthermore, theoretical calculations can provide significant insight into organic reaction mechanisms, which can be applied to study highly sensitive detection in chromatography, such as bromate and chemiluminescence detection. As a consequence, combining chromatography and computational chemistry offers new possibilities in developing a quantitative description of molecule interactions relevant to analytical separations. Furthermore, a combination of quantitative molecular recognition analysis and electron transfer, can permit the study of enzyme reaction mechanisms.

In this book, I propose and describe one approach to combining these methods, and illustrate the power of this strategy in biological applications.

For example, this method reveals a high correlation coefficient between measured capacity ratios and the sum of theoretically calculated molecular interaction energy and molecular property values, opening the possibility for quantitative analysis of chromatographic retention mechanisms.

This book is a pebble thrown in a pond. I hope the 'ripples' created will stimulate new research into questions about the basic phenomena of chromatographic separation and, perhaps, improve our understanding of enzyme reaction mechanisms.

Toshihiko Hanai
Yokohama

Acknowledgements

I appreciate Professor Rudolf Kaiser and the late Professor Jack Cazes who encouraged me in my challenge to study basic chromatographic phenomena. I thank the former CAChe™ group members, Dr George Purvis, David Gallagher, Munetaka Sawada, Ichiro Sakuma and Shimpei Hatta who supported my handling of computational chemical software, and Professor Richard Niger who taught me protein conformation.

I thank Professor Roger Smith, of Loughborough University, who reviewed the original manuscript of this book and offered many helpful suggestions. I also deeply appreciate the support of the Royal Society of Chemistry for this publication.

I thank the professors who have supported my research, the late Professor Wataru Funasaka, the late Professor Hiroyuki Hatano, Professor Shoji Hara, Professor Barry Karger, the late Professor Harold Walton, Professor Robert Sievers, Professor Joseph Hubert, the late Professor Toshio Kinoshita, and Professor Hiroshi Homma. I thank all the students and staff who prepared the chromatography data. I also give thanks to my family members.

RSC Chromatography Monographs No. 19
Quantitative *In Silico* Chromatography: Computational Modelling of Molecular Interactions
By Toshihiko Hanai
© Toshihiko Hanai, 2014
Published by the Royal Society of Chemistry, www.rsc.org

Contents

RSC Chromatography Monographs No. 19
Quantitative *In Silico* Chromatography: Computational Modelling of Molecular Interactions
By Toshihiko Hanai
© Toshihiko Hanai, 2014
Published by the Royal Society of Chemistry, www.rsc.org

CHAPTER 1

Introduction

1.1 Fundamental Phenomena in Chromatography

The quantitative analysis of molecular interactions is of fundamental interest, and the development of computer software has made it easy to calculate the theoretical properties of molecules. Feasibility can be demonstrated using simple, small molecules. Alkanes have demonstrated van der Waals energy contribution, and alkanols demonstrated the additional hydrogen-bonding energy contribution. Ion–ion interactions were related to the electrostatic energy contribution, and amino acids demonstrated the contribution of steric hindrance.

Chromatography is one technique that is used to measure molecular interaction strengths using model compounds, and an excellent technique for measuring the relative physico-chemical values of molecules in a short amount of time. Molecular recognition, the retention time difference, in chromatography can be quantitatively studied. Typical molecular interaction forces are clearly observed in different types of chromatography as the retention time differences of analytes. The individual molecular interaction forces are solubility factors. Chromatographic retention is based on the combination of solubility factors. Consistent with the concept of "like dissolves like" proposed by Henry Freiser, the retention mechanisms of chromatography are the same. Different types of chromatography demonstrate the typical molecular interaction forces, as summarized in Figure 1.1.[1] If we can reconstruct quantitatively obtained solubility factors, we can quantitatively analyze the chromatographic retention time.

Computational chemical analysis methods provide the molecular interaction energy as the sum of mainly van der Waals, hydrogen bonding, and electrostatic energy values. The van der Waals energy is related to molecular size, hence, the contact surface area between an analyte and an adsorbent contributes to the molecular interaction energy. When hydrogen bonding

RSC Chromatography Monographs No. 19
Quantitative *In Silico* Chromatography: Computational Modelling of Molecular Interactions
By Toshihiko Hanai
© Toshihiko Hanai, 2014
Published by the Royal Society of Chemistry, www.rsc.org

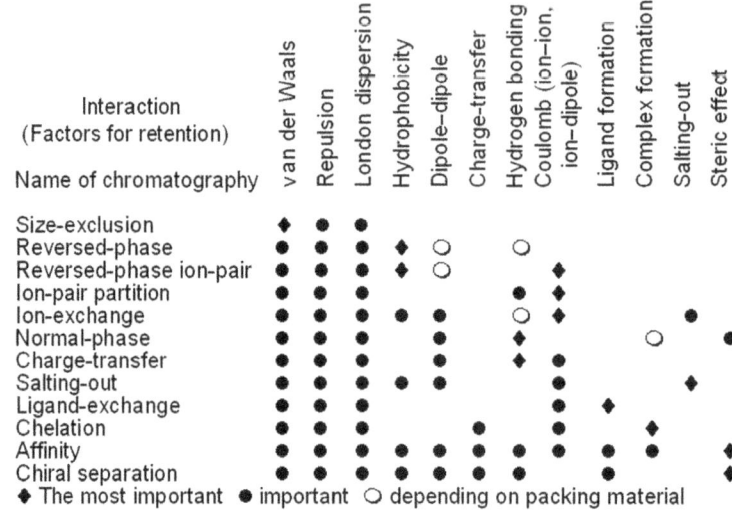

Figure 1.1 Classification in liquid chromatographic methods. Reproduced from ref. 1.

exists between an analyte and an adsorbent, hydrogen-bonding energy contributes to the molecular interaction energy. When ion–ion interactions exist, electrostatic energy contributes to the molecular interaction energy.

The measurement of direct interactions reveals the different strengths of molecular interactions between an analyte and the packing material surface or liquid phase. In gas chromatography, the retained compounds are vaporized and moved toward the column outlet. The analyte's volatility in the carrier gas affects the retention time.

In special cases, polar gases such as ammonia, formic acid and water are doped into the carrier gas to improve the analyte's solubility in the carrier gas. In both supercritical fluid and liquid chromatography, the analyte solubility in the carrier liquid affects the retention time. The carrier liquid is called the *eluent* and/or the *mobile phase*. The prediction of retention times in liquid chromatography is very difficult due to the lack of a solubility prediction method. However, the retention can be predicted by computational chemical methods using model phases.[2]

General computational chemical analysis of liquid chromatographic retention is performed without solvents in the calculation. Generally, mixed solvents with and without pH-controlled ions are present as the eluent components in liquid chromatography. At present, these solvent systems cannot be handled by computational chemical calculations. The measurement of direct interactions, however, reveals the different strengths of molecular interactions between an analyte and the packing material surface. The difference in molecular interaction energy values can be used as a relative retention time.

A typical example of a van der Waals energy contribution was observed in the analysis of chromatographic retention using a graphitized carbon column. The graphitized carbon column was used in gas chromatography, and in both normal- and reversed-phase liquid chromatography. The design of the model phase is important for the computational chemical analysis. A large polycyclic aromatic hydrocarbon was used to study the retention mechanisms of polar and non-polar compounds. The retention of a variety of compounds was quantitatively related to their van der Waals energy change before and after analyte adsorption on a model graphitized carbon phase. The ionic interaction between the model phase and an ion was explained using computational chemical methods. Carbohydrates are retained on a model graphitized carbon phase by hydrogen bonding at the edge of the model phase. Hydrocarbons are retained at the center of the model phase by van der Waals forces. The selective interaction was quantitatively analyzed with energy values calculated using a molecular mechanics (MM2) program.[3]

In gas chromatography, the retention time of compounds is determined by a combination of molecular interaction and vaporization energy values. The molecular interaction (adsorption) value was calculated as the molecular interaction energy between an analyte and a model stationary phase. Vaporization is related to the molecular properties of the analytes and was calculated as the optimized energy of a pair of analytes. This approach requires only a limited number of standard compounds for a column calibration. The standard chemicals for column calibration are alkanes, benzene, naphthalene, and anthracene for both methylsilicone and polyethyleneglycol phases. The properties were calculated *in silico* using a molecular mechanics program. This method can be used to predict retention times on both non-polar methylsilicone and polar polyethyleneglycol phases with high correlation coefficients between the logarithm of the capacity ratio (log k) values and the sum of the molecular interaction and vaporization energy values.[4] Inductive effects derived from Highest Occupied Molecular Orbital (HOMO) and Lowest Unoccupied Molecular Orbital (LUMO) energy values improved the precision.

The retention mechanisms in liquid and gas chromatography are the same, but the desorption mechanisms are different. Vaporization occurs in gas chromatography, and solvation is necessary in liquid chromatography. In normal-phase liquid chromatography, the adsorption strength can be quantitatively analyzed *in silico*, but the desorption strength cannot be quantitatively analyzed. Although a variety of organic solvents can be used as eluent components, the solvent strength does not directly relate to the elution order. The total solubility factor does not relate to solvent strength in normal-phase liquid chromatography, and no solubility prediction methods exist. The addition of acids and bases to the eluent modifies the silica gel surface and affects the ionization of the analyte. Therefore, quantitative analysis of retention in normal-phase liquid chromatography, other than for enantiomer separation, remains problematic. The approach can be used to predict enantiomer separation, but the bonded phases still have a mixed functional character, the silanol effect cannot be eliminated from present bonded-phase silica gels.[5–8]

In reversed-phase liquid chromatography, the octanol–water partition coefficient (log P) was demonstrated to be a quantitative molecular property of analytes, and later, log *P* values were used for the quantitative analysis of chromatographic optimization. Log *P* is a property of the molecular form, but not the ionized form of compounds. Therefore, the use of log *P* introduces errors into the predicted retention times of ionized compounds. If the direct interaction between an analyte and the surface of the packing materials or capillary tube surface is considered to be the predominant retention force, the retention mechanisms can be quantitatively analyzed, and taking the effect of the solvent into account should improve the precision of the analysis. The hydrophobic interactions between an alkyl phase and an analyte are related to the van der Waals energy value. The alkyl phase in reversed-phase liquid chromatography is hydrophobic, and therefore should reject adsorption of water molecules. An organic modifier may support the molecular interaction between an alkyl phase and an analyte, but works mainly to replace the analyte on the surface of the alkyl phase. If there are water molecules at the site of interaction, the contribution of hydrogen bonding decreases, but the contribution of water molecules is neglected in both the alkyl-bonded silica gels, and in the MM2 programs currently available. The design of the model phase is important for the quantitative analysis of retention mechanisms.

The correlation between log *P* and log *k* of standard compounds obtained for the molecular form of the analyte can be used to predict the maximum capacity ratios, and for the ionized form can be used to predict the minimum capacity ratios. Furthermore, these capacity ratios have been used to predict capacity ratios in eluents of different pH values. Drug analysis requires a three-dimensional model phase to take into account the contribution of the van der Waals energy, which is related to the contact surface area between the analyte and the model phase.

The dissociation constant, pK_a, controls the pH effect of the retention of ionizable compounds. Dissociation constants are measured by titration, and micro-volume flow titration has been developed. Dissociation constants can also be predicted from the atomic partial charge. An extended study of dissociation constants has been carried out, and a new method was proposed similar to a modified Hammett's equation. The dissociation constants used in this book were predicted by this new method from the atomic partial charge.[9]

This new approach has been examined for reversed-phase liquid chromatography of various compounds, *i.e.*, phenolic compounds, benzoic acid derivatives, acidic drugs, and basic drugs. The correlation coefficient between the measured and predicted capacity ratios was equivalent or better than that obtained using the log *P* system. In particular, the log *P* system could not handle, with good precision, analytes whose fragment's log *P* values were not established. This new method can be applied to estimate the capacity ratio of newly designed compounds, because this system does not require experimental data to predict the elution order.[10-13] This approach works well for groups of similar compounds. However, the solvation mechanisms for groups containing varied compounds differ; therefore, the effect of the organic modifier has to be considered in order to improve the precision. Alkanes were used in liquid chromatography as standard compounds, such as to calculate Kováts retention indices. The solvation was analyzed as a molecular interaction between a model phase and the analytes. The combined molecular interaction and solvation energy values improved the precision.[14]

These results indicate that the direct calculation of molecular interaction energies using the MM2 program was a new and practical approach to determining the quantitative structure–retention relationship in chromatography. The model phase used for the quantitative analysis of molecular interactions in reversed-phase liquid chromatography was modified for analysis of the ion-exchange mechanism. Ionized acidic and basic drugs demonstrated a strong contribution of electrostatic energy to the retention in ion-exchange liquid chromatography. A theoretical approach provides a new dimension in which to study molecular interactions quantitatively, and for designing new phases.[15–18] Stereoselectivity, or enantiomer recognition, in chromatography has also been quantitatively analyzed using MM2 calculations.[19–21] Furthermore, the combined model of reversed-phase and ion-exchange chromatography permitted the study and prediction of drug–albumin binding affinities.

1.2 Human Serum Albumin–Drug Binding Affinity Based on Liquid Chromatography

The measurement and prediction of human serum albumin (HSA)–drug binding affinity to determine bioavailability is necessary in the drug-discovery process. Several experimental methods have been applied using HSA, however, these methods are time consuming and show poor reproducibility, as seen by the varied binding affinity values (log nK) available from different references. The protein binding affinity of drugs was determined using a physically bonded protein-coated octadecyl-bonded silica gel (ODS) column and a chemically bonded bovine serum albumin column. The immobilized protein column method is simple but the columns are not stable, and the capacity ratios did not correlate well with HSA–drug binding affinities measured by free solution methods. The method requires a specific standard for the measurement of the pharmacokinetics of new chemicals, and the active sites are probably buried by the binding reaction. These are fundamental problems in protein–drug binding measurements.

Hummel–Dreyer and Frontal analyses have been used to measure protein–drug binding affinity by liquid chromatography. A new liquid chromatographic system was developed to measure protein–drug binding affinity indirectly, without albumin, and was evaluated using the log nK values of drugs measured by a modified Hummel–Dreyer method using purified HSA. The guanidino- and carboxyl-bonded silica gels were developed as mimic ion exchangers of HSA. The capacity ratios of acidic and basic drugs were measured by reversed-phase and ion-exchange liquid chromatography in a pH 7.40 eluent at 37 °C. The combined capacity ratios correlated well with the log nK

values measured by a modified Hummer–Dreyer method. Furthermore, a single-column method was developed. This liquid chromatographic method was reproducible and faster than the ordinary methods. The rapid analysis was further developed using a computational chemical method. The calculated molecular interaction energies using the new model phases correlated well with those measured using the Hummel–Dreyer method.[22,23]

1.3 Proteins as Affinity Phases

Affinity separation was first used in 1968, in biospecific adsorption studies to purify a number of enzymes.[24] It is important to note, however, that affinity separation does not guarantee 100% selectivity, since the use of two different affinity methods for the separation of the same protein can give different results. This is illustrated by the affinity separation of glycosylated human serum albumin (GHSA). The immunoassay method that uses an enzyme-linked immunosorbent (ELISA) gives relatively low values of protein, while a liquid chromatographic method with an affinity stationary phase resulted in higher values of protein. The percentage values of GHSA determined for healthy subjects and diabetic patients were 2.4% and 4.5% by ELISA and 16.1% and 39.9% by the liquid chromatographic method, respectively.[25] Affinity chromatography has also been used to separate GHSA and pure HSA from other proteins in blood serum.[26] However, such differences cannot be quantitatively analyzed by computational chemical methods, because the environmental conditions are not known. However, ligand density and spacer effects are very important factors in the specific interactions of macromolecules. Consequently, in certain cases, the molecular recognition of proteins can be quantitatively analyzed using computational chemical methods. This is possible because of the availability of stereostructures for a variety of proteins, as well as the capability of mutant construction. In Chapter 10, chromatographic phase monoamine oxidase selectivity, and the modification of related proteins are described.

The enzyme activity of monoamine oxidase (MAO) is inhibited by amphetamine and ephedrine. The immobilized enzyme as a chromatographic column is chirally selective. The enantiomer selectivity, affinity, can be analyzed by MM2 calculations. Furthermore, enzyme activity has been confirmed by the atomic partial charge change of the substrates, amino acids, by Molecular Orbital PACkage (MOPAC) calculations. MAO catalyzes the oxidative de-amination of amines, as D-amino acid oxidase selectively oxidizes D-amino acids. The enzyme exists in two forms. The MAO-A and MAO-B immobilized enzyme reactors demonstrated this selectivity. The enzyme reaction was inhibited by several compounds. The substrates of downloaded MAO-A and MAD-B were replaced by target compounds and the stereo-structure of complexes was optimized using molecular mechanics calculations. The atomic partial charge of atoms directly involved in the oxidation reaction was calculated using the MOPAC PM5 program. The difference in molecular interaction energy values demonstrated this selective inhibition.

The atomic partial charge indicated the poor oxidation reactions of MAO-A and MAO-B.[27]

Specifically, a mutant design process for D-amino acid oxidase was described based on downloaded stereostructures of this chirally selective enzyme. The evaluation method for the process was also presented.[28]

1.4 Mechanisms of Highly Sensitive Detection

Computational chemistry, which can predict the spectra of a variety of compounds that cannot be obtained in their pure form, was used to study the highly sensitive detection of bromate in ion chromatography. Several possible ions, molecules and their complexes were constructed by a molecular editor, and optimized by MM2 and MOPAC (PM3) calculations. Their possible electronic spectra were then obtained by the Zerner's Intermediate Neglect of Differential Overlap (ZINDO) (INDO)-Visualyzer in the CAChe™ program. The λ_{max} of the spectra and the transition dipoles were calculated using the ProjectLeader program. Comparison of the experimental and predicted results indicated that Br_3^- was the probable reaction product, and that NO_2^- and ClO^- accelerated the reaction.[29]

Computational chemistry permits the study of organic reaction mechanisms. The method was applied to study highly sensitive chemiluminescence detection. The efficiency of a chemiluminescence reaction can be expressed as the number of light-emitting molecules relative to the number of excited molecules. Peroxyoxalate luminescence is used to assay hydrogen peroxide or the number of fluorophores. Organic reducing compounds, including reducing sugars, ascorbic acid, uric acid, phenacyl alcohol derivatives, and steroids, are detected with the chemiluminescence method using lucigenin and luminol. The reaction process is considered the same for similar compounds, but the chemiluminescence sensitivity is thought to be structure-dependent. The intensity of chemiluminescence was quantitatively analyzed using computational chemical calculations based on a radical reaction mechanism in which a *keto–enol* rearrangement produced superoxide, and the superoxide reacted with luminol or lucigenin to produce the chemiluminescence. The atomic partial charge changed significantly and strongly correlated with the relative intensity of the chemiluminescence. This computational chemical analytical method can be used to determine the relative sensitivity of the chemiluminescence reaction. Furthermore, the chemiluminescence intensity was related to the toxicity of the analytes.[30]

Derivatization is performed to enable highly sensitive analyses. Selection of the derivatizing reagent depends on the desired sensitivity, selectivity, and reactivity. A variety of derivatizing reagents have been developed for trace amino acid analysis in biological samples. In particular, the analysis of trace amounts of (R)-amino acids in high-quantity (S)-amino acids is a subject of interest, because low levels of (R)-amino acids have been discovered in mammals. Fluorescent derivatives are preferred because of their higher sensitivity and selectivity and the elimination of the solvent background.

The MOPAC/ZINDO and MO-S (a molecular orbital package to calculate spectroscopic properties) programs can also be used to construct electronic (absorption) spectra. Although the absolute spectrum and intensity cannot be obtained, their relative values indicate the wavelength and intensity differences between compounds. At present, fluorescence spectra cannot be derived by computational chemical methods, but the absorption wavelength and intensity are related to the excitation wavelength and fluorescence intensity, respectively. Therefore, the calculated absorption spectra can be used to study sensitivity based on molar absorptivity values. However, a major limitation of this method is that solvent effects cannot be included for the calculation, and no program can produce a reliable spectrum or an absolute intensity. The relative sensitivities of the derivatized amino acids can be demonstrated by computational chemical techniques, and used to understand current reagent properties, and to design new reagents. Basic data used for the calculations are added as tables in the Appendix for personal study.

References

1. T. Hanai, *HPLC: A Practical Guide*, RSC Publishing, Cambridge, 1999.
2. T. Hanai, H. Hatano, N. Nimura and T. Kinoshita, Computer-aided analysis of molecular recognition in chromatography, *Analyst*, 1993, **118**, 1371–1374.
3. T. Hanai, Quantitative *in silico* analysis of the specificity of graphitized (graphitic) carbons, *Adv. Chromatogr.*, 2011, **49**, 251–284.
4. T. Hanai, Quantitative *in silico* analysis of retention time on methylsilicone and polyethyleneglycol phases in capillary gas chromatography. http://www.internet-chromatography.com/html/toshihikbeitrage.html.
5. T. Hanai, Quantitative in silico analysis of retention in normal-phase liquid chromatography, *J. Liq. Chromatogr., Relat. Technol.*, 2010, **33**, 297–304.
6. T. Hanai, Quantitative *in silico* analysis of retention in normal-phase liquid chromatography. http://www.internet-chromatography.com/html/toshihikbeitrage.html.
7. T. Hanai, Synthesis and properties of stable bonded silica gel packings and the performance, in *Advances in Liquid Chromatography*, ed. H. Hatano and T. Hanai, World Scientific, Singapore, 1996, pp. 307–327.
8. T. Hanai, New developments in liquid-chromatographic stationary phases, in *Advances in Chromatography*, ed. P. R. Brown and E. Grushka, Springer, New York, 2000, vol. 40, pp. 315–357.
9. T. Hanai, K. Koizumi, T. Kinoshita, R. Arora and F. Ahmed, Prediction of pK_a values of phenolic and nitrogen-containing compounds by computational chemical analysis compared to those measured by liquid chromatography, *J. Chromatogr., A*, 1997, **762**, 55–61.
10. T. Hanai, Chromatography *in silico*, basic concept in reversed-phase liquid chromatography, *Anal. Bioanal. Chem.*, 2005, **382**, 708–717.

11. T. Hanai, Chromatography *in silico*, quantitative analysis of retention of aromatic acid derivatives, *J. Chromatogr. Sci.*, 2006, **44**, 247–252.
12. T. Hanai, Chromatography *in silico* for basic drugs, *J. Liq. Chromatogr. Relat. Technol.*, 2005, **28**, 2163–2177.
13. T. Hanai, R. Miyazaki, A. Koseki and T. Kinoshita, Computational chemical analysis of the retention of acidic drugs on a pentyl-bonded silica gel in reversed-phase liquid chromatography, *J. Chromatogr. Sci.*, 2004, **42**, 354–360.
14. T. Hanai, Quantitative *in silico* analysis of organic modifier effect on retention in reversed-phase liquid chromatography, *J. Chromatogr. Sci.*, 2014, **52**, 75–80.
15. T. Hanai, Molecular modeling for quantitative analysis of molecular interaction, *Lett. Drug Des. Discovery*, 2005, **2**, 232–238.
16. T. Hanai, Quantitative *in silico* analysis of ion exchange from chromatography to protein, *J. Liq. Chromatogr. Relat. Technol.*, 2007, **30**, 1251–1275.
17. T. Hanai and H. Homma, Chromatography *in silico*: retention of acidic drugs on a guanidino ion-exchanger, *J. Liq. Chromatogr. Relat. Technol.*, 2007, **30**, 1723–1731.
18. T. Hanai, Y. Masuda and H. Homma, Chromatography *in silico*; retention of basic compounds on a carboxyl ion exchanger, *J. Liq. Chromatogr. Relat. Technol.*, 2005, **28**, 3087–3097.
19. T. Hanai, H. Hatano, N. Nimura and T. Kinoshita, Computational chemical analysis of chiral recognition in liquid chromatography, selectivity of *N*-(*R*)-1-(α-naphthyl)ethylaminocarbonyl-(*R* or *S*)-valine and *N*-(*S*)-1-(α-naphthyl)ethylaminocarbonyl-(*R* or *S*)-valine bonded aminopropyl silica gels, *Anal. Chim. Acta*, 1996, **332**, 213–224.
20. T. Hanai, Computational chemical analysis of enantiomer separations of derivatized amino acids in reversed-phase liquid chromatography, *Internet Electron. J. Mol. Des.*, 2004, **3**, 379–386.
21. F. Tazerouti, A. Y. Badjah-Hadj-Ahmed and T. Hanai, Analysis of the mechanism of retention on a modified β-cyclodextrin/silica chiral stationary phase using a computational chemical method, *J. Liq. Chromatogr. Relat. Technol.*, 2007, **30**, 3043–3057.
22. T. Hanai, log n*K* Chromatography and computational chemical analysis for drug discovery, *Curr. Med. Chem.*, 2005, **12**, 501–525.
23. T. Hanai, Evaluation of measuring methods of human serum albumin–drug binding affinity, *Curr. Pharm. Anal.*, 2007, **3**, 205–212.
24. P. Cuatrecasas, M. Willehek and C. B. Anfimeen, Selective enzyme purification by affinity chromatography, *Proc. Natl. Acad. Sci. U. S. A.*, 1968, **61**, 636–643.
25. K. Koizumi, C. Ikeda, M. Ito, J. Suzuki, T. Kinoshita, K. Yasukawa and T. Hanai, Influence of glycosylation on the drug binding of human serum albumin, *Biomed. Chromatogr.*, 1998, **12**, 203–210.
26. T. Hanai, M. Uchida, M. Minematsu, H. Homma, T. Kinoshita and G. Matsumoto, Fast, selective analysis of glycated albumin in HSA, *J. Liq. Chromatogr. Relat. Technol.*, 2002, **25**, 275–286.

27. T. Hanai, Quantitative *in silico* analysis of enzyme reactions: comparison of D-amino acid oxidase and monoamine oxidase, *Am. Biotechnol. Lab.*, 2007, **25**, 8–15.

28. T. Hanai, Quantitative *in silico* analysis of molecular recognition and reactivity of D-amino acid oxidase, *Internet Electron. J. Mol. Des.*, 2006, **5**, 247–259.

29. T. Hanai, Y. Inoue, T. Sakai and H. Kumagai, Computational chemical analysis of the highly sensitive detection of bromate in ion chromatography, *J. Chem. Inf. Comput. Sci.*, 1998, **38**, 885–888.

30. T. Hanai and T. Tachikawa, Quantitative analysis of chemiluminescence intensity and toxicity *in silico*, in *Bioluminescence & Chemiluminescence: Progress and Perspectives*, ed. A. Tsuji, World Scientific, Singapore, 2005, pp. 397–400.

CHAPTER 2

Basic Concepts of Molecular Interaction Energy Values

2.1 Introduction

Molecular interaction forces are a combination of solubility factors, and can be obtained as van der Waals, hydrogen-bonding and electrostatic energy values after molecular mechanics (MM2) calculations.[1] Simple studies can be carried out using small molecules. For example, the molecular interaction energy between alkanes is the van der Waals energy, and that between short-chain alcohols is mainly the hydrogen-bonding energy. Longer chain alcohols interact together with hydrogen-bonding and van der Waals energies. Polar opposite ion interactions can be observed as electrostatic energy value changes. (R)- and (S)-Amino acids form a compact complex, but (R)- and (R)- or (S)- and (S)-amino acids form larger size complexes.[2] The molecular interaction (MI) energy value can be calculated using following equations.

MIFS = FS(molecule A) + FS(molecule B) − FS(molecule A − molecule B complex),

MIHB = HB(molecule A) + HB(molecule B) − HB(molecule A − molecule B complex),

MIES = ES(molecule A) + ES(molecule B) − ES(molecule A − molecule B complex),

MIVW = VW(molecule A) + VW(molecule B) − VW(analyte A − molecule B complex),

where FS is the energy value of the final (optimized) structure, HB is the energy value of the hydrogen bonding, ES is the electrostatic energy value, and VW is the energy value of the van der Waals forces.

RSC Chromatography Monographs No. 19
Quantitative *In Silico* Chromatography: Computational Modelling of Molecular Interactions
By Toshihiko Hanai
© Toshihiko Hanai, 2014
Published by the Royal Society of Chemistry, www.rsc.org

When these equations are applied, one of molecules is the model phase and the target molecules are the analytes. The following equations can be used to study chromatographic retention in different types of chromatography.

$$\text{MIFS} = \text{FS(analyte)} + \text{FS(model phase)} - \text{FS(analyte} - \text{model phase complex)},$$

$$\text{MIHB} = \text{HB(analyte)} + \text{HB(model phase)} - \text{HB(analyte} - \text{model phase complex)},$$

$$\text{MIES} = \text{ES(analyte)} + \text{ES(model phase)} - \text{ES(analyte} - \text{model phase complex)},$$

$$\text{MIVW} = \text{VW(analyte)} + \text{VW(model phase)} - \text{VW(analyte} - \text{model phase complex)},$$

Simple examples follow below, and chromatographic retention analyses are demonstrated in later chapters.

2.2 Hydrophobic Interactions (van der Waals Forces)

Alkanes are completely saturated molecules having no specific physico-chemical properties except van der Waals volume. They interact together using van der Waals forces. Alkanes from methane to octadecane have been constructed and individual and identical pair energy values obtained after optimizing the structures using MM2 calculations. The calculated values are summarized in Table 2.1, and examples of complexes are shown Figure 2.1. The final (optimized) structure energy values are smaller than the sum of the individual energy values. The difference is considered to be the molecular interaction energy value. These pairs only demonstrated changes in van der Waals energy. That is, these molecules interact *via* hydrophobicity.

Table 2.1 Molecular properties of alkanes calculated using the MM2 program. VWV represents the van der Waals volume ($\mathring{A}^3$). fs, hb, es, and vw represent the energy value of the final (optimized) structure, the hydrogen-bonding energy, the electrostatic energy, and the van der Waals energy of individual analytes (kcal mol^{-1}), respectively. FS, HB, ES and VW represent the same energy values for analyte pairs, respectively.

Alkane	VWV	fs	hb	es	vw	FS	HB	ES	VW
Methane	28.69	0.00	0	0	0.00	−0.72	0	0	−0.71
Ethane	95.51	0.82	0	0	0.68	0.39	0	0	0.12
Propane	62.33	1.50	0	0	1.19	0.64	0	0	−0.03
Butane	79.15	2.18	0	0	1.68	1.01	0	0	−0.03
Pentane	95.97	2.83	0	0	2.15	1.35	0	0	−0.06
Hexane	112.79	3.47	0	0	2.61	1.63	0	0	−0.14
Heptane	129.55	4.12	0	0	3.07	1.93	0	0	−0.20
Octane	146.41	4.76	0	0	3.53	2.20	0	0	−0.28
Nonane	160.23	5.40	0	0	4.00	2.50	0	0	−0.32
Decane	180.01	6.04	0	0	4.46	2.79	0	0	−0.41

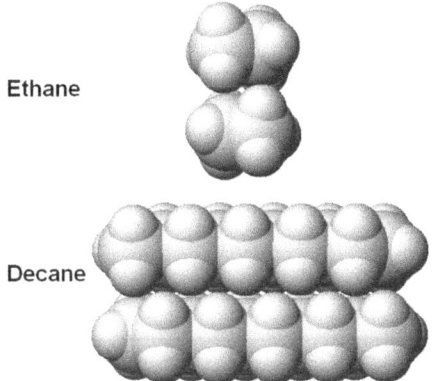

Ethane

Decane

Figure 2.1 Complexes of pairs of ethane and decane. Small white and light-gray balls represent hydrogen and carbon, respectively.

2.3 Hydrogen Bonding

The hydrogen bonding of alkyl alcohols depends on the alkyl chain length, and up to four methylene units can affect the hydrogen bonding. Alkyl alcohols from methanol to nonylalcohol have been constructed, and individual and identical pair energy values obtained after optimizing the structures using MM2 calculations. The calculated values are summarized in Table 2.2, and examples of complexes are shown in Figure 2.2. The molecular interaction energy values are plotted against the carbon number of the alkyl alcohol chain, along with the results for alkanes, in Figure 2.3.

The final energy of a pair of identical alcohols was smaller than twice the energy of one alcohol. This indicates that the energy values calculated using the MM2 program can explain the degree of molecular interaction. Methanol, ethanol, and propanol had similar values; their hydrogen-bonding energy value was about -2.9 kcal mol^{-1}, and their electrostatic energy value was about -1.3 kcal mol^{-1}. *n*-Butanol, *n*-pentanol and *n*-hexanol also showed similar energy values; their hydrogen-bonding energy values being very small (of the order of -0.1 kcal mol^{-1}), and their electrostatic energy values were about 0.7 kcal mol^{-1}. Their van der Waals energy values indicated their complexed form. Methanol, and hexanol formed pairs, as shown in Figure 2.2. The total energy values of the alkyl alcohols were corrected to eliminate the molecular size effect by using data from *n*-alkanes. As seen in Figure 2.3, their corrected final structure energy values increased dramatically with increasing alkyl chain length, then the value leveled off to a constant. Up to three carbons, there was a linear increase in the hydrogen bonding. This means hydrogen bonding contributes to molecular interactions for alcohols up to and including three carbon atoms, and van der Waals energy is the main energy for alkyl alcohols with longer chains. However, the corrected energies did not bear a linear relationship to their calculated hydrogen-bonding energies as given in Table 2.2. This result can

Table 2.2   Molecular properties of alkyl alcohols calculated using the MM2 program.

Alkane	VWV	fs	hb	es	vw	FS	HB	ES	VW
Methanol	36.85	0.06	0	0	−0.08	−3.69	−2.90	−1.21	0.13
Ethanol	53.73	0.80	0	0	0.58	−2.24	−1.61	−1.27	0.19
Propanol	70.57	1.47	0	0	1.09	0.25	−0.14	−0.07	−0.36
Butanol	87.41	2.13	0	0	1.57	0.60	−0.10	−0.05	−0.43
Pentanol	104.06	2.78	0	0	2.04	0.92	−0.08	−0.07	−0.48
Hexanol	120.92	3.42	0	0	2.50	1.21	−0.09	−0.06	−0.56
Heptanol	137.77	4.07	0	0	2.96	1.50	−0.08	−0.06	−0.62
Octanol	154.65	4.71	0	0	3.43	1.78	−0.08	−0.06	−0.70

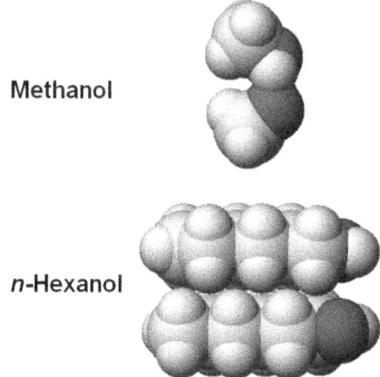

Methanol

n-Hexanol

Figure 2.2   Complexes of pairs of methanol and *n*-hexanol. Small white, light-gray, and black balls represent hydrogen, carbon, and oxygen, respectively. Reproduced by permission of Taylor and Francis, ref. 2.

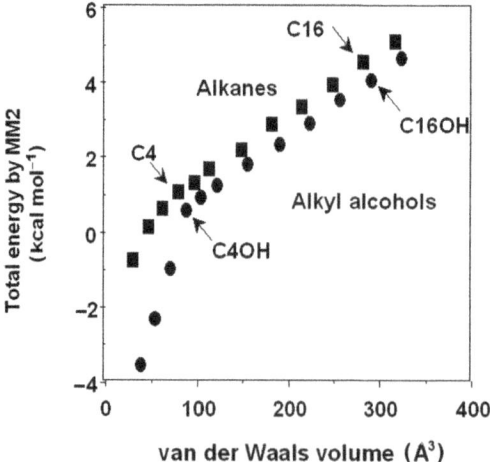

Figure 2.3   Final structure energy of *n*-alkanes and *n*-alkyl alcohol pairs correlated to their van der Waals volume.
Reproduced by permission of Taylor and Francis, ref. 2.

also be explained by the fact that the energy change for a chain of five or more carbon atoms was almost parallel to the energy change of alkanes.

2.4 Coulombic Forces (Ion–Ion Interactions)

The possibility of computational chemical analysis of molecular interactions can be understood from the analysis of simple model compounds. Ion–ion interactions were studied for the combination of an ammonium ion (cation) with an acetate ion (anion) by MM2 calculations. The calculated ion-pair formation energy values are as given in Table 2.3.

The difference (the MI energy value) after complex (ion-pair) formation indicates the energy contribution of the ion-pair formation. The electrostatic energy is the main contributor to the ion-pair formation between ammonium and acetate ions. The structure of the ammonium and acetate ions, and the ion-pair, are shown in Figure 2.4, where the atomic distance of the ammonium hydrogens has been expanded to 1.3 Å, and their atomic partial charge calculated using the MOPAC PM5 program is also indicated, where the atomic partial charge of the ammonium hydrogens has been doubled. The retention mechanisms of these cations and anions were then studied using a model graphitized carbon-phase (PAH14). The optimized complex

Table 2.3 Calculated ion-pair formation energies for the ammonium and acetic acid ions (kcal mol^{-1}).

	NH_4^+ +	CH_3COO^- =	CH_3COONH_4	Difference
Stretch	0.000	0.010	0.012	0.002
Stretch bend	0.000	0.003	0.003	0.000
Improper torsion	0.000	0.000	0.001	0.001
Electrostatic	0.000	0.000	−5.671	−5.671
Angle	1.526	0.038	1.729	0.165
Dihedral angle	0.000	−0.802	−0.803	−0.001
van der Waals	0.000	0.346	0.224	−0.122
Hydrogen bonding	0.000	0.000	0.000	0.000
Final structure (optimized)	1.626	−0.404	−5.505	−4.283

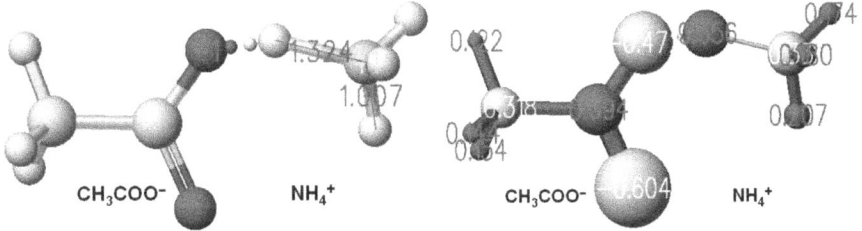

Figure 2.4 Structure of the ion-pair and the atomic partial charges. White, light-gray, black, and dark-gray balls represent hydrogen, carbon, oxygen, and nitrogen, respectively.
Reproduced by permission of Taylor and Francis, ref. 3.

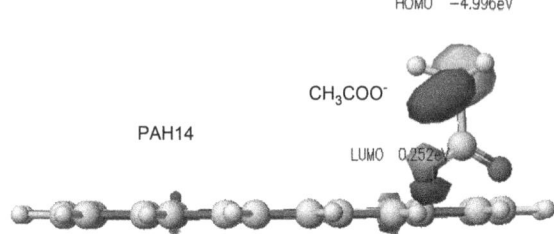

Figure 2.5 Electrostatic interaction between the acetate ion and a model graphitized carbon.
Reproduced by permission of Taylor and Francis, ref. 3.

form is shown in Figure 2.5, where the anions were retained by ion–ion
interactions at the model graphitized carbon edge, but the cations did not
show an ion–ion interaction. The HOMO and LUMO also indicate the
interaction.[3]

2.5   Steric Hindrance (Enantiomer Recognition)

Steric hindrance cannot be directly calculated, but a lower MI energy value
indicates lower steric hindrance in a complex. This phenomenon can be
observed in enantiomer recognition and protein–ligand interactions, with
the latter also being known as *affinity*. Such phenomena can be studied *via*
amino acid enantiomer complexes. The optimized energy values of (*R*)- and
(*S*)-alanine are the same, as summarized in Table 2.4.

   (*R*)- and (*S*)-Alanine complexes showed lower van der Waals energy values
than those of (*R*)- and (*R*)- or (*S*)- and (*S*)-alanine complexes. (*R*)- and (*R*)-
Alanine complexes form four hydrogen bonds, as do (*S*)- and (*S*)-alanine
complexes. (*R*)- and (*S*)-Alanine complexes form two hydrogen bonds as
shown in Figure 2.6. The hydrogen-bonding energy values of (*R*)- and (*R*)- (*S*-
and *S*-) complexes are lower than those of (*R*)- and (*S*)-complexes. The van der
Waals energy values of (*R*)- and (*S*)-alanine complexes are lower than those of
(*R*)- and (*R*)- (*S*- and *S*-) complexes. When steric hindrance is low, two molecules interact with less steric hindrance. That is, the change in van der
Waals energy value indicates the contribution of complex formation. The
lowest final structure and hydrogen-bonding energy values were obtained for
the complex formed between two hydroxyl groups (the first (*S*)- and (*R*)-ala-
nine complex in Table 2.4). The structure is like a long dimer, and typical
steric hindrance is observed. But this conformation is not suitable for discussing steric hindrance. The results indicate that if a correct model of chiral
recognition was available, the optimized form of the molecule could be
demonstrated from the different energies of the enantiomers. This type of
study could be used to demonstrate the affinity in the molecular recognition
of proteins.

Table 2.4 Molecular interaction energy values of alanine complexes calculated using the MM2 program.

Alanine	fs	hb	es	vw
(S)-Alanine	4.2502	−3.612	7.756	1.126
(R)-Alanine	4.2763	−3.591	7.781	1.113
2×(S)-Alanine complex	−0.1157	−13.768	14.938	0.691
2×(R)-Alanine complex	−0.1168	−13.774	14.939	0.693
(S)- and (R)-Alanine complex[a]	0.1152	−14.639	14.928	0.798
(S)- and (R)-Alanine complex[ab]	−2.8919	−18.207	14.191	2.860
(S)- and (R)-Alanine complex[c]	4.0125	−8.850	14.619	−0.084

[a]Mirror complex.
[b]Head − head complex.
[c]Complex between carboxyl and amino groups.

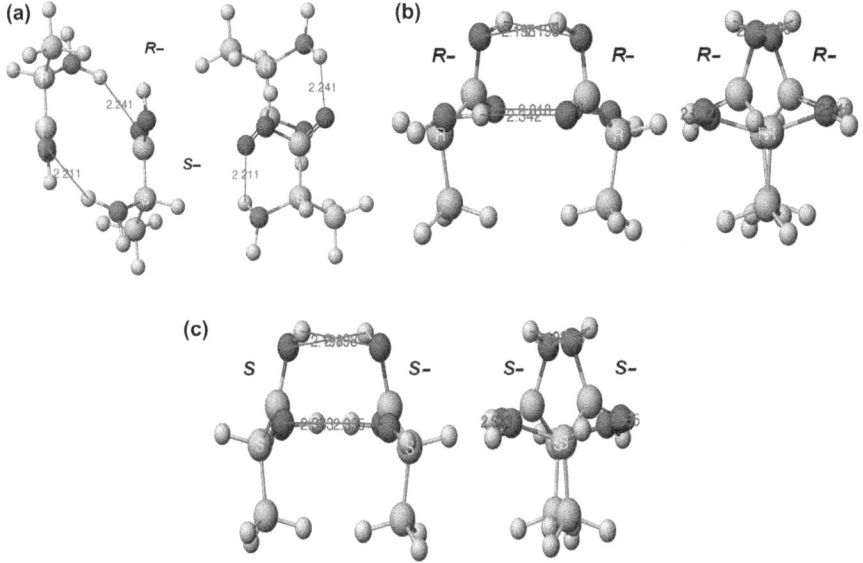

Figure 2.6 (a) Conformation of an (R)- and (S)-alanine complex. (b) Conformation of an (R)- and (R)-alanine complex. (c) Conformation of an (S)- and (S)-alanine complex. White, light-gray, black, and dark-gray balls represent hydrogen, carbon, oxygen, and nitrogen, respectively.

References

1. T. Hanai, H. Hatano, N. Nimura and T. Kinoshita, Computer-aided analysis of molecular recognition in chromatography, *Analyst*, 1993, **118**, 1371–1374.
2. T. Hanai, H. Hatano, N. Nimura and T. Kinoshita, Molecular recognition in chromatography aided by computational chemistry, *Supramol. Chem.*, 1994, **3**, 243–247.
3. T. Hanai, Quantitative *in silico* analysis of the specificity of graphitized (graphitic) carbons, *Adv. Chromatogr.*, 2011, **49**, 257–290.

The Design of Model Phases for Chromatography

3.1 Graphitized (Graphite) Carbon Phases

Graphite is a polycyclic aromatic hydrocarbon (PAH) comprising more than 10^5 carbons. The number of atoms in graphitized carbon is estimated from electric conductivity studies. The surface acts as a Lewis base towards polar solutes, and has Lewis acid–base and dispersive interactions with aromatic solutes. Based on theoretical calculations, Hosoya[1] reported that 100 carbon atoms are adequate to demonstrate graphitic properties. Based on theoretical calculations, the electrons localize at the edge of graphitized materials. The calculation capacity of a computer, however, is limited. Several PAHs were constructed, the structures were optimized using the CAChe™ molecular mechanics program (MM2), and the calculated energy values and the center and outer average atomic partial charges were calculated using the MOPAC PM5 program, as summarized in Table 3.1. The electron density, atomic partial charge, and the HOMO–LUMO density gradient of coronene (PAH7) are shown in Figure 3.1(a) and (b).

The final structure energy of PAH7 decreases about 18 kcal, and the van der Waals energy decreases by about 18 kcal after pair formation (PAH7×2). The hydrogen-bonding and electrostatic energy values are zero. The average atomic partial charge of the center carbons is -0.013 au and that of the outer carbons is -0.104 au. These values change after pair formation in graphite; the atomic partial charges of the center and outer carbons are -0.007 and -0.103, respectively.

RSC Chromatography Monographs No. 19
Quantitative *In Silico* Chromatography: Computational Modelling of Molecular Interactions
By Toshihiko Hanai
Published by the Royal Society of Chemistry, www.rsc.org

Table 3.1 Molecular properties of model PAHs. Cn represents the carbon number, fs, hb, es, and vw represent the energy value for the final (optimized) structure, the hydrogen-bonding energy, the electrostatic energy, and the van der Waals energy (kcal mol⁻¹), respectively. apcc, and apco represent the average atomic partial charge of the center and outer carbons (au).

Model	Cn	fs	hb	es	vw	apcc	apco
PAH7	24	−51.6412	0	0	16.176	−0.013	−0.104
PAH7×2	48	−121.0014	0	0	14.663	−0.007	−0.103
PAH14	42	−92.6645	0	0	29.590	−0.006	−0.096
PAH19	54	−119.1088	0	0	38.658	−0.005	−0.092
PAH22	62	−77.9879	0	0	44.325	−0.000	0.009
PAH37	96	−134.1108	0	0	68.046	0.002	0.001
PAH61	150	−225.0938	0	0	109.175	0.002	−0.012
PAH64	160	−86.465	0	0	164.359	−0.001	−0.064
PAH196	435	−953.3515	0	0	340.003	—	—

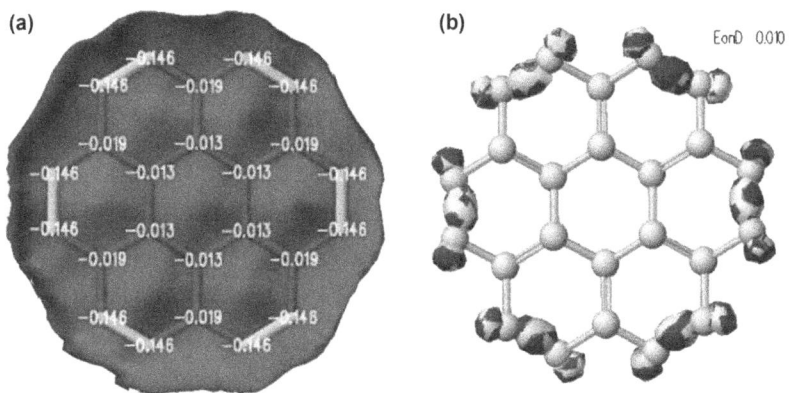

(a)

(b)

Atomic partial charge Electronic isopotential

Figure 3.1 (a) The electrostatic potential, and (b) the atomic partial charge of coronene.

In general, graphitized carbon comprises multiple layers of large PAHs. An increase in the molecular size decreases the final structure energy values, and the chemical stability increases with an increase in the size of the structure. The final heat of formation of a pair of PAHs is increased compared to that of a single layer. How does it affect the size of a single layer? The average atomic partial charge values of the center and outer atoms of PAH19 are −0.005 and −0.092; those of PAH37 are 0.002 and 0.001 au, respectively. Increasing the number of rings decreases the atomic partial charge value of the center atoms. The value is affected by the symmetry of the molecule. The value is −0.000 au for PAH22 and 0.002 au for PAH37, and depends on the length and width ratio of the PAH. The electrostatic

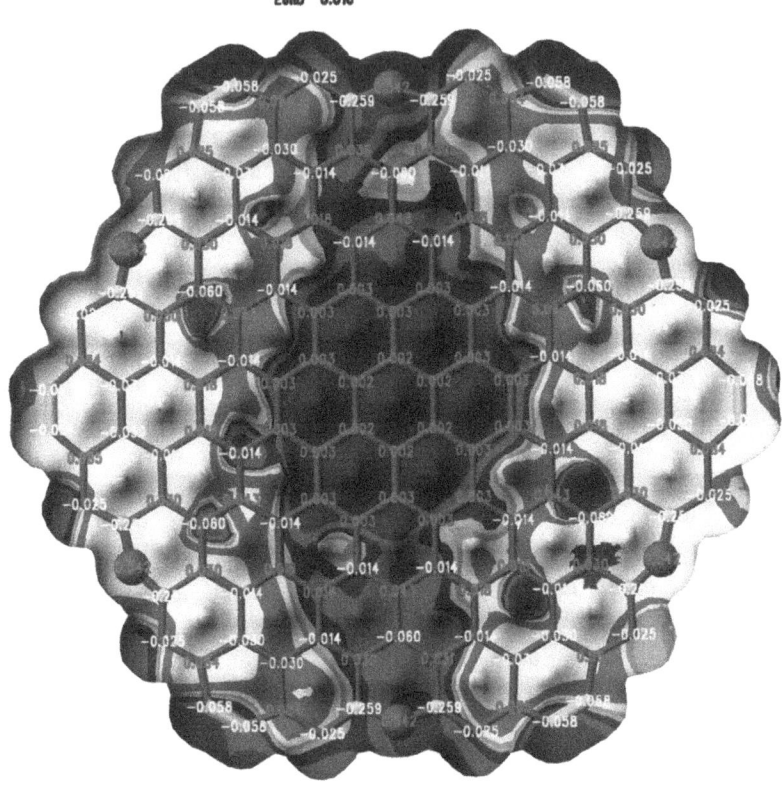

EonD 0.010

Figure 3.2   The electrostatic potential and atomic partial charge of PAH61.

potentials of PAH61 are shown in Figure 3.2, where the center and outer atomic partial charge indicates the electron concentration difference.

The structure allows for two types of molecular interactions; a hydrophobic interaction at the center of the large molecules and an electrostatic interaction at the edge of the graphitized carbon. Electron density is low at the center, and high at the edge of the molecule. The electron charge of the center atoms of larger molecules is lower than that of smaller molecules. These observations suggest that the electron charge of the center of graphitized carbon is close to zero and neutral. Similar phenomena are also observed in a saturated carbon-layer of 22 rings, which has a structure similar to that of PAH22 with a more homogeneous net atomic charge distribution. The atomic partial charge of the center carbon is − 0.104 and that of the outer carbon is − 0.148 au. Therefore, a PAH may be used as a model for a graphitized carbon phase. PAH196 was constructed for studying the molecular interaction of larger analytes such as fatty acids. The structure is shown in Figure 3.3, however the electrostatic potential is not given, due to computer capability.[2]

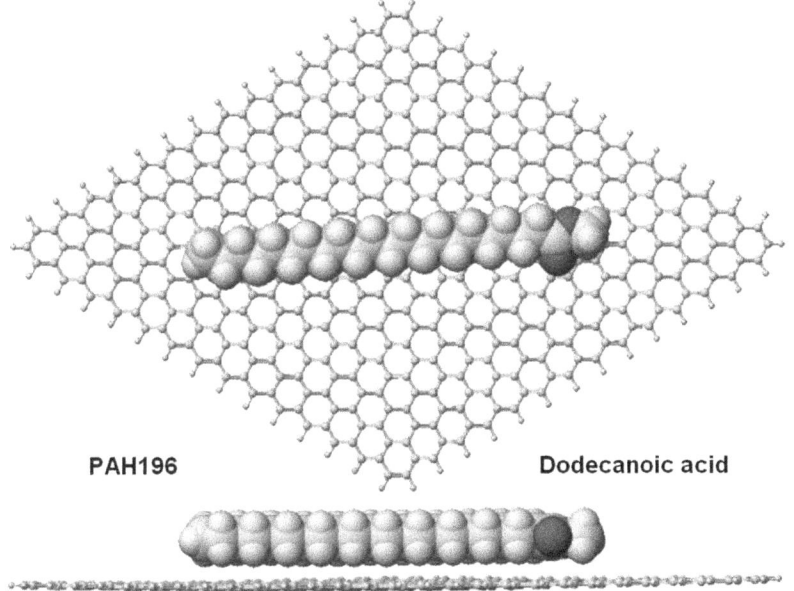

Figure 3.3   Adsorption of dodecanoic acid methylester on a PAH196 graphitized carbon model.

3.2   Methyl and Hydroxyl Groups of Organic Phases

A graphite-like layer is constructed, and an even number of carbon atoms in two model adsorbents are bonded tightly to diminish the flexibility of the model adsorbent, then the surface is saturated with hydrogen. The hydrocarbon layer, whose structure is shown in Figure 3.4, is constructed on the computer with 628 atoms (368 carbons and 260 hydrogens), 866 bonds and 1732 connectors, and the relative molecular mass is 4676. The white and gray balls represent hydrogen and carbon atoms, respectively. Part of a double-layer molecule is shown in Figure 3.5 for easy understanding of the bonded sites. It is designed as a simple model of a hydrophobic adsorbent such as an alkyl-bonded vinyl alcohol copolymer of a silica gel, the surface of which has to be completely covered with alkyl groups, and whose unreacted hydroxyl groups and matrix can not affect the solute retention. If the bonded alkyl chain length is short, unreacted hydroxyl groups and the matrix usually affect the retention in chromatography.

The model molecule demonstrates the selectivity of a methylsilicone phase in gas chromatography that theoretically works as a hydrophobic phase. One nonanol molecule was put at the center of this adsorbent, then the adsorption position was optimized by MM2 calculations. The result is shown in Figure 3.6, where a slightly bent nonanol molecule is laid on the adsorbent, and the black ball is an oxygen atom. The adsorbent and nonanol appear to be isolated in Figure 3.6, because the van der Waals radius is 20% of the actual size.

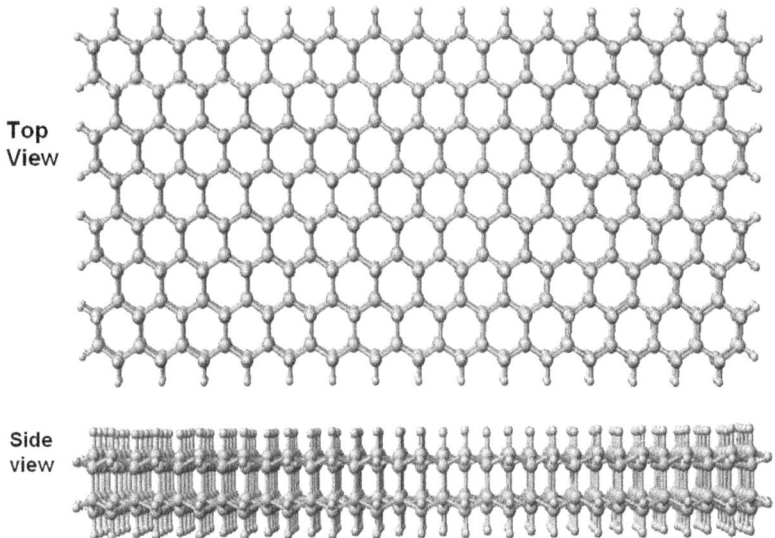

Figure 3.4   A model hydrocarbon layer. White and light-gray balls represent hydro-
                 gen and carbon, respectively.
                 Reproduced with permission from ref. 11.

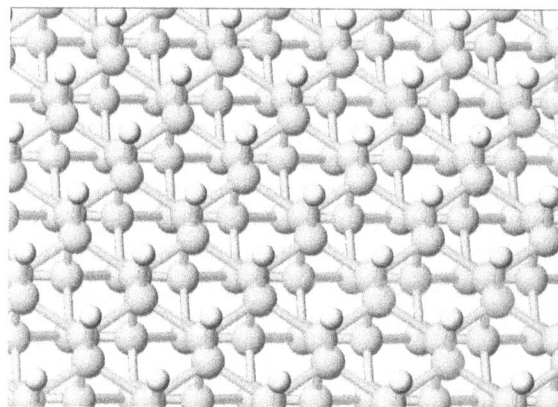

Figure 3.5   The structure of a model hydrocarbon layer. White and light-gray balls
                 represent hydrogen and carbon, respectively.
                 Reproduced with permission from ref. 11.

When the actual van der Waals radius is used, two molecules are completely adsorbed. One surface of the model phase was further modified to
study selectivity. In one case the surface was methylated and in another it
was hydroxylated, as shown in Figures 3.7 and 3.8, where decane and anthracene are adsorbed, respectively. The relative molecular masses of the
methylated and hydroxylated adsorbents are 5096 and 5156, respectively.
The methylated adsorbent is bent by the saturated methyl groups.

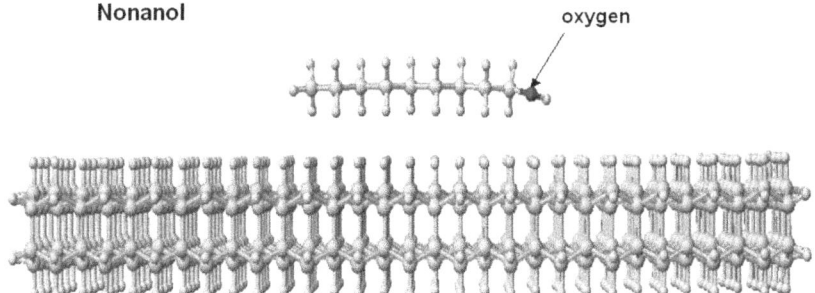

Figure 3.6 An adsorbed nonanol molecule on a model hydrocarbon phase. White, light-gray, and black balls represent hydrogen, carbon, and oxygen, respectively.
Reproduced with permission from ref. 11.

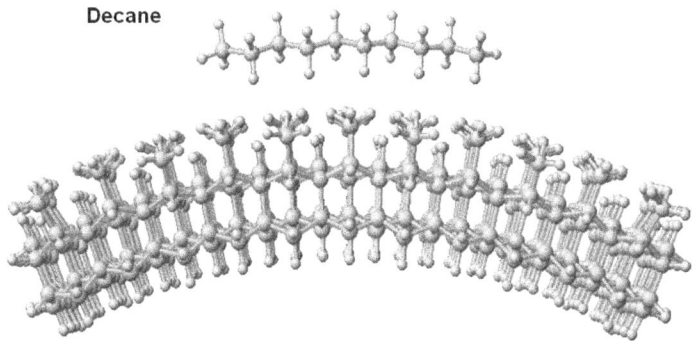

Figure 3.7 An adsorbed decane molecule on a model methylated hydrocarbon phase. White and light-gray balls represent hydrogen and carbon, respectively.
Reproduced with permission from ref. 11.

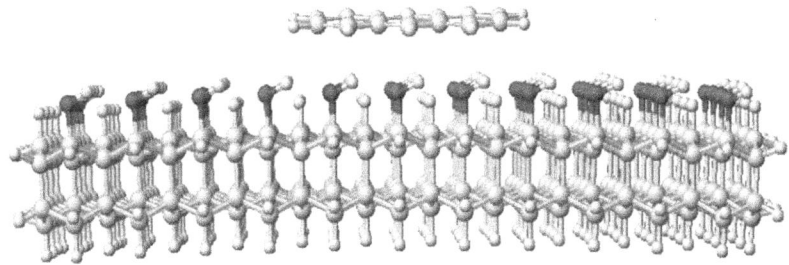

Figure 3.8 An adsorbed anthracene molecule on a model hydroxylated hydro-carbon phase. White, light-gray, and black balls represent hydrogen, carbon, and oxygen, respectively.
Reproduced with permission from ref. 11.

Table 3.2 Thc molecular properties of various alkanes, alkyl alcohols and cyclohydrocarbons. VWV represents the van der Waals volume ($\mathring{A}^3$).

Analyte	VWV	fs	hb	es	vw
Benzene	83.79	−8.077	0.000	0.000	3.006
Naphthalene	127.60	−18.688	0.000	0.000	5.766
Anthracene	171.49	−29.360	0.000	0.000	8.491
n-Hexane	112.79	3.472	0.000	0.000	2.614
n-Decane	180.01	6.039	0.000	0.000	4.461
n-Tetradecane	247.26	8.608	0.000	0.000	6.313
Pentanol	104.06	2.779	0.000	0.000	2.036
Nonanol	171.44	5.355	0.000	0.000	3.886
Tridecanol	238.62	7.917	0.000	0.000	5.744
Cyclohexane	101.40	6.558	0.000	0.000	3.634
Perhydronaphthalene	157.27	11.436	0.000	0.000	5.632
Perhydroanthracene	212.88	30.274	0.000	0.000	11.772
Basic adsorbent	—	1270.650	0.000	0.000	276.393
Methylated adsorbent	—	2175.051	0.000	0.000	625.216
Hydroxylated adsorbent	—	1325.913	−65.081	−67.809	349.294

Table 3.3 Molecular interaction energies for the analytes in Table 3.2 on different adsorbents. MIFS, MIVW, MIHB, and MIES represent the molecular interaction energy value of the final (optimized) structure, the van der Waals energy, the hydrogen-bonding energy, and the electrostatic energy (kcal mol^{-1}), respectively.

Analyte	Basic adsorbent		Methylated adsorbent		Hydroxylated adsorbent			
	MIFS	MIVW	MIFS	MIVW	MIFS	MIHB	MIES	MIVW
Benzene	10.019	9.872	9.805	9.254	19.918	−14.188	0.605	6.541
Naphthalene	16.261	16.033	13.748	12.923	32.110	−23.323	0.550	9.806
Anthracene	21.908	21.595	17.421	16.041	44.116	−32.441	0.721	13.079
n-Hexane	12.181	12.114	10.938	10.210	12.343	0.040	−0.040	12.230
n-Decane	19.577	19.417	18.407	17.426	19.855	0.071	−0.070	19.666
n-Tetradecane	26.918	26.686	25.144	24.025	27.340	0.097	−0.091	27.066
Pentanol	11.106	11.097	9.588	8.930	12.682	−2.431	−0.544	10.665
Nonanol	18.521	18.441	17.364	16.480	20.088	−3.044	−0.473	17.739
Tridecanol	25.859	25.698	23.826	22.586	27.550	−2.355	−0.529	25.520
Cyclohexane	8.239	8.181	8.845	8.521	12.682	−2.431	−0.544	10.665
Perhydronaphthalene	13.799	13.697	14.003	13.179	20.088	−3.044	−0.473	17.739
Perhydroanthracene	17.850	17.702	19.133	18.204	27.550	−2.355	−0.529	25.520

The optimized final energies and van der Waals energies obtained by MM2 calculations for these adsorbents are given in Table 3.2 with the properties of the solutes. The properties of three PAHs [benzene (Bz), naphthalene (Na) and anthracene (An)], which are model compounds for studying the effect of aromaticity, three alkanes (Al) [hexane (C_6), decane (C_{10}) and tetradecane (C_{14})], which are model compounds for studying the aliphatic effect, three alkyl alcohols (AlOH) [pentanol (C_5OH), nonanol (C_9OH) and tridecanol ($C_{13}OH$), which are model compounds for studying the hydrogen-bonding

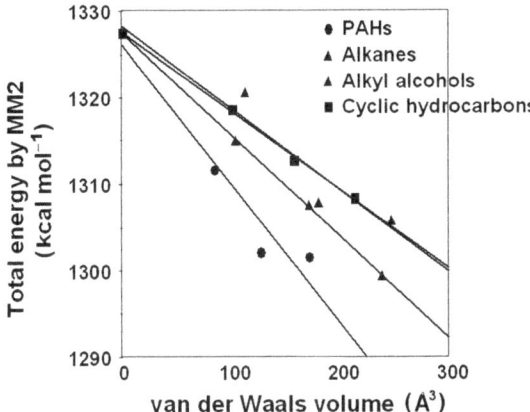

Figure 3.9   The selectivity of model hydroxylated phases.
Reproduced with permission from ref. 11.

effect, and three saturated cyclohydrocarbons (cyCH) [cyclohexane (cyC$_6$), perhydronaphthalene (cyC$_{10}$) and perhydroanthracene (cyC$_{14}$)] which are model compounds to study steric effects on the adsorbent, were calculated using the MM2 and MOPAC-BlogP programs.

The basic properties of the above 12 compounds were first analyzed based on their van der Waals volumes and enthalpies, and this indicated that these four groups of compounds have individual relationships, but are classified into two groups, one containing alkanes and alkyl alcohols and the other containing aromatic hydrocarbons and saturated cyclic hydrocarbons. A similar result was obtained for the relationship between their enthalpies and surface areas.[3]

The molecular interaction energy given by the MM2 calculation of each complex (given in Table 3.3) was analyzed after being corrected for individual properties. The slope between the van der Waals volume, calculated using the MOPAC-BlogP program, and the molecular interaction energy indicates the selectivity of the adsorbent. The slopes between the van der Waals volume and the final energy of a homologous series of compounds on the basic adsorbent are shown in Figure 3.9. The values obtained on different adsorbents are given in Table 3.4. The slopes for alkanes and alkyl alcohols obtained on the basic adsorbent are about the same, at 0.11. The slope for aromatic hydrocarbons is about 1.2 times of that for alkanes, at 0.14. The slope for saturated cyclic hydrocarbons differed from those for the other groups, being 0.09. The contribution of van der Waals energy in the MM2 calculation of each complex was also analyzed. The slope between the van der Waals volume, calculated using the MOPAC-BlogP program, and the van der Waals energy also indicates the selectivity of the adsorbent. The slope change after adsorption was 0.11 for alkanes and alkyl alcohols, 0.09 for saturated cyclic hydrocarbons compounds and 0.14 for aromatic hydrocarbons. This means that the size effects of the solutes are similar for

Table 3.4   The selectivity of the adsorbents. The selectivity is obtained from the slopes of plots of the van der Waals volume of solutes and their molecular interaction energy values. BA represents the basic adsorbent, MA represents the methylated adsorbent, OH represents the hydroxylated adsorbent. Cyclic HCs represents cyclic hydrocarbons.

Analyte	PAHs	Alkanes	Alkyl alcohols	Cyclic HCs
BA MIFS	0.136	0.110	0.110	0.086
BA MIVW	0.134	0.108	0.109	0.085
MA MIFS	0.087	0.106	0.106	0.092
MA MIVW	0.077	0.103	0.101	0.087
OH MIFS	0.276	0.112	0.110	0.133
OH MIHB	−0.208	0.000	0.001	0.001
OH MIES	−0.001	−0.000	0.000	0.000
OH MIVW	0.075	0.110	0.110	0.133

such molecular interactions except for aromatic hydrocarbons. The model adsorbent therefore appears to be selective for PAHs.

After methylation of this basic adsorbent, the slopes are changed, but the selectivity is similar to that obtained on the basic adsorbent. This is due to the similar properties of the adsorbents. Both are hydrocarbons, and the difference is in the smoothness of their surfaces. There is no indication that a small molecule stuck to the methyl groups. Such a phenomenon is expected on an alkyl-bonded phase due to atomic distance between the bonded alkyl groups.

The selectivity of the hydroxylated adsorbent is different from that of the hydrophobic adsorbents. The selectivity for aromatic hydrocarbons is dramatically changed, especially the slope relating to the final energy. Considering the slopes obtained for the final energies, the slope for aromatic hydrocarbons is 0.14 and 0.09 on the hydrophobic adsorbents, and 0.28 on the hydroxylated adsorbent. Such selectivity can also be obtained from their hydrogen-bonding energies in Table 3.3, where the hydrogen-bonding energies of other compounds do not demonstrate a significant difference, and only the hydrogen-bonding energy of the aromatic hydrocarbons demonstrates a clear difference. The selectivity of the hydroxylated adsorbent supports the selectivity of vinyl alcohol copolymer gels in liquid chromatography, where the retention of PAHs was stronger than that expected from their van der Waals volumes.

3.3   The Structure of Silica Gels

3.3.1   Construction of Model Silica Gels

Silica can form different polymer shapes such as fibers, sheets, columns, and porous and non-porous materials. Asbestos and quartz are naturally occurring different shapes of silica. The many colorful jewels are a mixture of

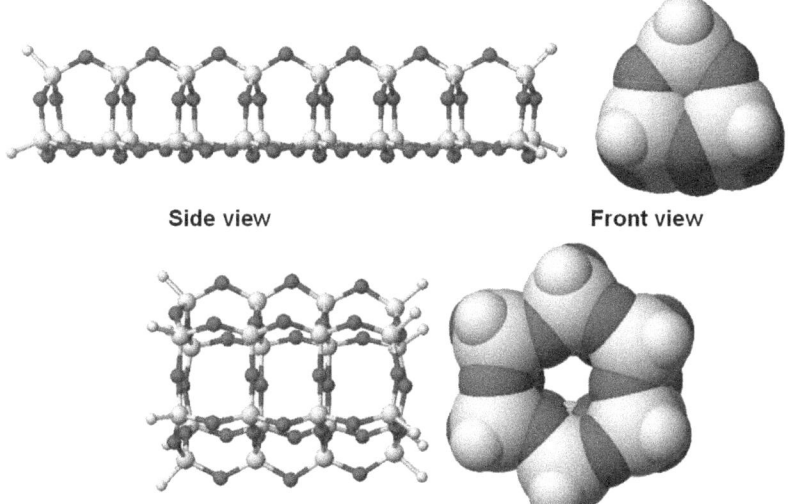

Side view   Front view

Figure 3.10 The structure of a polymerized silica, based on three and six units of SiO_2 groups. Small white, large white, and black balls represent hydrogen, silicon, and oxygen, respectively.

silica and other elements. The degree of homogeneity makes for different depths of color and physical strength. However, the structure of silica gel is not well understood. Therefore, several forms of silica gel were constructed by a computational chemical method, and the most suitable form was investigated. The monomer of silica, SiO_2, was polymerized to form a stable structure having a minimum of three SO_2 units. The three- and six-unit structures can be further polymerized as a fiber, as seen in Figure 3.10. However, the three-unit structure is not ideal for making the spherical balls that are suitable as a packing material. The polymerized fiber has silanol groups at each end. The six-ring unit can be three-dimensionally polymerized to make a large molecule as seen in Figure 3.11. The small ball is solid, except for a narrow channel. This may not be the final structure of packing material. These tiny silica sols can join together to form large ball-like materials for packing.[4]

3.3.2   Chemical Modification of Surface Silanol Groups

A maximum of 50% of the surface silicon atoms can be converted to silanol groups for further surface modification. The question is, how many silanol groups appear on the surface? Geometrical analysis of the surface of constructed silica gels has been undertaken. Six SiO_2 units can form a ring which can grow into a three-dimensional polymer. An example of a polymerized silica layer is shown in Figure 3.11, where one surface is saturated with silanol groups. In this large unit, all silicon atoms have two free valences to join together. Of the silicon atoms, 50% have free silanol groups,

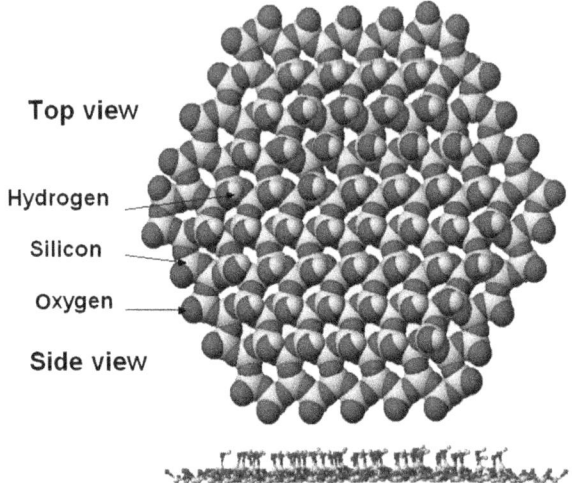

Figure 3.11 The structure of a polymerized silica gel based on six units of SiO_2 groups.

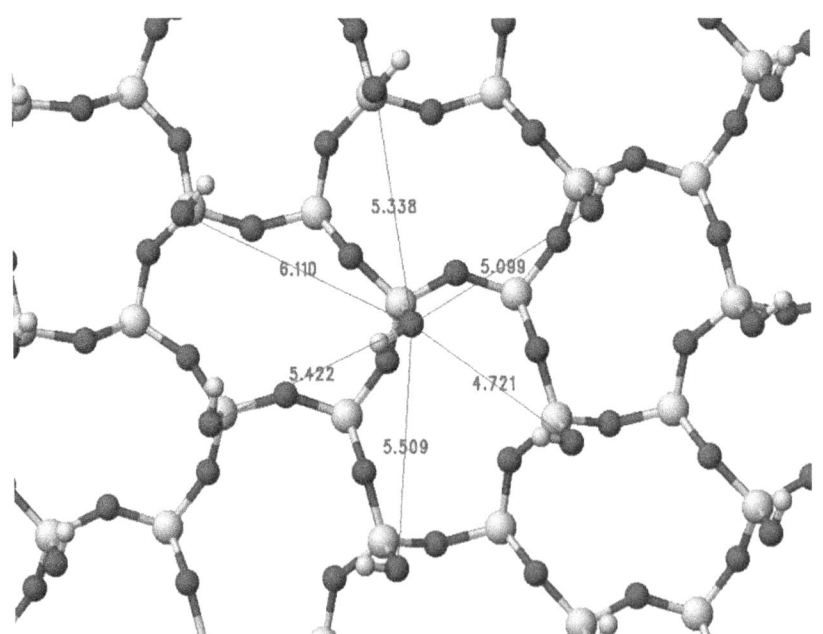

Figure 3.12 The atomic distances (Å) between oxygen atoms (the middle-sized circles) in a polymerized silica. Small white, large white, and black balls represent hydrogen, silicon, and oxygen, respectively.

and those on the surface of the polymerized structure can be further sila-nized. The distance between the oxygens of two silanol groups is about 5 Å, as Figure 3.12 shows. The shortest distance between the two chlorines of a silylating reagent is 3.43 Å similar to this distance. The number of silanol groups can be increased by cutting siloxane on the surface. However the atom distances in these silanol groups are not ideal for trichlorosilane. This means that three-way bonding cannot occur smoothly, and the cutting of many siloxanes will make the porous structure weaker.

In an ideal surface structure, one out of every two silicon atoms has a silanol group which can be silanized by either mono-, di- or trichlorosilane. If the surface of the silica matrix is completely silanized by the bonded organic phase, the organic phase can prevent the corrosion of the silica gel by acid or basic solutions. If the silanol groups are not completely silanized by the first reaction, a second silanization process, known as *end-capping*, is required. The carbon content did not significantly increase in this end-capping process.

When more than seven SiO_2 units form a large unit ring, the center hole is large enough to collect and keep a metal ion in it, but not large enough for total silanization of the inside, and the structure of polymerized silica may not be homogeneous. This means that the porous silica gel will be physically weak, and will not have an ideal surface for total silanization.

The real structures of porous silica gels are unknown. However electron microscopic analysis revealed a smooth surface. Bonded silica gels made from pure silica are very chemically stable compared to ordinary silica gels. This means that the purer the silica, the more homogeneous the structure, which improves the physical strength and the reactivity of silanization. The basic structure may be based on rings of six SiO_2 units.

According to the literature, all silicon atoms on the surface can be converted to silanol, and 50% of the silanols can be silanized. However, in the above newly designed silica gel, it appears that only 50% of silicon atoms on the surface are silanols, and all the silanols can be silanized. Although the surface coverage can be estimated from the carbon content, the surface area and the pore size, the following proposed inertness test is more practical for measuring the surface coverage than calculation from physical properties. The stability of bonded silica gels designed using computational chemical analysis was improved after more than 5000 h immersion in 0.1% trifluoroacetic acid solution. The batch-to-batch distribution of carbon content was less than 5% and the deviation of the capacity ratio of standard compounds was also less than 5%.[4]

3.4 Three-Dimensional Model Phases

Numerous reports have been published about the three-dimensional structure of proteins determined by X-ray crystallography and/or NMR, and by computational chemical calculation from the results of amino acid sequencing. An empirical approach to identifying catalytic sites, the location of metal ion and carbohydrate binding sites, and folding and unfolding, has been studied with molecular dynamics simulations. Once the binding site of a protein, the structure of the protein–small molecule interaction, is determined, the fitted small molecule is further modified, and a variety of drug candidates can be designed. A question is, how does a small molecule reach the center of the protein? No channel large enough for small molecules to pass through freely exists in a protein. Even water molecules cannot physically pass through a membrane. Agre and MacKinnon clarified how ions and water cross cell membranes where the existence of ion channels in a protein is important. Ions and water are driven through by electrolytic forces. The electronic movement is Arhenius' theory of electrolytic dissociation.[5] The active sites of some enzymes are nearly identical, but the catalytic reactions are different. One enzyme can have a different chemistry at the same active site. These observations indicate that a protein will catch different compounds *via* the same manner.[6] The correcting mechanism should be basically the same. It is not 100% affinity. The basic docking mechanism can be explained by a simple model experiment.

A fast analytical method is required to measure human serum albumin–drug binding affinity in drug discovery research. Drug–albumin binding sites have been studied, but albumin also functions as a scavenger. This indicates that the albumin structure has the flexibility to carry a variety of compounds and the affinity may not be specific. The main binding forces are hydrophobic interactions and ion–ion interactions, and specific steric effects may not be important. Previously, acidic drug–HSA and basic drug–HSA binding affinities were successfully determined by a combination of reversed-phase and ion-exchange liquid chromatography without albumin.[7–9] The guanidino groups of arginine should work as anion-exchange groups and the carboxyl groups of aspartic and glutamic acids should work as cation-exchange groups. The chromatographic behavior of acidic and basic drugs was studied using guanidino and carboxyl-phase columns, and their retention factors correlated well with their log nK values measured by the modified Hummel–Dreyer method. Using a computational chemical program to analyze liquid chromatographic data, the direct interaction between a model phase and a drug was calculated as energy values using MM2 calculations. Computational chemistry using a model adsorbent is a new method for quantitative analysis of the retention of acidic drugs on a guanidine phase used for ion-exchange liquid chromatography.

The model support consisted of 362 carbons and 248 hydrogens, 848 bonds and 3684 connectors, as shown in Figure 3.13. The molecular weight was 4592. The center hydrogen was replaced by a methyl group or a guanidyl group. The surrounding six hydrogens were replaced by methyl groups to make room for the analyte to contact the substituted group in the center. Twelve hydrogens from the second, and 18 hydrogens from the third circle were replaced by longer alkyl groups. Twenty-four hydrogens of the fourth circle were also replaced by longer alkyl groups if dense alkyl groups were necessary, as shown in Figure 3.14, where octyl groups were bonded.

The pentyl-bonded phase was constructed with pentyl groups that were open like petals of the flower chamomile, and the center of the bonded phase was completely open for the adsorption of a variety of compounds.

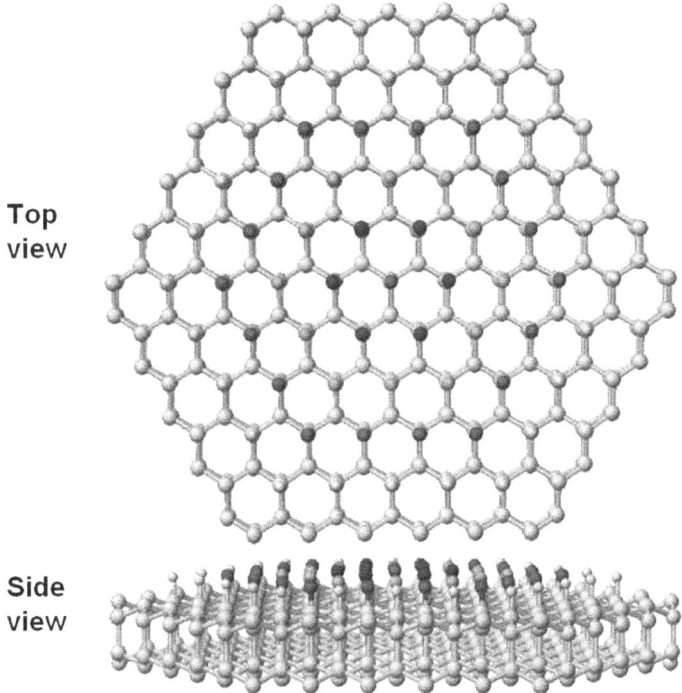

Figure 3.13 The three-dimensional structure of the model support.
Reproduced by permission of Bentham Sci.

Even the octyl-bonded phase was open for the adsorption of a small molecule, as shown in Figure 3.15, where a guanidyl group was located at the center for studying ion–ion interactions. The low-density guanidino phase is shown in Figure 3.16. Then, the octyl groups were replaced by dodecyl groups. The bonded phase was like a flower tulip, and a pocket remained where a small molecule could reach the methyl or ion-exchange group at the center of the bonded phase, like a bee entering the center of a tulip, as shown in Figure 3.17. These bonded phases were used to study the ion–ion interactions between the ionic substitute and the ion-exchange group located at the center of the pentyl-, octyl- or dodecyl-bonded phase, if ion–ion interactions were the predominant driving force.

The dodecyl groups were then replaced by octadecyl groups as Phase 4. The octadecyl groups were gathered together by hydrophobic interactions and the center pocket was covered, so no hole was observed like a flower bud where a small molecule could pass through and reach the surface of an ion-exchange or methyl group, as shown in Figure 3.18. Even benzoic acid could not reach the bottom and remained on the top of the octadecyl groups. Therefore, the octadecyl-bonded phase was not used for further experiments.

A computational chemical analysis using the MM2 program was applied to analyze the retention mechanism of benzoic acid on the guanidine phases. The methyl-bonded phase was used as the blank phase to measure

Phase 1 Me (C8)

Top view   Side view

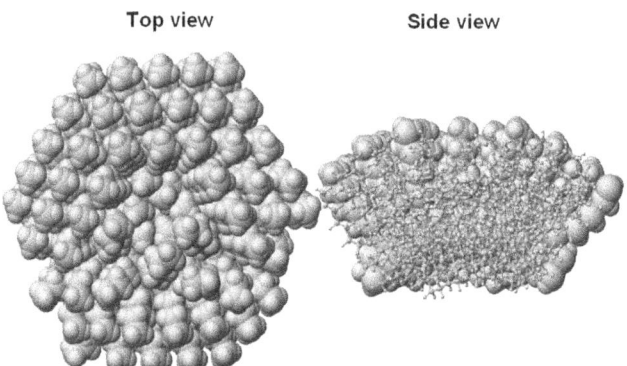

Figure 3.14 A Phase 1 Me (C8) structure used to study hydrophobic interactions. White and light-gray balls represent hydrogen and carbon, respectively.

Phase 1 GUA (C8)

Top view   Side view

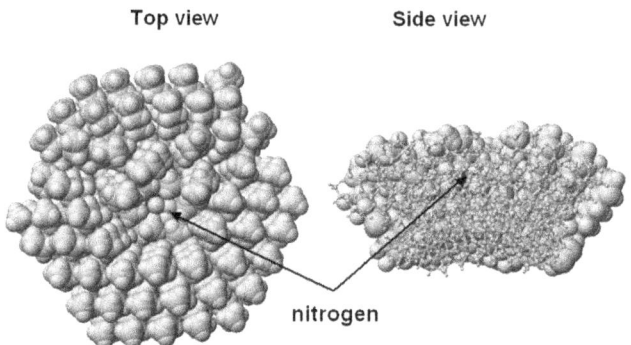

nitrogen

Figure 3.15 A Phase 1 GUA (C8) structure. White, light-gray, and dark-gray balls represent hydrogen, carbon, and nitrogen, respectively.
Reproduced by permission of Bentham Sci., ref. 10.

the hydrophobic interactions. The optimized energy value was less than 0.00001 kcal mol^{-1} by MM2 optimization. The calculated energy values are listed in Table 3.5. After subtraction of the individual energies of an analyte and a model bonded phase from the molecular interaction energy values, the substituted energy values were considered as the interaction energy values and are listed in Tables 3.6 as MIFS, MIHB, MIES and MIVW, and these MI were used for the analyses.

The molecular or ionized form of benzoic acid was placed outside the hole and faced toward the center methyl or guanidyl group at the bottom of the pocket before starting the docking as shown in Figure 3.19, then the complex structure was optimized to measure the direct hydrophobic and ion–ion interactions. (Me) and (GUA) indicate the center substituted groups; (Me)

Phase 2 GUA (C8)

Top view   Side view

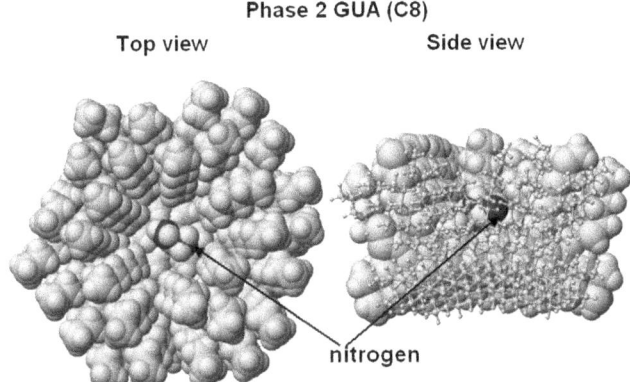

nitrogen

Figure 3.16 A Phase 2 GUA (C8) structure containing one guanidyl, six methyl, and 30 octyl groups. White, light-gray, and dark-gray balls represent hydrogen, carbon, and nitrogen, respectively
Reproduced by permission of Bentham Sci., ref. 10.

Phase 3 GUA (C12)

Top view   Side view

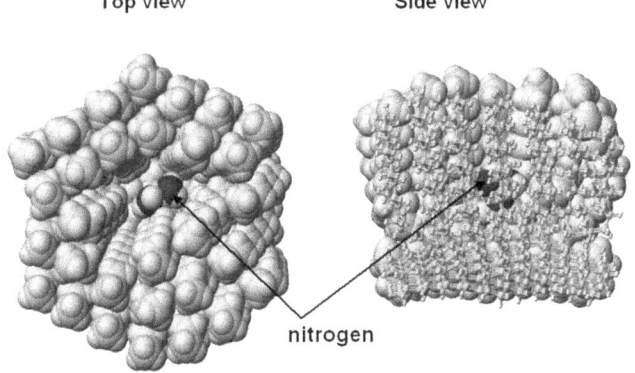

nitrogen

Figure 3.17 A Phase 3 GUA (C12) structure containing one guanidyl group, six methyl and 30 dodecyl groups. White, light-gray, and dark-gray balls represent hydrogen, carbon, and nitrogen, respectively.

represents a hydrophobic phase, (GUA) represents a guanidino phase, "phenyl" means the phenyl group of benzoic acid faced towards the bottom, and "COOH" and "COO⁻" mean these groups of benzoic acid faced towards the bottom before the optimization.

The pocket size of Phase 1 was about 9.5 Å inside diameter (i.d.) at the entrance, 7.5 Å at the bottom and 5 Å deep. That of Phase 2 was about 14 Å i.d. at the entrance, 5.0 Å at the bottom and 7.5 Å deep. That of Phase 3 was 8.8 Å i.d. at the entrance, 4.4 Å at the bottom and 12.5 Å deep. According to these pocket sizes, benzoic acid may reach the bottom without encountering a physical barrier. The docked molecular and ionized benzoic acids in Phase 1 (Me), Phase 1 (GUA), and Phase 3 (GUA) are shown in Figures 3.20–3.22.

Phase 4 GUA (C18)

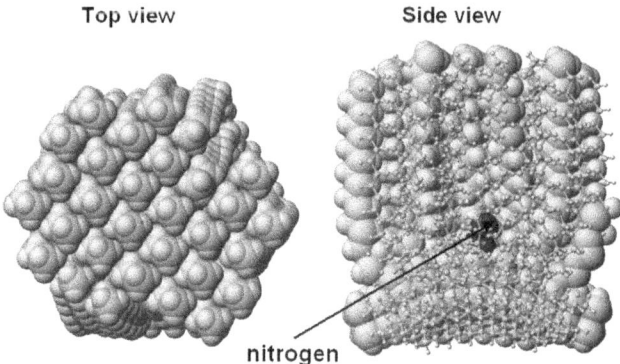

Top view   Side view

nitrogen

Figure 3.18 A Phase 4 (C18) GUA structure. White, light-gray, and dark-gray balls represent hydrogen, carbon, and nitrogen, respectively.

Table 3.5 The molecular properties of various model phases and analytes. M and I represent the molecular and ionized forms, and GUA and Me represent the guanidino and methyl phases. Reproduced by permission of Bentham Sci., ref. 10.

Analyte	fs	hb	es	vw
Benzoic acid (M)	−13.9182	−3.458	−6.671	4.876
Benzoic acid (I)	−2.5494	0.000	0.000	4.743
Phase 1 (Me)	6083.2002	0.000	0.000	1567.181
Phase 1 (GUA)	6136.1074	−0.683	−20.262	1587.223
Phase 2 (Me)	3641.5884	0.000	0.000	947.116
Phase 2 (GUA)	3615.2891	−6.684	−19.773	940.814
Phase 3 (Me)	3636.3321	0.000	0.000	831.617
Phase 3 (GUA)	3681.1654	−0.809	−21.050	835.088

The molecular interaction energy values of benzoic acid and the model phases are summarized in Table 3.6. The MIES between benzoic acid and a guanidine phase demonstrated the existence of ion–ion interactions between a carboxyl ion and a guanidino group. No such energy value changes were observed for the hydrophobic phases. Even the phenyl group was placed toward the guanidino group; a significant energy change was not obtained because benzoic acid was trapped by hydrophobic interactions before it reached the bottom of Phase 3, as shown in Figure 3.22. The molecular form of benzoic acid formed a hydrogen bond with the guanidyl group, but the energy change was less than that of the electrostatic energy of the ionized form. A hydrogen-bonding energy change was also observed on the hydrophobic phase whose pocket size was large enough for the free access of a small molecule to reach the bottom.

The above results demonstrate that ion–ion interactions are strong enough to drive a distance of more than 10 Å and push aside an alkyl wall

Table 3.6 Molecular interaction energy values (kcal mol^{-1}). (Me) and (GUA) – phenyl COOH: phenyl group faces methyl (Me) or guanidyl (GUA) group, HOOC phenyl and OOC phenyl: polar group faces methyl (Me) or guanidyl (GUA) group.

Complex form	MIFS	MIHB	MIES	MIVW
Phase 1 (Me) – phenyl COOH	12.20	−0.00	−0.00	11.27
Phase 1 (Me) – HOOC phenyl	10.64	−0.01	−0.01	9.70
Phase 1 (Me) – phenyl COO$^-$	16.59	0.00	0.00	15.02
Phase 1 (Me) – OOC phenyl	11.32	0.00	0.00	11.22
Phase 1 (GUA) – phenyl COOH	20.63	3.57	−0.74	16.24
Phase 1 (GUA) – HOOC phenyl	33.90	1.13	1.27	29.26
Phase 1 (GUA) – phenyl COO$^-$	21.33	3.61	0.92	15.38
Phase 1 (GUA) – OOC phenyl	35.49	−0.39	7.72	26.02
Phase 2 (Me) – phenyl COOH	21.11	−0.08	−0.05	22.09
Phase 2 (Me) – HOOC phenyl	14.26	0.00	−0.01	14.11
Phase 2 (Me) – phenyl COO$^-$	14.67	0.00	0.00	16.77
Phase 2 (Me) – OOC phenyl	14.53	0.00	0.00	15.11
Phase 2 (GUA) – phenyl COOH	13.91	1.64	0.73	11.25
Phase 2 (GUA) – HOOC phenyl	18.38	5.05	1.59	12.14
Phase 2 (GUA) – phenyl COO$^-$	14.69	−1.50	0.73	12.94
Phase 2 (GUA) – OOC phenyl	18.14	0.13	8.13	10.97
Phase 3 (Me) – phenyl COOH	14.85	0.00	0.00	13.70
Phase 3 (Me) – HOOC phenyl	14.53	−0.02	−0.01	13.44
Phase 3 (Me) – phenyl COO$^-$	20.02	0.00	0.00	19.30
Phase 3 (Me) – OOC phenyl	17.90	0.00	0.00	17.78
Phase 3 (GUA) – phenyl COOH	16.80	1.11	0.06	15.03
Phase 3 (GUA) – HOOC phenyl	18.11	2.27	−0.42	16.62
Phase 3 (GUA) – phenyl COO$^-$	17.28	1.47	0.87	14.96
Phase 3 (GUA) – OOC phenyl	19.23	0.13	5.78	12.45

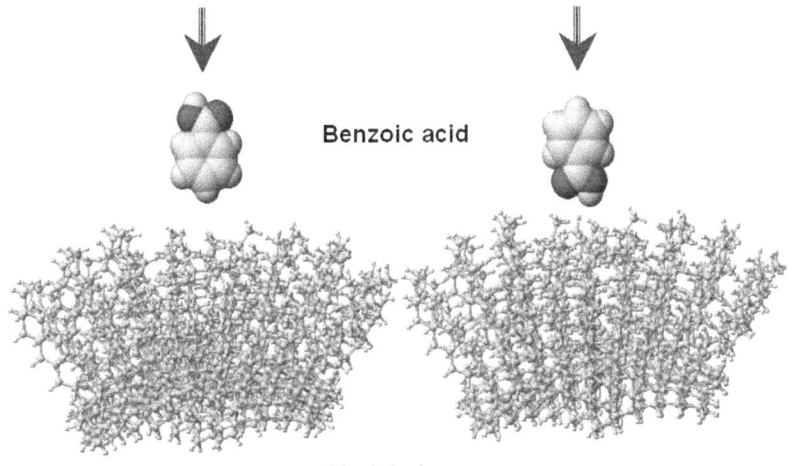

Figure 3.19 The docking process of benzoic acid into Phase 1 (Me).

Benzoic acid

molecular form

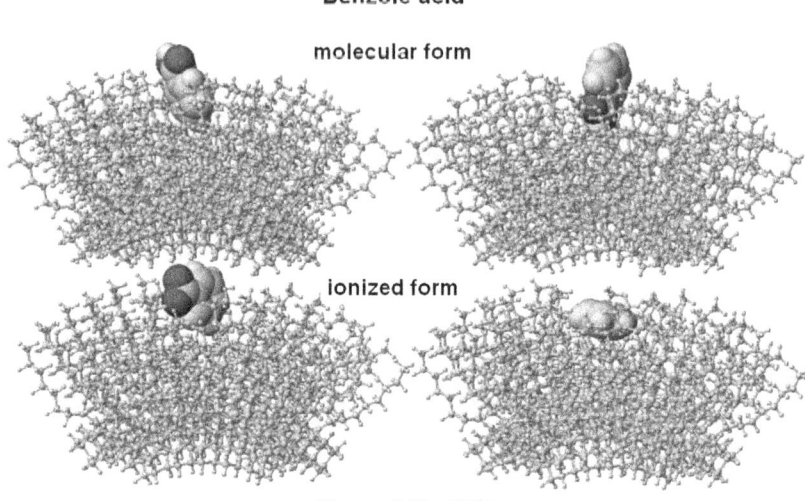

Phase 1 Me (C8)

Figure 3.20   A docked benzoic acid molecule in Phase 1 Me (C8).

Benzoic acid

molecular form

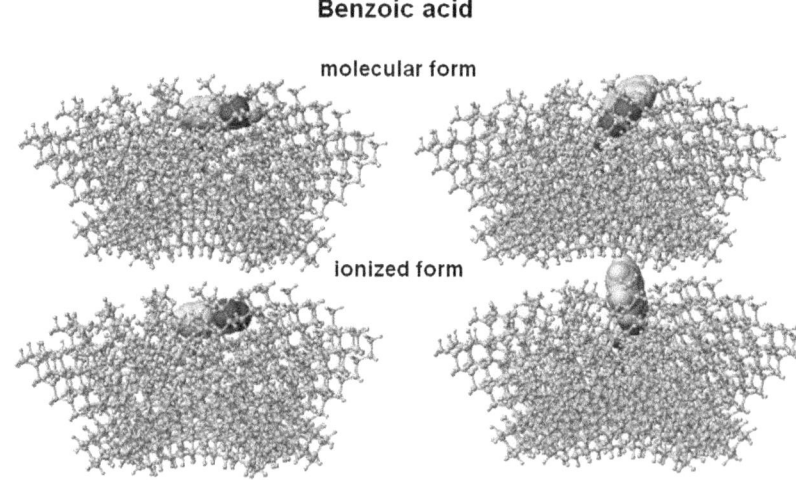

Phase 1 GUA (C8)

Figure 3.21   A docked benzoic acid molecule in Phase 1 GUA (C8).

whose hydrophobicity is about log $P = 5$. This means there may be a hole where a large molecule enters into a protein using ion–ion interactions. A flat guanidino phase was suitable for docking and demonstrated the highest correlation coefficient between the energy values calculated using molecular mechanics and the binding affinity measured using liquid chromatography. A model phase with a narrow channel demonstrated that a large ionized acid pushed aside the hydrophobic wall and reached the

Ionized benzoic acid

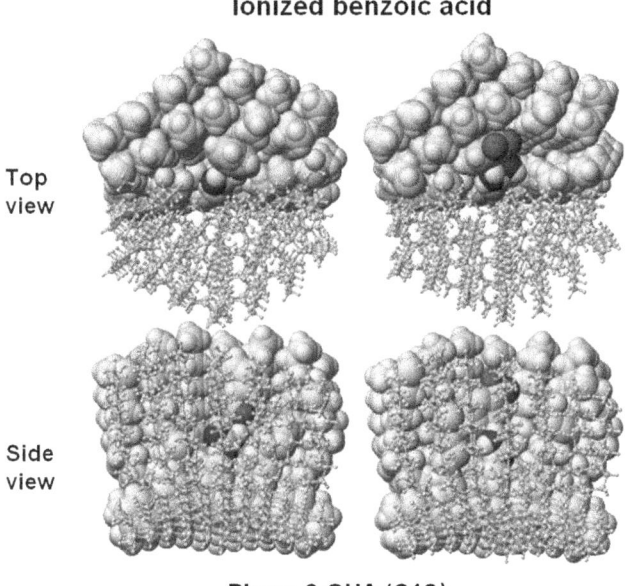

Phase 3 GUA (C12)

Figure 3.22 A docked ionized benzoic acid molecule in Phase 3 GUA (C12).

guanidyl group by ion–ion interactions. Ion–ion interactions should be the predominant driving force by which a molecule reaches the reaction site of a protein.[10]

References

1. H. Hosoya, Application of graph theory for chemical physics, *Kotaibutsuri*, 1988, **33**, 181–189.
2. T. Hanai, Quantitative *in silico* analysis of the specificity of graphitized (graphitic) carbons, *Adv. Chromatogr.*, 2011, **49**, 251–284.
3. T. Hanai, H. Hatano, N. Nimura and T. Kinoshita, Molecular recognition in chromatography aided by computational chemistry, *Supramol. Chem.*, 1994, **3**, 243–247.
4. T. Hanai, Synthesis and properties of stable bonded silica gel packings and the performance, in *Advances in Liquid Chromatography*, ed. H. Hatano and T. Hanai, World Scientific, Singapore, 1996, pp. 307–327.
5. A. Yarnell, Renaissance men, *Chem. Eng. News*, 2003, **81**, 35–38.
6. A. Yarnell, The power of promiscuity, *Chem. Eng. News*, 2003, **81**, 33–35.
7. T. Hanai, R. Miyazaki and T. Kinoshita, Quantitative analysis of human serum albumin–drug interactions using reversed-phase and ion-exchange liquid chromatography, *Anal. Chim. Acta*, 1999, **378**, 77–82.
8. T. Hanai, A. Koseki, R. Yoshikawa, M. Ueno, T. Kinoshita and H. Homma, Prediction of human serum albumin–Drug binding affinity without albumin, *Anal. Chim. Acta*, 2002, **454**, 101–108.

9. R. Miyazaki, T. Hanai, J. Suzuki and T. Kinoshita, Study of ion–ion interaction for protein–drug binding using a newly developed guanidino-bonded phase in liquid chromatography, *J. Liq. Chromatogr. Relat. Technol.*, 1998, **21**, 2887–2895.

10. T. Hanai, Molecular modeling for quantitative analysis of molecular interaction, *Lett. Drug Des. Discovery*, 2005, **2**, 232–238.

11. T. Hanai, H. Hatano, N. Nimura and T. Kinoshita, Computer-aided analysis of molecular recognition in chromatography, *Analyst*, 1993, **118**, 1371–1374.

CHAPTER 4

Retention in Gas Chromatography

4.1 Introduction

Since its development, gas chromatography has been used to measure molecular properties, such as boiling point, enthalpy, entropy, and free energy. Conversely, these molecular properties are used to calculate the retention index in gas chromatography. Boiling point has been used as a retention index since 1958. The retention time in gas chromatography using a non-polar phase has been related to the boiling points of a variety of compounds, including low-boiling organic compounds,[1] polycyclic aromatic hydrocarbons (PAHs) and biphenyls,[2] alkylbenzenes,[3] dialkyl-*N*-nitrosamines, dialkyldiimides, trialkyl phosphates and alkyl-*N*-methylpyrasols[4]; and O-, N-, and S-containing homologous series.[5] The boiling points of PAHs are determined from the contributions of molecular fragments, and correlate well with retention times.[6] The enthalpy of vaporization ($\Delta_{vap}H$) is also used as a retention index. The $\Delta_{vap}H$ of organic compounds is calculated using empirical methods.[7] The gas chromatographic retention of organochlorine pesticides and nitrochlorine analytes has been simulated using the enthalpy and entropy of vaporization.[8] The enthalpy of vaporization has been determined from gas chromatographic retention, for alkyl benzenes,[9] 80-branched esters,[10,11] polybrominated diphenyl ethers,[12] PAHs,[13] benzyl halides and benzylethers,[14] polychlorinated dibenzodioxins and polychlorinated dibenzofurans,[15] toxaphene congeners,[16] C21 to C30 and C31 to C38 *n*-alkanes,[17,18] linear aliphatic nitriles,[19] dicarboxylic acids,[20] cyclic organic peroxides,[21] and aliphatic alcohols.[22] Boiling is a process of evaporation. The molecular density can be computationally chemically calculated and is related to $\Delta_{vap}H$. The $\Delta_{vap}H$ of PAHs can be determined by gas

RSC Chromatography Monographs No. 19
Quantitative *In Silico* Chromatography: Computational Modelling of Molecular Interactions
By Toshihiko Hanai
Published by the Royal Society of Chemistry, www.rsc.org

chromatography using Kováts retention indices.[23] The properties estimated from the gas chromatography retention times are, however, dependent on the specificity of the stationary phase. The varied enthalpy measured on different stationary phases indicates that there is a molecular interaction between the stationary phase and the analyte.[24,25]

Since Kováts proposed the use of a homologous series of *n*-alkanes as standard compounds for calibrating columns, Kováts retention indices[26] have been used to analyze the specificity of stationary phases. Kováts retention indices are applicable to the chromatography of non-polar phases, but further modification is necessary for determining the molecular properties of analytes on a variety of stationary phases.[27] The temperature effect on the retention indices has been studied using transferable potentials for phase equilibrium force fields, and the vapor–liquid coexistence densities and saturated vapor pressures were calculated using configurational-bias Monte Carlo, which improves the precision of the predicted boiling point. This approach was applied to determine the gas–liquid partitioning for a model chromatographic system, and also improved the accuracy of the retention indices. Furthermore, the molecular simulations allowed for the direct determination of both the Gibbs free energy and the enthalpy of transfer using standard thermodynamic relations.[28] The method was applied to chromatography on a polyethylene oxide phase, and the varied results indicated the selectivity of the analytes.[29] Gas chromatographic retention times have been calculated using the quantitative structure–retention relationship based on molecular structure.[30] Kováts retention indices for both linear and non-linear temperature gas chromatography of a homologous series of compounds, have been studied using various stationary phases. The methylene unit value is affected by the electron density and the molecular structure.[31] The analytes are retained on the stationary phase by molecular interactions. Unfortunately, a quantitative method has not yet been developed to analyze these molecular interactions.

Using these methods is similar to reconstructing a puzzle. How the retention and vaporization mechanisms can be quantitatively analyzed, and the predicted retention times improved, based on molecular properties calculated *in silico*, are fundamental questions in chromatography. In gas chromatography, no solvent is used except in special cases where water vapor and ionic gas are mixed with the carrier gas. The basic retention mechanisms depend on the strength of the molecular interaction with the stationary phase, and the vaporization mechanism depends on the properties of the analytes.

4.2   Retention of Volatile Compounds on Graphitized Carbon

There is very little data in the literature concerning the retention of organic volatile compounds, during gas chromatography with a graphitized carbon column. In general, unsaturated compounds elute faster than saturated compounds. The retention time of ethylene is shorter than that of ethane,[32] and benzene elutes before *n*-hexane.[33] The retention order of the following four-carbon hydrocarbons is isobutane, 1-butene, *n*-butane, and *trans*-2-butene.[34] The retention mechanism was studied using model standard compounds. The model phase was a 22-ring PAH. The molecular interaction energy value was calculated using the MM2 program. No retention time was measured under isocratic conditions, therefore the correlation coefficient was not calculated. The final structure energy appears to be related to the retention time, and the van der Waals energy contributes to the retention, whereas the other energy values do not contribute to the retention, according to the MM2 calculations, as summarized in Table 4.1.

MOPAC calculations do not clearly indicate the forces contributing to molecular interactions. They demonstrate the location of the molecular interaction center from an electron density map, using the tabulator of the CAChe™ program, and calculate the change in the atomic partial charge before and after optimization of the complex form. These hydrocarbons interact at the center of the PAH molecule, as shown in Figure 4.1, where *trans*-2-butene is located at the center of PAH22.

The atomic partial charge of the model phase and hydrocarbon does not change significantly after optimization of the complex form, while the electron potential is slightly shifted toward the molecular interaction side. These results clearly indicate the existence of different retention mechanisms on graphitized carbon phases, a hydrophobic interaction and hydrogen bonding.

Table 4.1 Molecular interaction energies calculated using the MM2 program. MIFS, MIVW, MIHB, and MIES represent the molecular interaction energy value of the final (optimized) structure, the van der Waals energy, the hydrogen-bonding energy, and the electrostatic energy (kcal mol^{-1}), respectively. Reproduced by permission of Taylor and Francis, ref. 40.

Analyte	MIFS	MIVW	MIHB	MIES
Isobutane	6.689	6.696	0	0.000
n-Butane	7.923	7.965	0	0.000
1-Butene	7.695	6.673	0	0.000
trans-2-Butene	11.382	7.022	0	0.099
Cyclohexane	8.235	8.268	0	0.000
Benzene	8.591	8.601	0	0.000
1,3,5-Hexene	9.868	9.883	0	0.000
1,3-Hexene	10.390	9.835	0	0.000
1-Hexene	11.069	9.777	0	0.000
n-Hexane	11.288	11.341	0	0.000

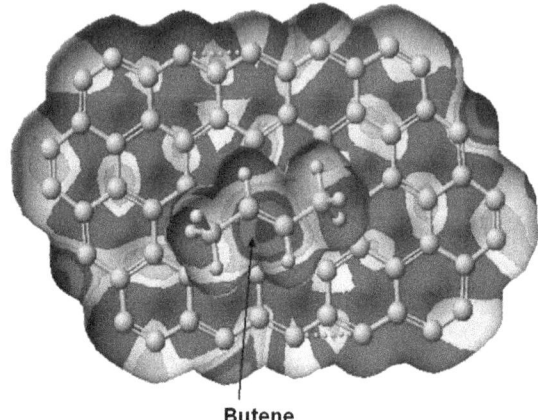

Butene

Figure 4.1   Electron density map of a *trans*-2-butene and PAH22 complex.
Reproduced by permission of Taylor and Francis, ref. 40.

4.3   Retention on Methylsilicone

Boiling point is sometimes related to the retention time measured on a non-polar phase. If the boiling point is available, the retention time can be predicted. The boiling point of a homologous series of compounds was, therefore, calculated based on the relationship between a reference boiling point[35] and the van der Waals volume. The calculated and reference boiling points are given in Table 1 of the Appendix (p. 278). The capacity ratios of five groups of compounds measured using two methylsilicone capillary columns under isocratic conditions[36] correlated well to the boiling point. The correlations between the boiling points and log k values at 240 °C were 0.992 and 0.988 ($n = 48$) for DB1 and CPSil5 columns, respectively. The relationship is shown in Figure 4.2.

The boiling points of peaks identified by mass spectrometry can be estimated from the boiling points of standard compounds, and enthalpies can be obtained from retention times measured at different temperatures. While the measured enthalpy correlates well between each group of logarithm of capacity ratio (log k) values, the relationship is not consistent,[25] as shown in Figure 4.3.

Retention time in gas chromatography is related to a combination of retention and volatility, similar to solubility in liquid chromatography. Predicting volatility is as difficult as predicting solubility. Volatility has been explained as the enthalpy of vaporization ($\Delta_{vap}H$), and a method for predicting volatility has been proposed.[4] If the $\Delta_{vap}H$ values are available, it may be possible to predict retention time. Unknown $\Delta_{vap}H$ values have been calculated from the relationship between the van der Waals volume and reference $\Delta_{vap}H$ values.[7,23] The values are summarized with the corresponding reference values in Table 1 of the Appendix (p. 278). Values of $\Delta_{vap}H$ have also been related to capacity ratios. The correlation coefficients were 0.896 and 0.852 ($n = 48$) for DB1 and CPSil5 columns, respectively, which appear to be acceptable correlation coefficients, except that the relationship for alkyl alcohols deviated from those of other compounds as seen Figure 4.4.

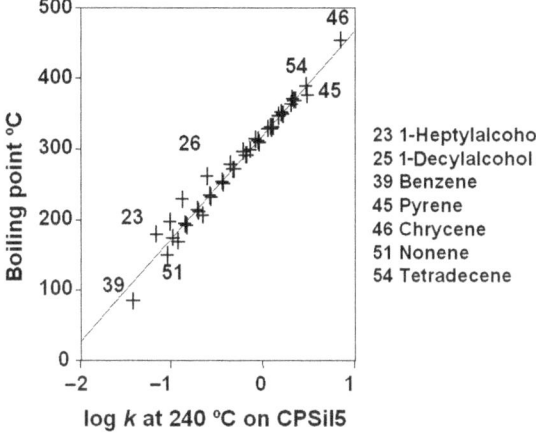

Figure 4.2 The relationship between log k (240 °C) and boiling point on CPSil5.

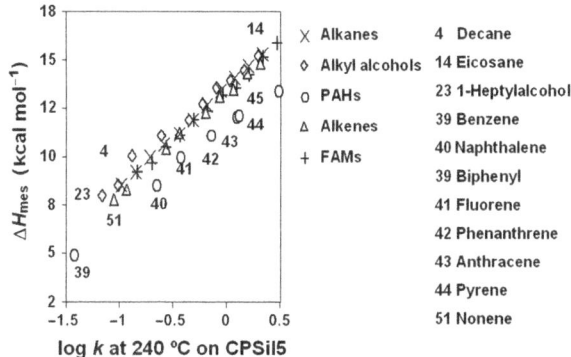

Figure 4.3 The relationship between log k (240 °C) and the measured enthalpy (ΔH_{mes}) on CPSil5.

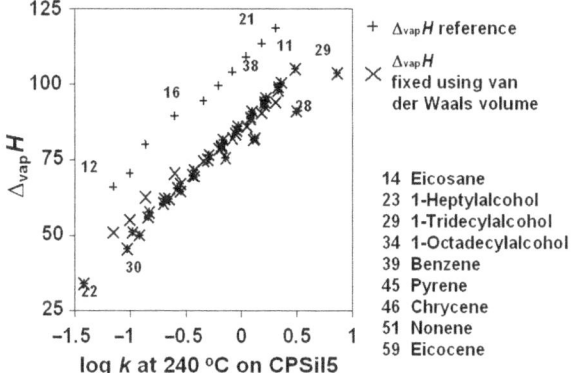

Figure 4.4 The relationship between log k (240 °C) and the enthalpy of vaporization ($\Delta_{vap}H$) on CPSil5.

The higher $\Delta_{vap}H$ of alkyl alcohols indicates that they require more energy for evaporation. When the alkyl alcohol $\Delta_{vap}H$ was fixed based on the boiling points of other aliphatic compounds, the correlation coefficients improved to 0.992 and 0.978 ($n = 48$) for the DB1 and CPSil5 columns, respectively. If the experimental and/or predicted boiling point and $\Delta_{vap}H$ values are available, it is possible to predict the retention time using the properties of alkanes as the standard.

In general, the above molecular properties (boiling point and $\Delta_{vap}H$) are not available. Computational chemical methods, however, can be used to calculate a variety of molecular properties. The calculated molecular interaction energy and the analyte van der Waals volume, enthalpy, and optimized energy values can be used to further quantitatively study the retention mechanism.

The surface of a well-treated fused silica capillary column is very inert, therefore, a saturated alkyl phase was first constructed from a model graphitized carbon phase, and the capacity ratios of five groups of compounds, alkanes, alkenes, alkyl alcohols, fatty acid methylesters (FAMs), and PAHs, measured on a methylsilicone phase,[36] were analyzed *in silico* based on the molecular interaction energy values calculated using the MM2 module of the CACheᵀᴹ program. The docking of individual molecules on such a simple model phase is straightforward. The molecular interaction energy value is calculated by subtracting the energy value of the complex from the sum of the individual energy values and the model phase. The molecular interaction energy values of alkanes were used as the standard, similar to Kováts retention index method. The energy value change after complex formation is considered to be the molecular interaction energy value and comprises the final (optimized) structure (MIFS), and the hydrogen-bonding (MIHB), electrostatic (MIES), and van der Waals (MIVW) forces. MIHB and MIVW indicate the contribution of the hydrogen bonding and molecular size effects, respectively. The calculated molecular interaction energy values were correlated with the log k values of these compounds. For example, to predict retention times using MIFS $= a \times$ log $k + b$, the slope a and constant b of all groups should be the same for an ideal system.

The molecular interaction energy values of each individual group correlated well with their log k values, but the molecular interaction energy values did not correlate well with their log k values on a saturated alkyl phase. Alkyl alcohol molecular interaction energy values were lower than expected based on their log k values, compared to those of alkanes and alkenes. This finding might be due to the low concentration of methyl groups, because the liquid phase of DB1 is methylsilicone. The C–H hydrogens were converted to methyl groups. The model phase was bent, and not suitable for studying the adsorption of larger molecules. The basic support carbons were therefore changed to silicons to increase the atomic distance and to build a flat alkyl-bonded phase. A model 100% methylsilicone phase is shown in Figure 4.5 with adsorbed octadecane.

The correlation between the molecular interaction energy values and their log k values was the same as that obtained using the saturated alkyl phase.

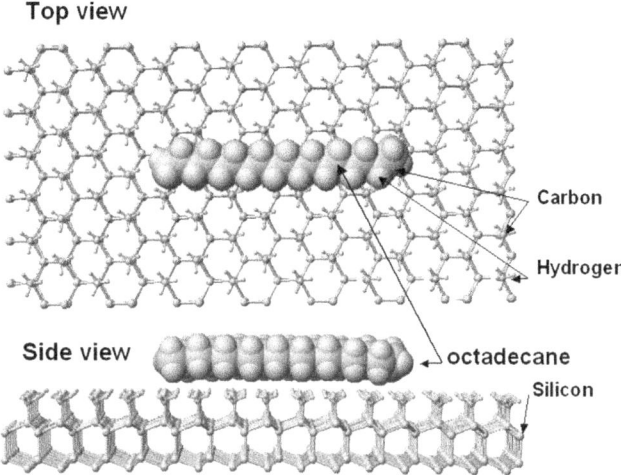

Figure 4.5 Adsorption of octadecane on a model methylsilicone phase. Small white, large white, and light-gray balls represent hydrogen, silicon, and carbon, respectively. The atomic size of the model phase is 20% of the analyte. Reproduced by permission of Institute for Chromatography, ref. 38.

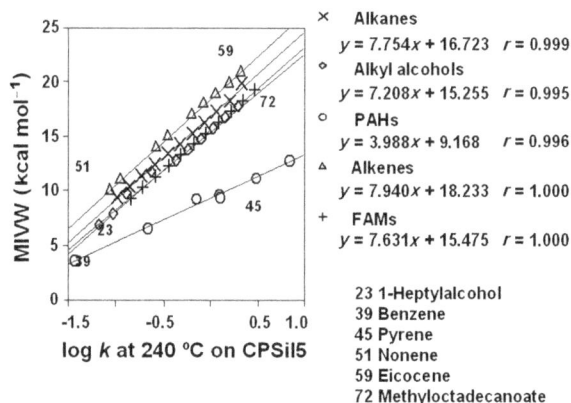

Figure 4.6 The relationship between log k (240 °C) and the molecular interaction energy (MIVW) on CPSil5.

The correlation coefficient of individual groups was close to 1. The order of the MIFS values was alkanes, alkenes > FAMs > alkyl alcohols ≫ PAHs, as seen in Figure 4.6.

The correlation between the van der Waals volume and the MIFS values demonstrated an excellent linear relation, excluding PAHs. The result indicated that the contact surface area for PAHs is important for obtaining higher MIFS values on an alkyl phase. Some of the methyl groups were therefore grown as hexyl groups to hold the analyte inside the model phase. If the relationship between alkane log k values and MIFS values is considered the standard for *in silico* retention analysis, and the retention is based on a hydrophobic interaction, the retention of other types of compounds should follow the relationship of the alkanes. The retention strength was as follows: PAHs > alkyl alcohols > alkanes > alkenes > FAMs, based on MIFS values. The longer alkyl phase did not support the strong retention of PAHs, and the MIFS values were lower than those of other compounds. The molecular interaction energy values depended on the contact surface area. The van der Waals energy values support the strength of hydrophobic retention. If their retention mechanisms are the same, all compounds should demonstrate the same linear relationship between their log k and the molecular interaction energy values observed in liquid chromatography. The chromatographic peaks of alkyl alcohols and amines showed tailing among the standard column test compounds (*e.g.*, Grob's mixture) on the DB1 methylsilicone phase. On the other hand, the molecular interaction energy values of PAHs and a model polyethylene glycol phase were high. The details are reported below. A similar silanol effect was observed in liquid chromatography using a graphitized carbon column. The findings using a graphitized carbon column synthesized from 100% organic materials, supported a theoretical approach for the quantitative analysis of the retention mechanism, but graphitized carbon synthesized from a porous silica gel demonstrated a silanol effect.[37] A model methylsilicone phase with free silanol groups was therefore constructed, similar to the silanol pinhole in the alkylsilane phase.

One silanol group produced MIHB, but the MIHB was not strong enough to increase the molecular interaction energy values of the PAHs for this retention prediction system. The slopes for alkenes, FAMs, alkanes, and alkyl alcohols were parallel (12.1–12.4), but the slope for PAHs was 5.0. The presence of only three silanol groups was not sufficient to obtain high hydrogen-bonding energy values for PAHs, even when the relationships between MIFS and log k values for alkenes and FAMs were similar, and the slopes for alkenes, FAMs, alkanes, and alkyl alcohols were parallel (11.0–12.5), and the slope for PAHs was low (5.8). Using a seven-silanol model phase, the correlation coefficient between the van der Waals volume and the van der Waals energy was excellent, $r = 0.995$ ($n = 49$), and the slopes for alkenes, FAMs, alkanes, and alkyl alcohols were 10.4–11.9, but the slope for PAHs was 7.7. Increasing the number of silanols increased the MIHB of the PAHs, thereby increasing the molecular interaction energy values. The measured enthalpy related well to the log k values, but the enthalpy of PAHs was lower than that of the other groups.[25] The measured enthalpy correlated well with each group of molecular interaction energy values. The molecular interaction energy values of PAHs were smaller than those of the other compounds, indicating that more PAHs were retained than expected, based on the measured enthalpy and calculated molecular interaction energy values. Therefore, a model phase with a valley of silanols was constructed. The calculated molecular interaction energy values are summarized in Table 1 of the Appendix (p. 278). The high correlation coefficient, $r = 0.996$ ($n = 49$), between the van der Waals volume and the van der Waals energy indicated that these compounds were in good contact with the model phase. The hydrogen-bonding energy values of PAHs increased, and the slope against log k became similar to those of other aliphatic compounds. The following relations were obtained:

MIFS of alkanes $= 10.035 \times (\log k) + 26.847, n = 11, r = 0.999,$

MIFS of alkyl alcohols $= 10.194 \times (\log k) + 25.517, n = 10, r = 0.999,$

MIFS of PAHs $= 9.941 \times (\log k) + 33.002, n = 8, r = 0.978,$

MIFS of alkenes $= 10.924 \times (\log k) + 30.116, n = 9, r = 0.998,$

MIFS of FAMs $= 10.887 \times (\log k) + 36.877, n = 11, r = 0.999.$

The slope for PAHs was approximately parallel with those of the other compounds. This result indicates that PAHs must contact the silanol groups, and their contact with alkyl groups was not well correlated with their log k values. This model phase performed better than the others, but the contribution of the hydrogen-bonding energy was a little low, and required an additional constant. The relationship between van der Waals energy and log k values indicated that the contribution of the hydrogen-bonding energy was an important factor in the retention, hence the existence of silanols clearly affected retention on the methylsilicone phase. The selectivity, especially the silanol effect, of a model phase can be easily observed from the molecular interaction energy values of PAHs. The molecular interaction energy values were relatively lower than those of aliphatic compounds on inert and hydrophobic phases, and high in the presence of silanol groups. The values depend on the contact surface area with the silanol groups. This tendency was observed for alkyl alcohols, but the energy value differed little from that of alkanes.

The relationship between van der Waals volume and van der Waals energy indicated that the slopes of each group were identical, and these compounds had the same relationship. The difference in the hydrogen-bonding energy values demonstrated a parallel relationship between the MIFS and van der Waals volume, except for PAHs whose slope was higher than the other compounds due to their hydrogen-bonding energy values. The relationships between enthalpy and log k values were similar to the MIFS and MIVW and log k values. Another methylsilicone phase, CPSil5, produced similar results, but there was little selectivity between alkyl alcohols and alkanes on the CPSil5 phase. If the adsorption mechanisms of a variety of compounds are the same, many silanol groups are required to obtain a good correlation with the retention time. Generally, the silanol groups are inactivated during the column manufacturing process. The vaporization mechanisms were further studied using a computational chemical method. In the above model system, retention time may correspond with the molecular interaction. The retention of PAHs requires many silanol groups. This requirement is unrealistic, even though there are silanol groups on the surface of fused silica tubes, polymethylsilicones, and their degradation products. The difference in molecular interaction energy of a variety of compounds from that of standard compound alkanes can be considered to be due to the volatility of the analyte (the stability of the analyte). *Volatility* refers to a reduction of the molecular density. Generally, molecules are clustered in solution or exist as small drops. It is difficult to calculate volatility directly from a liquid model. The simplest liquid model is a pair of molecules, therefore, a pair of molecules was constructed as a complex, and the structure was optimized using the MM2 program. The calculated final (optimized) structure, hydrogen-bonding, electrostatic, and van der Waals energies are summarized as *fsp*, *hbp*, *esp*, and *vwp* in Table 1 of the Appendix (p. 278).

The relationship between log k and the original enthalpy values[25] was similar to that between log k and the MIFS calculated *in silico* using different model hydrophobic phases such as a hydrocarbon phase, a methylhydrocarbon phase, and a methylsilicone phase. There was no specific selectivity between ΔH calculated using PM5 and AM1. The van der Waals energy was the main contributor to the molecular interaction. The smallest enthalpy and MIFS were obtained for PAHs, and the values increased as follows: PAHs $\ll$ alkyl alcohols (OHs) $\leq$ FAMs $<$ alkenes (Aks) $\approx$ alkanes (Als). The relationship between van der Waals volume and MIFS calculated for different hydrophobic phases such as a hydrocarbon phase, a methylhydrocarbon phase, and a methylsilicone phase, was similar. The enthalpy of PAHs was lowest, and for the others OHs $\geq$ FAMs $>$ Aks $\approx$ Als $\gg$ PAHs.

The retention (molecular interaction) force on methylsilicone phases should be a van der Waals force because methylsilicone phases are non-polar. The hydrogen-bonding (MIHB) and electrostatic (MIES) energy values contribute little to the molecular interaction energy value. The van der Waals energy values (MIVW) were therefore used for the quantitative analysis of retention time on methylsilicone phases instead of the optimized (final structure) energy values (MIFS). The van der Waals energy values (MIVW) are nearly equal to the final structure energy values (MIFS) on model methylsilicone phases. The following equations were obtained between log k and the MIVW values of alkanes at different column temperatures for CPSil5 and DB1 columns:

$$\text{MIVW} = 14.178 \times (\log k_{280\,°C}) + 34.366, \ r = 0.999, \ n = 17, \ \text{for CPSil5,}$$

$$\text{MIVW} = 13.517 \times (\log k_{260\,°C}) + 31.610, \ r = 1.000, \ n = 17, \ \text{for CPSil5,}$$

$$\text{MIVW} = 12.140 \times (\log k_{240\,°C}) + 28.603, \ r = 0.999, \ n = 18, \ \text{for CPSil5,}$$

$$\text{MIVW} = 10.209 \times (\log k_{220\,°C}) + 25.112, \ r = 0.997, \ n = 16, \ \text{for CPSil5,}$$

$$\text{MIVW} = 10.431 \times (\log k_{200\,°C}) + 23.214, \ r = 0.999, \ n = 14, \ \text{for CPSil5,}$$

$$\text{MIVW} = 8.747 \times (\log k_{180\,°C}) + 21.764, \ r = 0.999, \ n = 12, \ \text{for CPSil5,}$$

$$\text{MIVW} = 8.052 \times (\log k_{160\,°C}) + 19.753, \ r = 0.999, \ n = 11, \ \text{for CPSil5,}$$

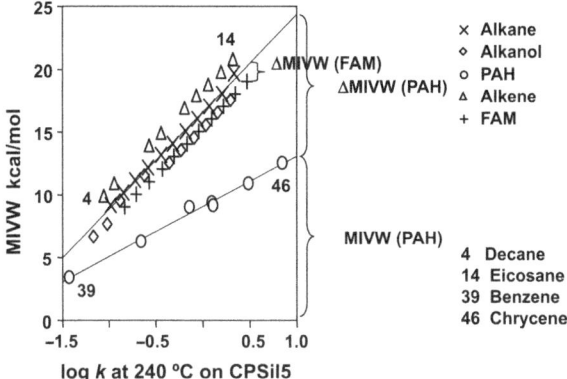

Figure 4.7 The relationship between log k (240 °C) and the molecular interaction energy (MIVW) on CPSil5. The difference in MIVW is indicated as ΔMIVW.

$$\text{MIVW} = 14.726 \times (\log k_{280\,°C}) + 38.158, \; r = 0.999, \; n = 16, \text{ for DB1,}$$

$$\text{MIVW} = 12.142 \times (\log k_{240\,°C}) + 31.901, \; r = 1.000, \; n = 16, \text{ for DB1,}$$

$$\text{MIVW} = 9.829 \times (\log k_{200\,°C}) + 26.604, \; r = 0.999, \; n = 15, \text{ for DB1,}$$

$$\text{MIVW} = 7.970 \times (\log k_{160\,°C}) + 21.989, \; r = 0.999, \; n = 11, \text{ for DB1.}$$

The difference (ΔMIVW) between the alkanes and the selected PAH (benzene, naphthalene or anthracene) was considered to be the volatility of the analyte, as shown in Figure 4.7.

ΔMIVW was defined as the vaporization energy (Δ_{vap}), and related to the stability of a pair of analytes, which was considered to be the simplest cluster model. That is, the energy necessary to disperse the individual molecules from the cluster. The final structure energy value of a pair of analytes, *fsp*, was used to obtain Δ_{vap}, but the Δ_{vap} of alkyl alcohols was small. Generally, cleaving a hydrogen bond requires more energy than the van der Waals force. The Δ_{vap} values of FAMs were slightly smaller than the expected values. The boiling points of molecules with hydrogen donor substituents are relatively high in comparison to their van der Waals volumes. A typical example is water, with two hydrogens, which has a high boiling point. The boiling points of molecules with hydrogen acceptors are a little higher. The contribution of the hydrogen-bonding and electrostatic energy values were therefore considered to increase Δ_{vap}. The Δ_{vap} values were obtained as a combination of *fsp*, $200 \times hbp$, and *esp*, instead of *fsp*.

$$\Delta\text{MIVW}_{280\,°C} = -0.175 \times (\Delta_{vap}) + 0.849, \; r = 0.989, \; n = 20, \text{ for CPSil5,}$$

$$\Delta\text{MIVW}_{260\,°C} = -0.173 \times (\Delta_{vap}) + 0.873, \; r = 0.995, \; n = 20, \text{ for CPSil5,}$$

$$\Delta\text{MIVW}_{240\,°C} = -0.160 \times (\Delta_{vap}) + 0.759, \; r = 0.990, \; n = 21, \text{ for CPSil5,}$$

$$\Delta\text{MIVW}_{220\,°C} = -0.146 \times (\Delta_{vap}) + 0.644, \; r = 0.983, \; n = 19, \text{ for CPSil5,}$$

$$\Delta\text{MIVW}_{200\,°C} = -0.146 \times (\Delta_{vap}) + 0.518, \; r = 0.983, \; n = 17, \text{ for CPSil5,}$$

$$\Delta\text{MIVW}_{180\,°C} = -0.152 \times (\Delta_{vap}) + 0.567, \; r = 0.997, \; n = 15, \text{ for CPSil5,}$$

$$\Delta\text{MIVW}_{160\,°C} = -0.148 \times (\Delta_{vap}) + 0.475, \; r = 0.997, \; n = 14, \text{ for CPSil5,}$$

$$\Delta MIVW_{280\,°C} = -0.180 \times (\Delta_{vap}) + 0.966, \, r = 0.995, \, n = 19, \text{ for DB1},$$

$$\Delta MIVW_{240\,°C} = -0.168 \times (\Delta_{vap}) + 0.823, \, r = 0.996, \, n = 19, \text{ for DB1},$$

$$\Delta MIVW_{200\,°C} = -0.161 \times (\Delta_{vap}) + 0.679, \, r = 0.996, \, n = 18, \text{ for DB1},$$

$$\Delta MIVW_{160\,°C} = -0.150 \times (\Delta_{vap}) + 0.537, \, r = 0.998, \, n = 14, \text{ for DB1}.$$

The calculated vaporization energy values at different temperatures were added to the original interaction van der Waals energy values (MIVW). The sum of MIVW and ΔMIVW correlated well with the measured log k values. The relationships are shown in the following equations. The relationship between MIVW + ΔMIVW and log k at 240 °C on CPSil5 is shown in Figure 4.8.

$$\log k_{280\,°C} = 0.070 \times (MIVW + \Delta MIVW) - 2.359, r = 0.993, n = 65, \text{ for CPSil5},$$

$$\log k_{260\,°C} = 0.078 \times (MIVW + \Delta MIVW) - 2.421, r = 0.995, n = 65, \text{ for CPSil5},$$

$$\log k_{240\,°C} = 0.085 \times (MIVW + \Delta MIVW) - 2.417, r = 0.992, n = 67, \text{ for CPSil5},$$

$$\log k_{220\,°C} = 0.096 \times (MIVW + \Delta MIVW) - 2.445, r = 0.995, n = 61, \text{ for CPSil5},$$

$$\log k_{200\,°C} = 0.101 \times (MIVW + \Delta MIVW) - 2.310, r = 0.995, n = 55, \text{ for CPSil5},$$

$$\log k_{180\,°C} = 0.119 \times (MIVW + \Delta MIVW) - 2.564, r = 0.995, n = 47, \text{ for CPSil5},$$

$$\log k_{160\,°C} = 0.129 \times (MIVW + \Delta MIVW) - 2.497, r = 0.994, n = 42, \text{ for CPSil5},$$

$$\log k_{280\,°C} = 0.070 \times (MIVW + \Delta MIVW) - 2.630, \, r = 0.994, \, n = 61, \text{ for DB1},$$

$$\log k_{240\,°C} = 0.085 \times (MIVW + \Delta MIVW) - 2.671, \, r = 0.994, \, n = 59, \text{ for DB1},$$

$$\log k_{200\,°C} = 0.105 \times (MIVW + \Delta MIVW) - 2.752, \, r = 0.996, \, n = 54, \text{ for DB1},$$

$$\log k_{160\,°C} = 0.128 \times (MIVW + \Delta MIVW) - 2.770, \, r = 0.994, \, n = 43, \text{ for DB1}.$$

This new method demonstrated an excellent correlation with the measured log k values. The correlation coefficient was 0.99, for all temperatures

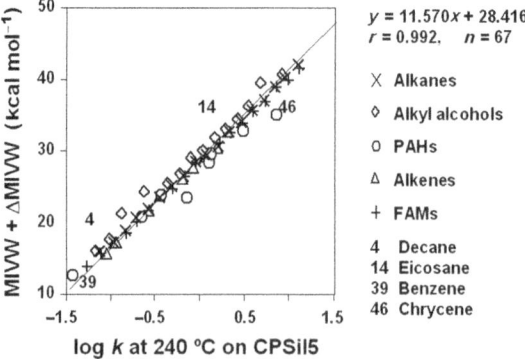

Figure 4.8 The relationship between log k (240 °C) and the calculated energy values (MIVW + ΔMIVW) on CPSil5.

on both CPSil5 and DB1 methylsilicone phases. Consideration of the vaporization mechanisms improves the results.

4.4   Retention on Carbowax™ (Polyethyleneglycol)

The construction of a polyethylene oxide phase was difficult, and the oxygen atoms were not homogenously distributed on the surface. A polyethylene oxide phase was therefore prepared as a model polyethyleneglycol phase. The model phase with anthracene is shown in Figure 4.9. The calculated molecular interaction energy values between the model polyethyleneglycol phase and the analytes are summarized in Table 2 of the Appendix (p. 286). The relationship between the log k values and the molecular interaction energy (MIFS) values was Als ≈ Aks > FAMs > OHs > PAHs. The slopes of the groups of the aliphatic compounds were parallel, and the slope of PAHs was the lowest. The MIFS did not relate simply to the log k values.

In general, boiling point and $\Delta_{vap}H$ cannot be predicted. The prediction of retention time in chromatography requires the use of predictable properties. The retention time can be calculated *in silico* using a model phase, but vaporization has not been quantitatively analyzed. In gas chromatography, vaporization may be related to analyte volatility. The retention on methylsilicone phases was quantitatively analyzed *in silico* as the molecular interaction energy values. In this model system, retention time may correspond with the molecular interaction as described for retention on the methylsilicone phase. The smallest MIFS was obtained for PAHs, and the values

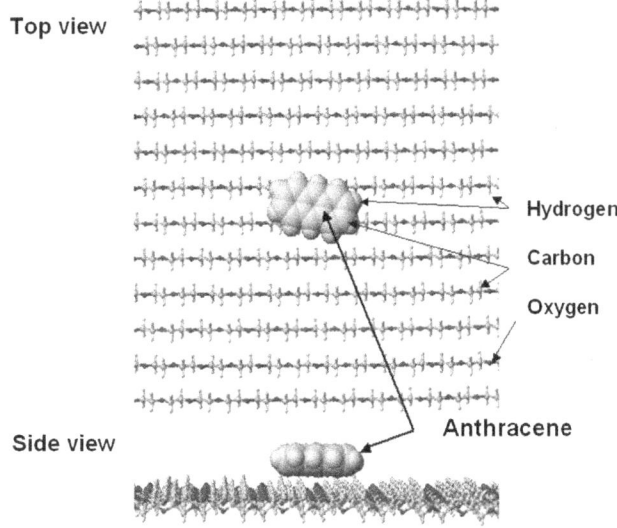

Figure 4.9   Adsorption of anthracene on a model polyethyleneglycol film. White, gray, and black balls represent hydrogen, carbon, and oxygen respectively. Reproduced by permission of Institute of Chromatography, ref. 38.

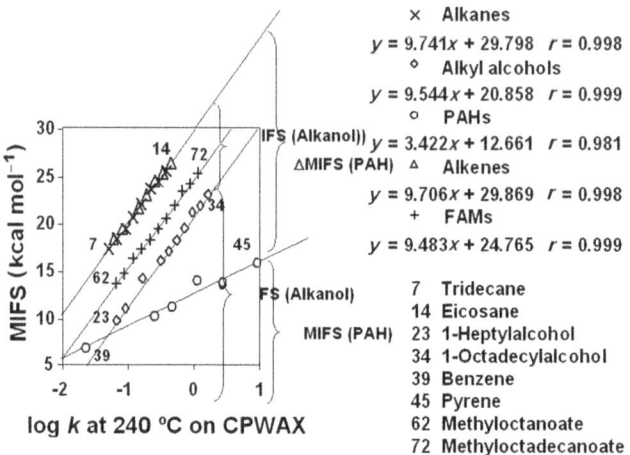

Figure 4.10 The relationship between $\log k$ (240 °C) and the calculated molecular interaction energy (MIFS) on CPWAX.

increased as follows: PAHs < OHs < FAMs < Aks ≈ Als, as shown in Figure 4.10. The comparison of Figures 4.7 and 4.10 indicates the selectivity of polyethyleneglycol phase.

However, the method of predicting retention time on polyethyleneglycol phases was somewhat complicated, therefore simplification of the prediction method was attempted. The stationary phase selectivity was included in the first correction factor by increasing the contribution of the hydrogen-bonding (*hbp*) and electrostatic (*esp*) energy values. The selectivity of the polar column was a little different, but the same constant was used to analyze the appropriateness of the new approach in the quantitative analysis of gas chromatographic retention times. Calculation of the retention mechanisms on polyethyleneglycol phases should include both hydrogen-bonding and electrostatic interactions. Therefore, MIFS was used instead of MIVW, which was used to evaluate chromatographic behavior on methylsilicone phases. The relationships obtained between the MIFS and $\log k$ values of alkanes are shown in the following equations:

$$\text{MIFS}_{240\,°C} = 10.757 \times (\log k) + 30.862, \; r = 0.999, \; n = 14, \text{ for CPWAX,}$$

$$\text{MIFS}_{200\,°C} = 8.829 \times (\log k) + 25.948, \; r = 0.999, \; n = 16, \text{ for CPWAX,}$$

$$\text{MIFS}_{240\,°C} = 10.695 \times (\log k) + 31.516, \; r = 0.998, \; n = 15, \text{ for DBWAX,}$$

$$\text{MIFS}_{200\,°C} = 9.055 \times (\log k) + 26.445, \; r = 0.999, \; n = 16, \text{ for DBWAX.}$$

The above methodology was applied to the chromatographic behavior of a polar phase, polyethyleneglycol (CPWAX, DVWAX), which demonstrates selectivity compared to non-polar methylsilicone phases. The difference in energy values (ΔMIFS) between the alkanes and the other groups was calculated using the following equations. The selected standard compounds

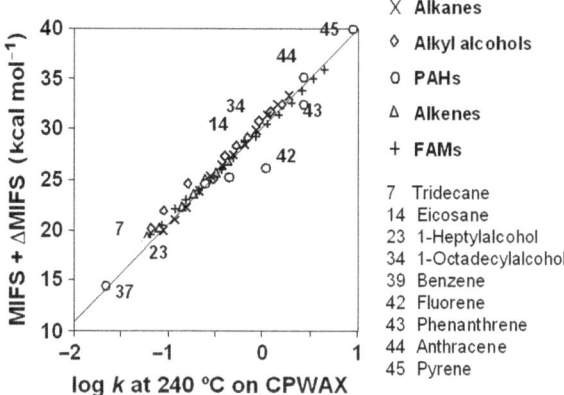

Figure 4.11 The relationship between log k (240 °C) and the calculated energy values (MIFS + ΔMIFS) on CPWAX.

were alkanes, benzene, naphthalene, and anthracene. The sum of *fsp*, 350*hbp*, and 2*esp* was used as Δ_{vap} for the calculation instead of *fsp*.

$$\Delta MIFS_{240\,°C} = -0.289 \times (\Delta_{vap}) + 1.557, r = 0.997, n = 17, \text{for CPWAX,}$$

$$\Delta MIFS_{200\,°C} = -0.290 \times (\Delta_{vap}) + 1.521, r = 0.993, n = 19, \text{for CPWAX,}$$

$$\Delta MIFS_{240\,°C} = -0.301 \times (\Delta_{vap}) + 1.561, r = 0.995, n = 18, \text{for DBWAX,}$$

$$\Delta MIFS_{200\,°C} = -0.287 \times (\Delta_{vap}) + 1.484, r = 0.997, n = 19, \text{for DBWAX.}$$

The calculated ΔMIFS values were added to MIFS values calculated using the model polyethyleneglycol phase. The following equations were obtained, and one example of a relationship is shown in Figure 4.11.

$$\log k_{240°C} = 0.105 \times (MIFS + \Delta MIFS) - 3.130, r = 0.986, n = 54, \text{for CPWAX,}$$

$$\log k_{200°C} = 0.127 \times (MIFS + \Delta MIFS) - 3.232, r = 0.988, n = 56, \text{for CPWAX,}$$

$$\log k_{240°C} = 0.106 \times (MIFS + \Delta MIFS) - 3.200, r = 0.984, n = 55, \text{for DBWAX,}$$

$$\log k_{200°C} = 0.128 \times (MIFS + \Delta MIFS) - 3.285, r = 0.981, n = 56, \text{for DBWAX.}$$

The total energy values demonstrated excellent correlation with the measured log k values. The above approach may therefore be applied to the chromatographic behavior of a variety of compounds on polar phases.[38]

4.5 Retention on a 50% Methylphenylsilicone Phase

The chromatographic behavior of a phenylsilicone phase (which is expected to demonstrate π–π interactions) was analyzed for its selectivity. A methylsilicone phase (OV1) was used as the reference column, and a 50%

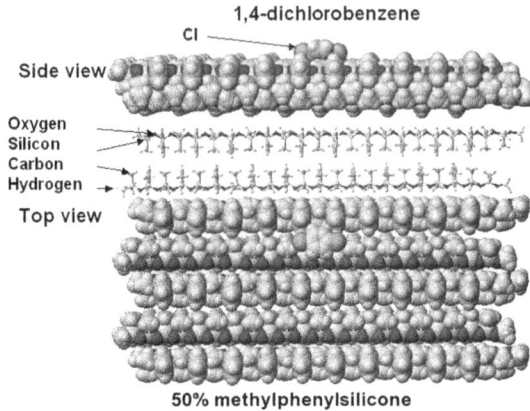

Figure 4.12   Adsorption of 1,4-dichlorobenzene on a model 50% methylphenylsilicone phase. Small white, large white, light-gray, black, and dark-gray
balls represent hydrogen, silicone, carbon, oxygen, and chlorine,
respectively.

methylphenylsilicone phase was the OV17. The analytes were alkanes, alkyl
alcohols, polymethylbenzenes (PMBZs), polychlorobenzenes (PCBZs), and
PAHs. The column temperature range was 140–230 °C and 160–230 °C for
the OV1 and OV17 columns, respectively.[39] PCBZs were selected to study π–π
interactions on OV17 phase.

A polyethylene oxide with 50% methyl and phenyl groups was prepared
and put side by side as a model 50% methylphenylsilicone phase (OV17).
The structure is shown in Figure 4.12, where 1,4-dichlorobenzene is adsorbed on the surface.

The calculated molecular interaction energy values between the model
phase and the analytes are summarized in Table 3 of the Appendix (p. 287).
The main retention (molecular interaction) forces were van der
Waals forces. The correlation between the log k values and the molecular
interaction energy values (MIVW) was closed to 1. The order of the
MIVW values was Als > OHs > alkylbenzenes (AlBZs) > PMBZs > PCBZs ≈ PAHs,
as seen in Figure 4.13. The behavior of PCBZs indicated that their MIVWs
were similar to those of PAHs, and those of PMBZs were similar to those
of AlBZs, as shown in Figure 4.13 where experimental data measured at 200 °C
is presented.

The alkane MIVW was used as the standard, and the difference of the
MIVW (ΔMIVW) for the other compounds was related to the optimized energy values of a pair of compounds. The ΔMIVW was defined as Δ_{vap}, and
was calculated as $fsp + 200hbp - 2esp$ where the *fsp*, *hbp*, and *esp* values are
summarized in Table 1 of the Appendix (p. 278) along with other properties.
The combined energy values of the molecular interaction (MIVW) and Δ_{vap}
were related to the log k values measured at 200 °C, as shown in Figure 4.14.
The results described in Sections 4.3 and 4.4 demonstrated an excellent
relationship between log k values and the combined energy values (MIVW

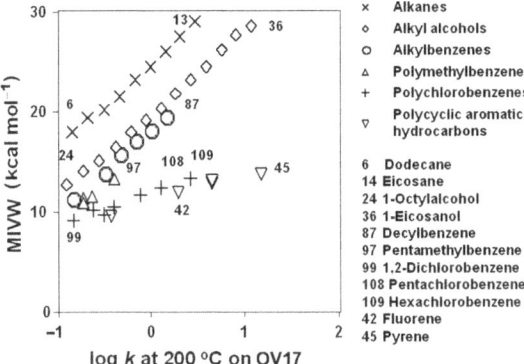

Figure 4.13 The relationship between log k (200 °C) and molecular interaction energy values (MIVW) on OV17.

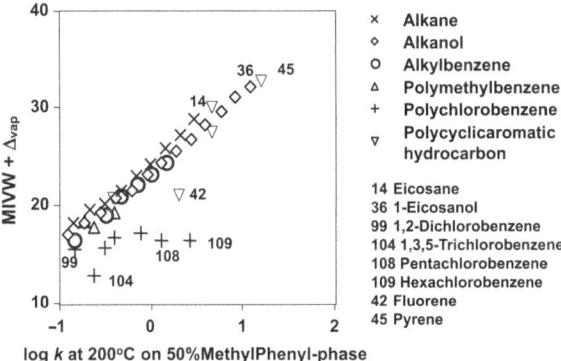

Figure 4.14 The relationship between log k (200 °C) and the calculated energy values (MIVW $+ \Delta_{\text{vap}}$) on OV17.

and Δ_{vap}) for Als, OHs, Aks, FAMs, and PAHs. However, although Als, OHs, AlBZs, PMBZs and PAHs demonstrated similar results, PCBZs demonstrated different behavior on a 50% methylphenylsilicone phase.

After surveying many molecular properties, the HOMO and LUMO energy calculated using the MOPAC PM5 program was found to be suitable for adjusting PCBZ specificity. The HOMO and LUMO energies of PCBZs demonstrated different behavior to those of PAHs. Finally, after the addition of the HOMO and LUMO energy values to the MIVW and Δ_{vap} as a secondary correction, the combined energy values correlated well with the measured log k values, as shown in Figure 4.15. The best combination of LUMO and HOMO was a 1 and 0.6 mixture.

The standard compounds used for calibration were Als, hexylalcohol, decylalcohol, butylbenzene, octylbenzene, 1,4-dichlorobenzene, penta-chlorobenzene, naphthalene and anthracene. These compounds were se-lected for use under a wide range of temperatures. However, the collection of

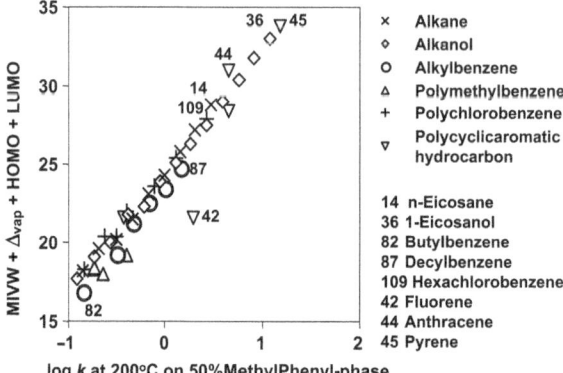

Figure 4.15   The relationship between log k (200 °C) and the sum of calculated energy values (MIVW $+ \Delta_{vap} +$ HOMO $+$ LUMO) on OV17.

all retention times under different temperatures was unnecessary. The combined energy value "TE" is the sum of molecular interaction energy (MIVW), the vaporization energy (Δ_{vap}), which is related to ($fsp + 200hsp - 2esp$), and the secondary correction related to (LUMO $+ 0.6 \times$ HOMO). The correlation is given in the following equations:

$$\text{TE}(160\,^\circ\text{C}) = 6.093 \times (\log k) + 20.404, r = 0.977, n = 53,$$

$$\text{TE}(170\,^\circ\text{C}) = 6.412 \times (\log k) + 21.139, r = 0.975, n = 46,$$

$$\text{TE}(180\,^\circ\text{C}) = 6.825 \times (\log k) + 22.273, r = 0.977, n = 47,$$

$$\text{TE}(190\,^\circ\text{C}) = 7.490 \times (\log k) + 23.237, r = 0.977, n = 43,$$

$$\text{TE}(200\,^\circ\text{C}) = 7.706 \times (\log k) + 23.240, r = 0.979, n = 44,$$

$$\text{TE}(210\,^\circ\text{C}) = 8.241 \times (\log k) + 26.593, r = 0.973, n = 36,$$

$$\text{TE}(220\,^\circ\text{C}) = 8.241 \times (\log k) + 26.593, r = 0.973, n = 36,$$

$$\text{TE}(230\,^\circ\text{C}) = 8.935 \times (\log k) + 27.492, r = 0.988, n = 32.$$

The retention times of these compounds can be predicted from the combination of TE and the linear relationship between log k and the MIVW of alkanes. The relationship between predicted log k (log k_{pred}) and measured log k (log k_{mes}) is given in the following equations:

$$\log k_{pred}(160\,^\circ\text{C}) = 0.908 \times (\log k_{mes}) - 0.036, r = 0.977, n = 52,$$

$$\log k_{pred}(170\,^\circ\text{C}) = 0.921 \times (\log k_{mes}) - 0.007, r = 0.987, n = 46,$$

$$\log k_{pred}(180\,^\circ\text{C}) = 0.916 \times (\log k_{mes}) - 0.079, r = 0.970, n = 47,$$

$$\log k_{pred}(190\,^\circ\text{C}) = 0.977 \times (\log k_{mes}) - 0.048, r = 0.982, n = 43,$$

$$\log k_{pred}(200\,^\circ\text{C}) = 0.935 \times (\log k_{mes}) - 0.080, r = 0.974, n = 44,$$

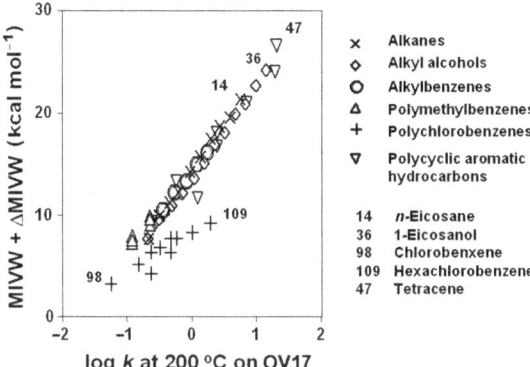

Figure 4.16 The relationship between log k (200 °C) and the calculated energy values (MIVW + ΔMIVW) on OV17.

$$\log k_{\text{pred}} (210\,°C) = 0.879 \times (\log k_{mes}) - 0.085, \ r = 0.960, \ n = 36,$$
$$\log k_{\text{pred}} (220\,°C) = 0.891 \times (\log k_{mes}) - 0.079, \ r = 0.965, \ n = 36,$$
$$\log k_{\text{pred}} (230\,°C) = 0.869 \times (\log k_{mes}) - 0.085, \ r = 0.987, \ n = 32,$$

PCBZs demonstrated different behavior from other compounds on the 50% methylphenylsilicone phase. The behavior of PCBZs, indicated by ΔMIVW, was similar to that of PAHs and that of PMBZs, and PMBZs were similar to alkyl alcohols, as shown in Figure 4.16, where experimental data measured at 200 °C is presented. The alkane MIVW was used as the standard, and the difference of the MIVW for other compounds was related to the optimized energy values for a pair of compounds. The combined energy values of molecular interaction (MIVW) and Δ$_{\text{vap}}$ were related to log k values measured at 200 °C, as shown in Figure 4.16.

A combination of the HOMO and LUMO energies was applied. The correlations at different temperatures are summarized in the following equations:

$$\text{TE} (140\,°C) = 6.608 \times (\log k) + 7.395, \ r = 0.979, \ n = 53,$$
$$\text{TE} (150\,°C) = 6.802 \times (\log k) + 8.455, \ r = 0.981, \ n = 55,$$
$$\text{TE} (160\,°C) = 7.204 \times (\log k) + 9.624, \ r = 0.980, \ n = 53,$$
$$\text{TE} (170\,°C) = 7.197 \times (\log k) + 10.880, \ r = 0.977, \ n = 52,$$
$$\text{TE} (180\,°C) = 7.200 \times (\log k) + 12.211, \ r = 0.984, \ n = 55,$$
$$\text{TE} (190\,°C) = 7.506 \times (\log k) + 11.716, \ r = 0.977, \ n = 57,$$
$$\text{TE} (200\,°C) = 8.563 \times (\log k) + 13.837, \ r = 0.987, \ n = 52,$$
$$\text{TE} (210\,°C) = 9.039 \times (\log k) + 14.744, \ r = 0.985, \ n = 48,$$
$$\text{TE} (220\,°C) = 9.901 \times (\log k) + 14.861, \ r = 0.981, \ n = 41,$$
$$\text{TE} (230\,°C) = 10.614 \times (\log k) + 17.148, \ r = 0.984, \ n = 39.$$

The retention times of these compounds can again be predicted by the combination of TE and the linear relationship between log k and the MIVW of alkanes. The relationship between predicted log k (log k_{pred}) and measured log k (log k) is given in the following equations:

$$\log k_{pred} (140\,°C) = 0.905 \times (\log k) - 0.005,\ r = 0.979,\ n = 53,$$
$$\log k_{pred} (150\,°C) = 0.898 \times (\log k) - 0.008,\ r = 0.981,\ n = 55,$$
$$\log k_{pred} (160\,°C) = 0.893 \times (\log k) + 0.016,\ r = 0.980,\ n = 53,$$
$$\log k_{pred} (170\,°C) = 0.943 \times (\log k) - 0.024,\ r = 0.977,\ n = 52,$$
$$\log k_{pred} (180\,°C) = 0.907 \times (\log k) - 0.004,\ r = 0.984,\ n = 55,$$
$$\log k_{pred} (190\,°C) = 0.870 \times (\log k) - 0.019,\ r = 0.979,\ n = 57,$$
$$\log k_{pred} (200\,°C) = 0.879 \times (\log k) - 0.159,\ r = 0.987,\ n = 52,$$
$$\log k_{pred} (210\,°C) = 0.922 \times (\log k) - 0.052,\ r = 0.985,\ n = 48,$$
$$\log k_{pred} (220\,°C) = 0.931 \times (\log k) - 0.136,\ r = 0.981,\ n = 41,$$
$$\log k_{pred} (230\,°C) = 0.945 \times (\log k) - 0.045,\ r = 0.984,\ n = 39.$$

The behavior of fluorene was different to the other polycyclic aromatic hydrocarbons. However, the behavior of MIVW was similar to that of the others. When the electrostatic energy was considered, a discrepancy appeared involving the HOMO and LUMO energies. It seems that the existence of an sp^3 orbital on a carbon in the aromatic carbon ring caused difficulties in calculating the resonance energy. The elimination of fluorene improved the correlation coefficient to 0.98. One error source was the measured log k values. Elimination of those log k values less than -1 improved the slopes and precision.

4.6   Classification of Gas Chromatography

The retention times of various compounds on methylsilicone, polyethyleneglycol, and 50% methylphenylsilicone phases can be predicted based on the combination of molecular interaction energies and the molecular properties of the analytes. The molecular interaction energy was calculated from a model phase and an analyte complex. The molecular property was the optimized structure energy value, calculated from a pair of compounds. The standard chemicals for a column calibration are alkanes, benzene, naphthalene, anthracene, hexyl- and decylalohols, butyl- and octylbenzenes, and 1,4-dichloro- and pentachlorobenzenes for polyethyleneglycol phases. The retention mechanisms in liquid and gas chromatography are the same, but the desorption mechanisms are different. Vaporization occurs in gas chromatography, and solvation is necessary in liquid chromatography. The adsorption forces are a combination of solubility factors based on the adsorption (stationary) phase and the analytes. van der Waals forces are predominant on non-polar phases. Hydrogen bonding and electrostatic forces contribute to polar phases. Ion–ion interactions should be considered on ionic phases. The final interaction type is represented by the electrostatic energy (Figure 4.17).

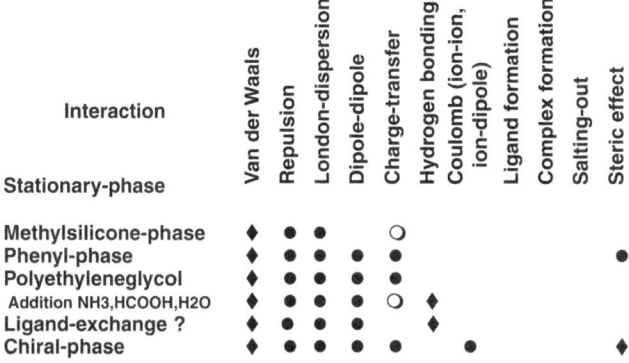

Chemistry in Gas Chromatography

Interaction / Stationary-phase	Van der Waals	Repulsion	London-dispersion	Dipole-dipole	Charge-transfer	Hydrogen bonding	Coulomb (ion-ion, ion-dipole)	Ligand formation	Complex formation	Salting-out	Steric effect
Methylsilicone-phase	◆	●	●			○					
Phenyl-phase	◆	●	●	●	●						●
Polyethyleneglycol	◆	●	●	●	●						
Addition NH3,HCOOH,H2O	◆	●	●	●		○		◆			
Ligand-exchange ?	◆	●	●	●				◆			
Chiral-phase	◆	●	●	●	●				●		◆

◆ The most imporant, ● important, ○ depending on phases.

Figure 4.17 Gas chromatography retention mechanisms based on solubility factors.

Gas chromatography retention mechanisms can be summarized in Figure 4.17 like that of liquid chromatography. The difference between gas and liquid chromatography is the desorption mechanisms. Desorption is evaporation in gas chromatography, and solvation in liquid chromatography. Evaporation is directly related to the analyte properties, and solvation depends on solvent properties that we cannot quantitatively obtain. Further study is therefore required to determine quantitative solvation mechanisms.

Relative retention times in gas chromatography using an inert gas as the mobile phase, can be quantitatively estimated as a combination of molecular interaction, evaporation, and inductive effects. The retention mechanisms in liquid and gas chromatography are the same, but the desorption mechanisms are different. However, solvent interaction in so-called "chemical gas chromatography" should be considered. The molecular interaction forces are a combination of solubility factors based on the adsorption (stationary) phase and the analytes, as well as on the analytes and components of the mobile phase. Although van der Waals forces are predominant in gas chromatography, hydrogen bonding contributes to the retention of alcoholic compounds, and electrostatic forces cannot be neglected for aromatic compounds. The balance of the HOMO and LUMO energies contributes to the inductive effects of substituents.

References

1. G. Zenkevich, New applications of the retention index concept in gas and high performance liquid chromatography, *Fresenius' J. Anal. Chem.*, 1999, **365**, 305–309.

2. K. D. Bartle, M. L. Lee and S. A. Wise, Factors affecting the retention of polycyclic aromatic hydrocarbons in gas chromatography, *Chromatographia*, 1981, **14**, 69–72.
3. H. Heberger and T. Kowalska, Thermodynamic properties of alkylbenzenes from retention boiling point correlations in gas chromatography, *Chromatographia*, 1997, **44**, 179–186.
4. I. G. Zenevich, Calculation of gas chromatographic retention indices of organic compounds from the boiling points of their structural analogs, *J. Struct. Chem.*, 1999, **40**, 101–107.
5. D. N. Grigoreva, R. V. Golovnya, L. A. Semina and A. L. Samusenko, Applicability of an universal equation in capillary GC for calculating the retention indexes and boiling points of members of O-, N-, and S-containing homologous series, *Izv. Akad. Nauk, Ser. Khim.*, 1989, **2**, 301–306.
6. I. B. Golovanov and S. M. Zhenodarova, Quantitative structure–property relationship: XXII. Polycyclic aromatic hydrocarbons, *Russ. J. Gen. Chem.*, 2005, **75**, 1790–1794.
7. E. V. Sagadeev and V. P. Barabanov, Calculations of the enthalpies of combustion of organic compounds by the additive scheme, *Russ. J. Phys. Chem.*, 2006, **80**, S152–S162.
8. E. C. Dose, Simulation of gas chromatographic retention and peak width using thermodynamic retention indexes, *Anal. Chem.*, 1987, **59**, 2414–2419.
9. K. Heberger and T. Kowalska, Determination of heats of vaporization and Gibbs free energies of alkylbenzenes on GC stationary phases of different polarity, *Chromatographia*, 1998, **48**, 89–94.
10. S. P. Verevkin and A. Heintz, Determination of vaporization enthalpies of the branched esters from correlation gas chromatography and transpiration methods, *J. Chem. Eng. Data*, 1999, **44**, 1240–1244.
11. S. P. Verevkin, E. L. Krasnykh, T. V. Vasiltsava and A. Heintz, Determination of ambient temperature vapor pressures and vaporization enthalpies of branched ethers, *J. Chem. Eng. Data*, 2003, **48**, 591–599.
12. A. Wong, Y. D. Lei, M. Alaee and F. Wania, Vapor pressures of the polybrominated diphenyl ethers, *J. Chem. Eng. Data*, 2001, **46**, 239–242.
13. Y. D. Lei, R. Chankalal, A. Chan and F. Wania, Supercooled liquid vapor pressures of the polycyclic aromatic hydrocarbons, *J. Chem. Eng. Data*, 2002, **47**, 801–806.
14. E. L. Krasnykh, T. V. Vasiltsova, S. P. Verevkin and A. Heintz, Vapor pressures and enthalpies of vaporization of benzyl halides and benzyl ethers, *J. Chem. Eng. Data*, 2002, **47**, 1372–1378.
15. B. T. Mader and J. F. Pankow, Vapor pressures of the polychlorinated dibenzodioxins (PCDDs) and the polychlorinated dibenzofurans (PCDFs), *Atomos. Environ.*, 2003, **37**, 3103–3114.
16. T. F. Bidieman, A. D. Leone and R. L. Falconer, Vapor pressures and enthalpies of vaporization for toxaphene congeners, *J. Chem. Eng. Data*, 2003, **48**, 1122–1127.

17. J. S. Chickos and W. Hanshaw, Vapour pressures and vaporization enthalpies of the *n*-alkanes from C_{21} to C_{30} at $T = 298.15$ K by correlation gas chromatography, *J. Chem. Eng. Data*, 2004, **49**, 77–85.

18. J. S. Chickos and W. Hanshaw, Vapour pressures and vaporization enthalpies of the *n*-alkanes from C_{31} to C_{38} at $T = 298.15$ K by correlation gas chromatography, *J. Chem. Eng. Data*, 2004, **49**, 620–630.

19. V. N. Emel'yanenco, S. P. Verevkin, B. Koutek and J. Doubsky, Vapour pressures and enthalpies of vaporization of a series of the linear aliphatic nitriles, *J. Chem. Thermodyn.*, 2005, **37**, 73–81.

20. M. V. Roux, M. Temprado and J. S. Chickos, Vaporization, fusion and sublimation enthalpies of the dicarboxylic acids from C_4 to C_{14} and C_{16}, *J. Chem. Thermodyn.*, 2005, **37**, 941–953.

21. A. I. Canizo, G. N. Eyler and G. P. Barreto, Determination of the enthalpies of vaporization of cyclic organic peroxides by correlation of changes in gas chromatographic retention times, *Chromatographia*, 2007, **65**, 31–34.

22. K. Ciazynska-Halarewicz, M. Helbin, P. Korzenecki and T. Kowaiska, Mathematical models of solute retention in gas chromatography as sources of thermodynamic data. Part IV. Aliphatic alcohols as the test analytes, *J. Chromatogr. Sci.*, 2007, **45**, 492–499.

23. J. J. H. Haftka, J. R. Parsons and H. A. J. Govers, Supercooled liquid vapour pressures and related thermodynamic properties of polycyclic aromatic hydrocarbons determined by gas chromatography, *J. Chromatogr., A*, 2006, **1135**, 91–100.

24. J. M. Santiuste, Contribution to the study of solute-stationary phase retention interactions in terms of activity coefficients obtained by gas-liquid chromatography, *Anal. Chim. Acta*, 2001, **441**, 63–72.

25. T. Hanai, Effect of enthalpy on structure-relation correlation in capillary gas chromatography, *J. High Resolut. Chromatogr.*, 1990, **13**, 178–181.

26. E. Kováts, Gas-chromatographische Charakterisierung Organischer Verbindungen, Teil 1; Retentionsindics aliphatischer Halogenide, alkohole, Aldehyde und Ketone, *Helv. Chim. Acta*, 1958, **41**, 1915–1932.

27. G. Castello, Retention index systems: alternatives to the *n*-alkanes as calibration standard, *J. Chromatogr., A*, 1999, **842**, 51–64.

28. C. D. Wick, J. I. Siepmann, W. L. Klotz and M. R. Schure, Temperature effects on the retention of *n*-alkanes and arenas in helium-squalane gas-liquid chromatography experiment and molecular simulation, *J. Chromatogr., A*, 2002, **954**, 181–190.

29. L. Sun, J. I. Siepmann, W. L. Klotz and M. R. Schure, Retention in gas-liquid chromatography with a polyethylene oxide stationary phase: Molecular simulation and experiment, *J. Chromatogr., A*, 2006, **1126**, 373–380.

30. N. E. Moustafa, Prediction of GC retention times of complex petroleum fractions based on quantitative structure-retention relationships, *Chromatographia*, 2008, **67**, 85–91.

31. C. T. Peng, Prediction of retention indices. VI: Isothermal and temperature-programmed retention indices, methylene value, functionality constant, electronic and steric effect, *J. Chromatogr., A*, 2010, **1217**, 3683–3694.

32. Spelco data from Fig. 37 of the *Superco GC Catalogue*, Sigma-Aldrich, Bellefont, PA, 1999, p. 29.

33. Spelco data from Fig. 39 of the *Superco Bulletin: Packed column GC Application Guide*, Sigma-Aldrich, Bellefont, PA, 1999, vol. 890A.

34. Spelco data from Fig. 94 of the *Superco GC Catalogue*, Sigma-Aldrich, Bellfonte, PA, 1999, p. 48.

35. D. R. Lide, *CRC Handbook of Chemistry and Physics*, CRC Press, Boca Raton, 75th edn, 1994.

36. T. Hanai and C. Hong, Structure-retention correlation in CGC, *J. High Resolut. Chromatogr.*, 1989, **12**, 327–332.

37. T. Hanai and H. Homma, Quantitative *in silico* analysis of the selectivity of graphitic carbon synthesized by different methods, *Anal. Bioanal. Chem.*, 2008, **390**, 369–375.

38. T. Hanai, Quantitative *in silico* analysis of retention time on methylsilicone and polyethyleneglycol phases in capillary gas chromatography. http://www.internet-chromatography.com/html/toshihikbeitrage.html.

39. Y. Yamane, K. Miyaji, K. Hanafusa, T. Hanai and H. Hatano, Structure-retention correlation on methylphenylpolysiloxane phases in capillary gas chromatography, *Bull. Chem. Soc. Jpn.*, 1993, **66**, 1881–1885.

40. T. Hanai, Quantitative *in silico* analysis of the specificity of graphitized (graphitic) carbons, *Adv. Chromatogr.*, 2011, **49**, 257–290.

CHAPTER 5

Retention in Normal-Phase Liquid Chromatography

5.1 Retention of Saccharides on Graphitized Carbon

Since Samuelson *et al.* separated saccharides using the combination of an ion-exchange resin and aqueous ethanol in 1965,[1] a variety of combined systems have been used for their liquid chromatographic separation. The retention was considered to depend on the nature and position of the substituents, and the degree of adsorption was, at least partially, related to the electron-withdrawing properties of the substituents. The results suggested that adsorption occurred by the formation of either charge transfer or hydrogen bonding between the analyte and the anion on the resin. A variety of packing materials, ion exchangers, bonded-phase silica gels, and organic polymers have been used for the chromatography of saccharides in organic solvent mixtures. The retention times on a graphitized phase measured in water were obtained from the literature[2] and are collected in Table 5.1. A 22-ring PAH was used as a model graphitized carbon phase, and the molecular interaction energy between this model phase and a standard compound was calculated using the MM2 program. The values are also summarized in Table 5.1. The optimized energies of the complexes calculated using the MOPAC-AM1 program are summarized in Table 5.2. The MM2 calculation of one pair of molecular interactions was complete within a few minutes, but the MOPAC calculation required longer to optimize the complex form.

The correlation coefficient between these retention times and the final structure energy (MIFS) was 0.906, $n = 6$. The correlation coefficients between the retention times and the hydrogen-bonding (MIHB) and electrostatic energies (MIES) were 0.742 and 0.752, respectively. However, the correlation

RSC Chromatography Monographs No. 19
Quantitative *In Silico* Chromatography: Computational Modelling of Molecular Interactions
By Toshihiko Hanai
© Toshihiko Hanai, 2014
Published by the Royal Society of Chemistry, www.rsc.org

Table 5.1 Molecular interaction energy values for saccharides on graphitized carbon calculated using the MM2 program. MIFS, MIVW, MIHB and MIES represent the molecular interaction energy value of the final (optimized) structure, the van der Waals energy, the hydrogen-bonding energy, and the electrostatic energy (kcal mol^{-1}), respectively. t_R represents the retention time. Reproduced by permission of Elsevier, ref. 24.

Saccharide	MIFS	MIVW	MIHB	MIES	t_R/min
β-Glucose	32.825	3.415	33.711	19.831	2.40
β-Galactose	39.577	6.903	40.627	17.348	2.60
α-Glucose	39.405	6.094	37.769	34.864	2.62
α-Galactose	43.389	5.434	39.738	33.195	3.05
Ribose	27.553	5.128	25.891	13.618	2.30
β-Mannose	35.033	6.515	30.271	19.071	2.50

Table 5.2 Contributions to the optimized energies of the complexes calculated using the MOPAC program. $\Delta_f H$ represents the heat of formation, TE the total energy, EE the electronic energy, CC the core–core repulsion and IP the ionization potential. Reproduced by permission of Elsevier, ref. 24.

Saccharide	$\Delta_f H$/kcal	ΔTE/eV	ΔEE/eV	ΔCC/eV	ΔIP
β-Glucose	−230.2	−10.0	31566.3	−31576.3	11.1
β-Galactose	−225.5	−9.8	31128.5	−31138.2	11.2
α-Glucose	−235.6	−10.2	33320.5	−33330.7	11.2
α-Galactose	−241.7	−10.5	24757.5	−24768.0	11.2
Ribose	−231.8	−10.0	20782.6	−20792.7	11.4
β-Mannose	−233.8	−10.1	30751.3	−30761.5	11.3

between the retention times and van der Waals energy (MIVW) was 0.171. These results indicate that hydrogen-bonding and electrostatic energy contributed to the retention of saccharides on a graphitized carbon in liquid chromatography, as shown in Figure 5.1. The van der Waals energy values were nearly equal and no contribution to retention was observed.

The MOPAC calculations did not indicate a specific contribution to the molecular interaction. However, the heat of formation ($\Delta_f H$) was correlated to the retention times of the saccharides. The correlation coefficient between the retention times and $\Delta_f H$ or the total energy was 0.684 ($n = 6$).

Molecular mechanics calculations are useful for studying retention order in chromatography. Although the MOPAC calculations did not clearly indicate the contributing force to the molecular interaction, they identified the molecular interaction center on an electron density map constructed by the tabulator of the CAChe™ program, and the change of atomic partial charge before and after the optimization of the complex form. The hydrocarbons

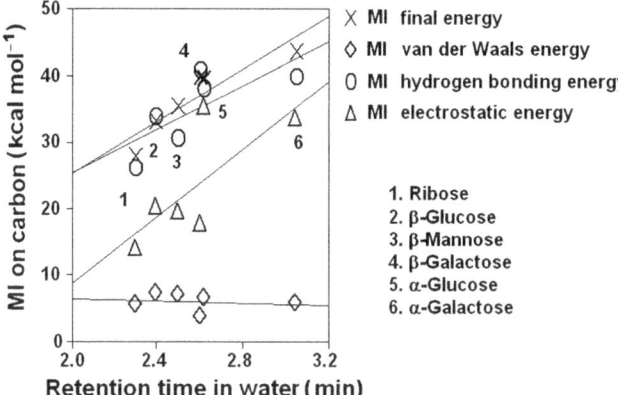

Figure 5.1 The relationship between retention time and molecular interaction energy values on graphitized carbon.
Reproduced by permission of Elsevier, ref. 24.

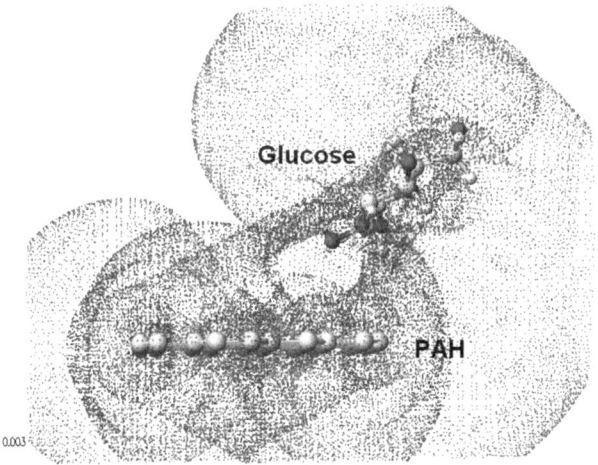

Figure 5.2 The electrostatic potential of a PAH–β-glucose complex.

interacted at the center of the PAH molecule, and the carbohydrates interacted at the edge of the PAH molecule *via* hydrogen bonding, as seen in Figure 5.2.

The atomic partial charge of the model phase and these hydrocarbons did not change significantly after optimization of the complex form, but the electron potential was slightly shifted toward the molecular interaction side. On the other hand, a carbohydrate was first put on the center of the model phase, but was shifted to the edge of model phase after the optimization. The result was the same for all compounds used. The atomic partial charge of atoms at hydrogen-bonding sites was changed by about 0.01 eV. These results indicated clearly the existence of different retention mechanisms on graphitic carbon phases *i.e.*, both hydrophobic and electrostatic mechanisms.

5.2 Retention of Aromatic Compounds on Graphitized Carbon

The chromatographic behavior of aromatic compounds on a graphitized carbon phase was studied using reference and measured retention times and a model graphitized carbon phase. The analytes used were simple aromatic compounds, similar to those used in a previous study,[3] and are summarized in Table 5.3. The eluent was *n*-heptane. *n*-Hexane was used at first, but the retention times of these compounds were very short, and handling of the eluent was complicated by its volatility. The retention times are summarized in Table 5.3. A model graphitic carbon phase was constructed using a molecular editor from the CAChe™ program. The molecular size of the model phase was decided based on the molecular size of the analytes. The phase was a 22-aromatic-ring PAH, as shown in Figure 5.3 where 2,3,4,6-tetrachlorophenol is adsorbed. The calculated energy values are summarized in Table 5.3.

The coefficient of correlation between the calculated individual molecular interaction (MI) energies and the logarithmic capacity ratio indicated the contribution of individual factors to the retention. MIVW was the main contributor to the interaction in reversed-phase liquid chromatography,[4] and MIES was the main contributor to the retention in ion-exchange liquid chromatography.[5] Steric hindrance affected the molecular interaction in enantiomeric separation.[6]

MI energy values were correlated with measured (log k_{mes}) and reference (log k_{ref}) log k values, and the results are given in the following equations:

$$\log k_{mes} = 0.239 \times (MIFS) - 4.000, \ r = 0.908, \ n = 14,$$
$$\log k_{mes} = 0.128 \times (MIHB) - 1.076, \ r = 0.504, \ n = 14,$$
$$\log k_{mes} = -92.767 \times (MIES) - 0.653, \ r = 0.384, \ n = 14,$$
$$\log k_{mes} = 0.168 \times (MIVW) - 2.296, \ r = 0.563, \ n = 14,$$

$$\log k_{ref} = 0.053 \times (MIFS) - 0.913, \ r = 0.362, \ n = 18,$$
$$\log k_{ref} = 0.056 \times (MIHB) - 0.411, \ r = 0.482, \ n = 18,$$
$$\log k_{ref} = -0.193 \times (MIES) - 0.162, \ r = 0.021, \ n = 18,$$
$$\log k_{ref} = -0.077 \times (MIVW) + 0.619, \ r = 0.296, \ n = 18.$$

The correlation coefficients for log k_{ref} were very poor compared with the results for log k_{mes}. MIFS contributed to the molecular interaction, but MIHB, MIES, and MIVW did not contribute to the main interaction force. The energy values of nitro-substituted compounds had to be optimized for

Table 5.3 Molecular properties and capacity ratios measured on different graphitized carbon columns. FS, HB, ES, and VW represent the energy value of the final (optimized) structure, the hydrogen-bonding energy, the electrostatic energy, and the van der Waals energy of the complexes (kcal mol^{-1}), respectively. fs, hb, es, and vw represent the same energies for each analyte. Reproduced by permission of Springer, ref. 25.

Analyte	FS	HB	ES	VW	fs	hb	es	vw	log k$_{ref}$[a]	log k$_{mes}$[b]
Acetophenone	18.5112	−9.147	−0.001	62.219	−4.5108	−1.459	0.000	5.846	0.283	—
Anisole	21.3212	0.000	0.000	57.841	−6.4954	0.000	0.000	3.301	−0.317	—
Benzene	22.2027	0.000	0.000	60.045	−8.0770	0.000	0.000	3.006	−1.060	−1.846
Benzylalcohol	16.8235	−13.144	−0.941	58.994	−9.0400	−0.606	−0.708	3.481	−0.211	—
Carvacrol	13.4350	−9.563	−0.495	61.248	−9.5419	−1.485	−0.500	4.146	−0.194	—
Chlorobenzene	20.8825	0.000	0.000	58.658	−7.8015	0.000	0.000	3.213	−0.606	—
p-Cresol	13.6417	−8.612	0.017	60.310	−10.7545	−1.459	0.017	2.974	0.456	−0.620
Ethylbenzene	20.1366	0.000	0.000	57.713	−6.7841	0.000	0.000	4.051	—	−1.557
Nitrobenzene	19.7300	0.000	0.000	59.149	−8.4983	0.000	0.000	4.161	0.799	−0.526
Phenol	15.0378	−10.801	0.000	64.145	−10.2088	−1.462	0.000	2.957	0.431	−1.076
2-tert-Butylphenol	7.2615	−10.224	−0.539	64.175	−5.1559	−1.537	−0.540	6.475	−0.419	—
4-tert-Butylphenol	15.4805	−9.053	0.017	60.855	−6.3768	−1.452	0.017	5.426	−0.022	—
4-thylphenol	13.9452	−8.532	0.017	59.832	−8.9665	−1.456	0.017	3.983	0.277	—
3,4-Dimethyl-phenol	12.3997	−7.992	0.171	58.725	−10.5499	−1.460	0.171	3.414	—	−0.024
2,4,6-Trimethyl-phenol	10.4488	−5.338	−1.376	56.221	−11.7192	−1.364	−1.379	4.006	—	−0.374
4-Propylphenol	12.9862	−8.328	0.017	58.357	−8.2767	−1.456	0.017	4.470	0.312	—
4-Chlorophenol	14.3741	−9.076	−0.088	61.057	−10.0237	−1.463	−0.087	3.153	—	−0.642
2,4-Dichloro-phenol	13.3258	−9.101	−0.178	59.968	−9.7063	−1.516	−0.177	3.567	—	−0.283
2,4,6-Trichloro-phenol	10.8876	−5.519	−3.198	57.343	−12.6811	−1.961	−3.207	4.095	—	−0.102
2,3,4,6-Tetrachloro-phenol	17.2990	−5.612	2.403	57.696	−4.5515	−1.980	2.393	6.099	—	0.456
4-Chloro-2-methyl-phenol	12.8184	−7.876	−0.066	58.884	−10.2495	−1.485	−0.069	3.403	—	0.058
4-Phenylphenol	1.3839	−8.571	0.000	61.004	−16.1472	−1.459	0.000	10.989	—	1.246
Thymol	17.2627	−4.993	−0.419	59.669	−7.5324	−1.302	−0.419	4.961	−0.148	—
Toluene	19.7250	0.000	0.000	58.153	−8.6003	0.000	0.000	3.036	−0.791	−1.721
o-Xylene	18.5017	0.013	0.000	57.130	−8.3949	0.000	0.130	3.498	−0.433	—
m-Xylene	17.1997	0.000	−0.125	56.343	−9.2696	0.000	−0.125	3.047	−0.666	—
p-Xylene	17.2987	0.000	0.024	56.298	−9.1330	0.000	0.024	3.042	−0.658	—
Carbon phase	—	—	—	—	39.0519	0.000	0.000	65.823	—	—

[a]Measured on a Hypercarb[3,4] column.
[b]Measured on a BTR column.

Side view   Top view

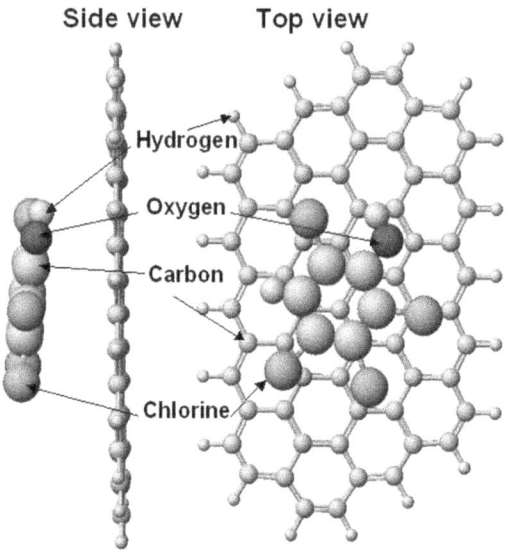

Figure 5.3   Adsorption of 2,3,4,6-tetrachlorophenol on a model graphitized carbon
phase. White, light-gray, dark-gray, and black balls represent hydrogen,
carbon, chlorine, and oxygen, respectively. The atomic size of 2,3,4,6-
tetrachlorophenol is 2.5 times the size of the model phase.
Reproduced by permission of Springer, ref. 25.

this approach,[7] therefore nitrobenzene was eliminated from the calculation.
The following results were obtained:

$$\log k_{mes} = 0.269 \times (\text{MIFS}) - 4.514, \; r = 0.961, \; n = 13,$$
$$\log k_{mes} = 0.149 \times (\text{MIHB}) - 1.222, \; r = 0.543, \; n = 13,$$
$$\log k_{mes} = -94.170 \times (\text{MIES}) - 0.665, \; r = 0.387, \; n = 13,$$
$$\log k_{mes} = 0.168 \times (\text{MIVW}) - 2.293, \; r = 0.564, \; n = 13.$$

$$\log k_{ref} = 0.074 \times (\text{MIFS}) - 1.280, \; r = 0.558, \; n = 17,$$
$$\log k_{ref} = 0.075 \times (\text{MIHB}) - 0.568, \; r = 0.706, \; n = 17,$$
$$\log k_{ref} = 0.059 \times (\text{MIES}) - 0.222, \; r = 0.007, \; n = 17,$$
$$\log k_{ref} = -0.089 \times (\text{MIVW}) + 0.680, \; r = 0.386, \; n = 17.$$

Even when nitrobenzene was eliminated from the calculation, the correlation coefficient of MIFS was high for $\log k_{mes}$, but poor for $\log k_{ref}$. The
results indicated these graphitized carbons have different properties. It
seemed to be based on the strong retention of phenol, and *p*-cresol, as
shown in Figure 5.4, compared with the results shown in Figure 5.5.
   The hydrogen-bonding energy seemed to contribute to the retention on the
graphitized carbon (Hypercarb™) column, but not to the retention on another
carbon BioTechnologyResearch (BTR™) column. Therefore, only alkyl-group-
substituted compounds were selected and their $\log k$ values were correlated with
MI values. The correlation coefficient was 0.718 for $\log k_{ref}$. The $\log k_{ref}$ values of
benzene, toluene, *o*-xylene, *m*-xylene, *p*-xylene, *o-tert*-butylphenol, 4-ethylphenol,

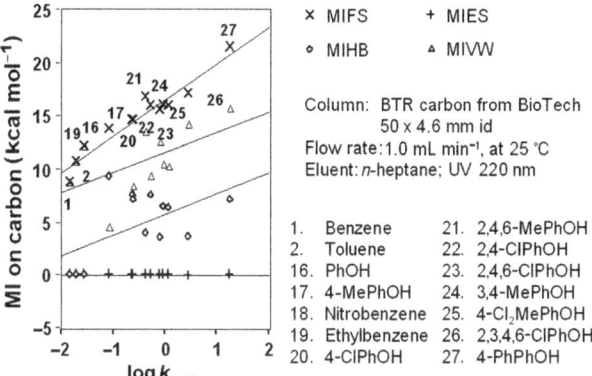

Figure 5.4 The relationship between log k_{mes} and molecular interaction energy values on BTR carbon.

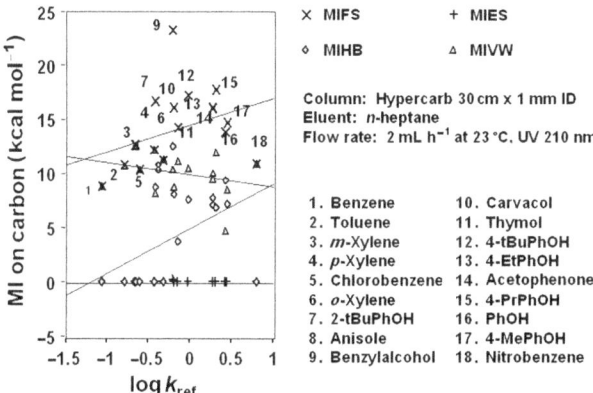

Figure 5.5 The relationship between log k_{ref} and molecular interaction energy values on Hypercarb™.

4-propylphenol, 4-*tert*-butylphenol, phenol and *p*-cresol were used for the calculation. The correlation coefficient increased to 0.889 ($n = 9$). The correlation coefficient for log k_{mes} was 0.945 without phenol and *p*-cresol. The log k_{mes} values of benzene, toluene, ethylbenzene, phenol, *p*-cresol, 3,4-dimethylphenol and 2,4,6-trimethylphenol were used for the correlation.

$$\log k_{mes} = 0.945 \times (\text{MIFS}) - 4.089, \ r = 0.945, \ n = 7,$$
$$\log k_{mes} = 0.126 \times (\text{MIHB}) - 1.518, \ r = 0.408, \ n = 7,$$
$$\log k_{mes} = -255.556 \times (\text{MIES}) - 1.141, \ r = 0.408, \ n = 7,$$
$$\log k_{mes} = 0.037 \times (\text{MIVW}) - 1.394, \ r = 0.394, \ n = 7.$$

$$\log k_{ref} = 0.134 \times (\text{MIFS}) - 2.128, \ r = 0.718, \ n = 11,$$
$$\log k_{ref} = 0.110 \times (\text{MIHB}) - 0.718, \ r = 0.827, \ n = 11,$$
$$\log k_{ref} = -4.543 \times (\text{MIES}) - 0.224, \ r = 0.542, \ n = 11,$$
$$\log k_{ref} = -0.080 \times (\text{MIVW}) + 0.566, \ r = 0.347, \ n = 11.$$

The correlation coefficients obtained for log k_{ref} were still poor based on the strong retention of phenol, and *p*-cresol. Why were these compounds retained strongly on the graphitic carbon column? According to Knox *et al.*[8] the graphitized carbon was synthesized by washing the silica from graphitized carbon using potassium hydroxide. Using this method, a high-porosity HPLC silica gel was impregnated with a phenol–formaldehyde resin. The resin was carbonized at 2000–2800 °C in nitrogen or argon, and the silica particles dissolved out with alkali. This process indicated the possibility of trace amounts of silica and metals in silica and potassium hydroxide in the final products. The possibility was expected due to the relatively high correlation coefficient related to the hydrogen-bonding energy values for polar compounds log k_{ref}.

The probability of silanol groups affecting the graphitized carbon was studied using a model silanol phase that was employed to study retention on silica gels.[4] The molecular interaction energy values were calculated using the model silanol phase shown in Figure 5.6, with adsorbed phenol. The calculated molecular interaction energy values are summarized in Table 5.4, and correlated with the polar compounds log k_{ref} values.

The correlation coefficients are given in the following equations:

$$\log k_{ref} = 0.098 \times (MIFS) - 2.485, \ r = 0.895, \ n = 11,$$
$$\log k_{ref} = 0.094 \times (MIHB) - 1.902, \ r = 0.947, \ n = 11,$$
$$\log k_{ref} = 0.526 \times (MIES) - 0.048, \ r = 0.713, \ n = 11,$$
$$\log k_{ref} = -0.116 \times (MIVW) + 0.517, \ r = 0.351, \ n = 11.$$

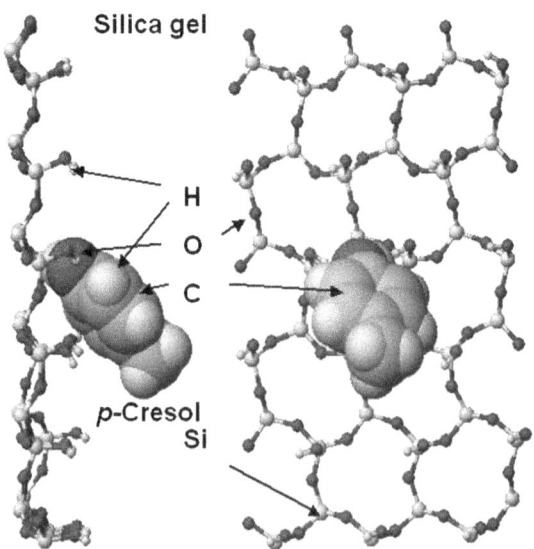

Figure 5.6   Retention of *p*-cresol on a model silanol phase. Small white, large white, light-gray, and black balls represent hydrogen, silicon, carbon, and oxygen, respectively. The atom size of *p*-cresol is five times the size of the model phase.
Reproduced by permission of Taylor and Francis, ref. 26.

Table 5.4 Optimized energy values of complexes and analytes (kcal mol^{-1}). Reproduced by permission of Springer, ref. 25.

Analyte	FS	HB	ES	VW	fs	hb	es	vw
Acetophenone	-873.0734	-59.870	-700.124	-251.212	-4.511	-1.459	0.000	5.846
Anisole	-868.7356	-48.342	-701.867	-253.950	-6.495	0.000	0.000	3.301
Benzene	-867.6150	-48.828	-700.373	-252.265	-8.077	0.000	0.000	3.006
Benzylalcohol	-872.2130	-50.747	-703.456	-252.270	-9.040	-0.606	-0.708	3.481
Carvacrol	-880.1182	-57.286	-702.529	-254.112	-9.542	-1.485	-0.500	4.146
Chlorobenzene	-874.8493	-55.106	-701.495	-252.578	-7.802	0.000	0.000	3.213
p-Cresol	-880.0758	-63.850	-701.228	-249.504	-10.754	-1.459	0.017	2.974
Phenol	-878.9949	-60.181	-702.662	-251.303	-10.209	-1.462	0.000	2.957
2-tert-Butylphenol	-872.1420	-53.834	-701.914	-251.970	-5.156	-1.537	-0.540	6.475
4-tert-Butylphenol	-876.5835	-58.231	-702.047	-251.773	-6.377	-1.452	0.017	5.426
4-Ethylphenol	-879.7051	-59.776	-701.163	-252.751	-8.966	-1.456	0.017	3.983
4-Propylphenol	-881.5762	-60.945	-702.530	-252.899	-8.277	-1.456	0.017	4.470
Thymol	-874.7628	-53.043	-703.176	-255.121	-7.532	-1.302	-0.419	4.961
Toluene	-869.9332	-48.892	-700.548	-253.810	-8.600	0.000	0.000	3.036
o-Xylene	-869.1134	-48.735	-700.798	-253.337	-8.395	0.000	0.130	3.498
m-Xylene	-870.9968	-48.845	-701.471	-253.384	-9.270	0.000	-0.125	3.047
p-Xylene	-870.0407	-47.748	-701.322	-253.590	-9.133	0.000	0.024	3.042
Silica phase	—	—	—	—	-842.7318	-36.059	-701.773	-249.742

The coefficient of the correlation between $\log k_{ref}$ and the hydrogen-bonding energy values was very high. The result was typical of normal-phase liquid chromatography using silica gels, even if the $\log k$ values were measured using a graphitized carbon column. On other hand, the coefficient of the correlation between $\log k_{mes}$ and MI values calculated using the model silica gel was not improved, but rather decreased. The results are given in the following equations:

$$\log k_{mes} = 0.150 \times (MIFS) - 4.511, \ r = 0.895, \ n = 7,$$
$$\log k_{mes} = 0.093 \times (MIHB) - 2.757, \ r = 0.803, \ n = 7,$$
$$\log k_{mes} = 0.378 \times (MIES) - 0.882, \ r = 0.552, \ n = 7,$$
$$\log k_{mes} = -0.166 \times (MIVW) - 0.077, \ r = 0.540, \ n = 7.$$

The graphitized (BTR) carbon used in this experiment was 100% organic.[9] The graphitized carbon (BTR) was made by an entirely different procedure. A roughly 50:50 mixture of low-molecular-mass pitch (viscous resin products from petroleum, MW $\sim$ 300) and a polymerizable monomer, such as styrene and/or divinylbenzene, along with a suitable initiator, was suspended in water and the monomer polymerized. The beads were then separated and heated in stages to 1100 °C, and finally to around 2800 °C. These results support the speculation about the silanol activity of one of the graphitized carbon columns. It seems that the matrix silica gel was not completely washed, and some silanol activity remained. An additional calculation was performed including the other compounds reported in ref. 3. The correlations between MI values and $\log k$ are given in the following equations, and the relations are shown in Figure 5.7.

$$\log k_{ref} = 0.085 \times (MIFS) - 2.211, \ r = 0.807, \ n = 17,$$
$$\log k_{ref} = 0.082 \times (MIHB) - 1.651, \ r = 0.856, \ n = 17,$$
$$\log k_{ref} = 0.208 \times (MIES) - 0.179, \ r = 0.373, \ n = 17,$$
$$\log k_{ref} = -0.051 \times (MIVW) + 0.130, \ r = 0.188, \ n = 17.$$

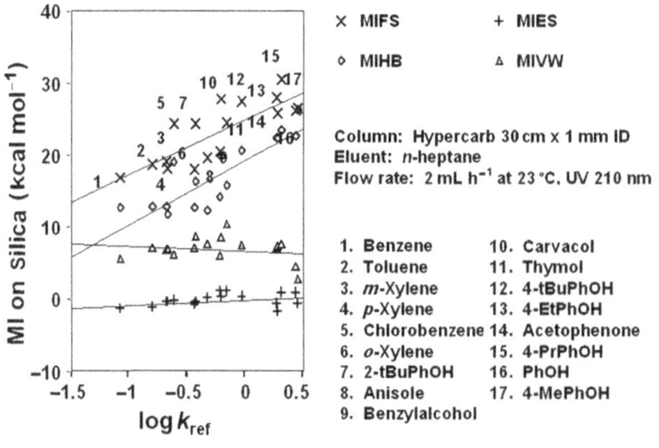

Figure 5.7 The relationship between $\log k_{ref}$ and molecular interaction energy values on silica.

The correlation coefficient for silica was significantly improved compared to the MI values calculated using the graphitized (Hypercarb™) carbon model. The coefficients for MIFS and MIHB were 0.807 and 0.856, compared to 0.558 and 0.706 obtained with the model carbon phase. No reasonable correlation was obtained for MIES and MIVW because these interactions are not the main contributors to retention in normal-phase liquid chromatography. In terms of chromatographic behavior, phenol and *p*-cresol were not outliers. Their strong retention on the graphitized carbon is partly supported by the silica gel model phase.

A semi-empirical molecular statistical theory of adsorption, based on an atom–atom approximation for the potential function for intermolecular adsorbate–adsorbent interactions, was studied to obtain Henry's constant. The study was based on gas chromatography data using graphitized thermal carbon, even with inhomogeneous porous adsorbents. A simple quantitative correlation of the thermodynamic characteristics of adsorption has been applied for liquid chromatography.[10] The graphitized carbon surface was considered to be homogeneous and described by one kind of adsorption site, after experimental measurement of the isotherms for one of the test solutes was performed, followed by non-linear model fitting.[11] Hence, a molecular mechanics calculation was performed to obtain Henry's constant using a flat model.[11] The latter method is a simple way of analyzing a variety of chromatographic data, however, it is not a simple way of synthesizing model phases, except graphitized carbon phases. Chromatographic phases are synthesized homogeneously, but the steric structure is inhomogeneous. The original computer software is used for the conformational analysis of proteins; therefore such an approach can be applied to the analysis of chromatographic retention if a prospective model is designed. The molecular interaction energy values calculated using the MM2 program support a set of data measured on a graphitized carbon column synthesized using 100% organic materials, but do not support the retention on a graphitized carbon column synthesized using a silica matrix. There is a difference in the chromatographic behavior of these graphitized carbon columns, as is clearly demonstrated by the above computational chemical approach. Knowledge about column properties is important for designing a suitable model phase for the quantitative analysis of retention mechanisms, and for retention time prediction.

5.3 Retention of Polycyclic Aromatic Hydrocarbons on Silica Gels

Normal-phase column liquid chromatography using unmodified silica gels has been used for the purification of both natural and synthetic compounds. It is a powerful tool for stereoisomer separations. The instability of unmodified silica gel means that its use for quantitative analysis in column liquid chromatography is avoided because of poor reproducibility. Normal-phase separation has, however, been performed using thin-layer liquid chromatography. The easy handling in thin-layer liquid chromatography permits a quick analysis, even if the sensitivity is not so good. The relative retention distance is used for qualitative analysis. The relative retention distance, R_f, can be converted to relative retention time, k, in column liquid chromatography using the following equation: $k = (1 - R_f)/R_f$, if the experimental conditions are well controlled.[12] The separation data of sedative (sleeping) medications measured using thin-layer liquid chromatography[13] was, therefore, used for the quantitative analysis of normal-phase liquid chromatography retention.

A model silanol phase was constructed with 150 silicon atoms, 288 oxygen atoms, and 48 hydrogen atoms using the CAChe™ molecular modeling program, and the structure was optimized using the MM2 program. A column liquid chromatogram of PAHs was analyzed to study the feasibility of quantitative *in silico* analysis of retention in normal-phase liquid chromatography. The molecular properties of benzene, naphthalene and anthracene were obtained from their optimized structures using the MM2 program. The optimized (final structure, fs), hydrogen-bonding (hb), electrostatic (es), and van der Waals (vw) energy values are summarized in Table 5.5. The energy values of the complexes are also summarized in Table 5.5 as final structure, hydrogen-bonding, electrostatic and van der Waals energy values. The complex form of anthracene and a model phase is shown in Figure 5.8.

The log k values were estimated from a chromatogram obtained using a Spherosil XOA 400 column (17.5×4 mm i.d.) with n-hexane.[14] The molecular interaction energy values (MIFS, MIHB, MIES, and MIVW) were obtained by subtraction of the complex energy value from sum of the analyte and the silanol phase energy values. The interaction energy values were correlated with the log k values. The correlation coefficients between log k and the MIFS, MIHB, MIES, and MIVW values were 1.000, 0.985, 0.508, and 0.992, respectively. The log k values correlate well to the hydrogen-bonding and van der Waals energy values. The hydrogen-bonding energy was demonstrated to be predominant compared to the van der Waals energy, by the higher slope of the former, as shown in Figure 5.9.

Table 5.5 Molecular properties of polycyclic aromatic hydrocarbons measured on a model silanol phase. Reproduced by permission of Taylor and Francis, ref. 17.

Chemicals	log k	fs	hb	es	vw
Benzene	−0.016	−8.077	0	0	3.006
Naphthalene	0.084	−18.6883	0	0	5.769
Anthracene	0.260	−29.3609	0	0	8.489
Silica phase	—	842.9634	−34.699	−699.649	−249.575
	FS	HB	ES	VW	
Benzene	−866.2433	−45.423	−699.207	−252.423	
Naphthalene	−882.1955	−47.960	−699.948	−250.884	
Anthracene	−901.7348	−57.069	−698.747	−249.469	

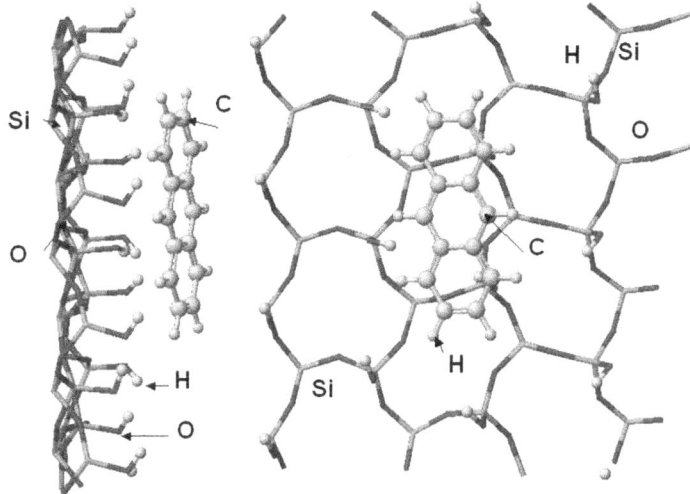

Figure 5.8 A complex of anthracene and the model silanol phase. White, light-gray, dark-gray, and black balls represent hydrogen, carbon, silicon, and oxygen, respectively.

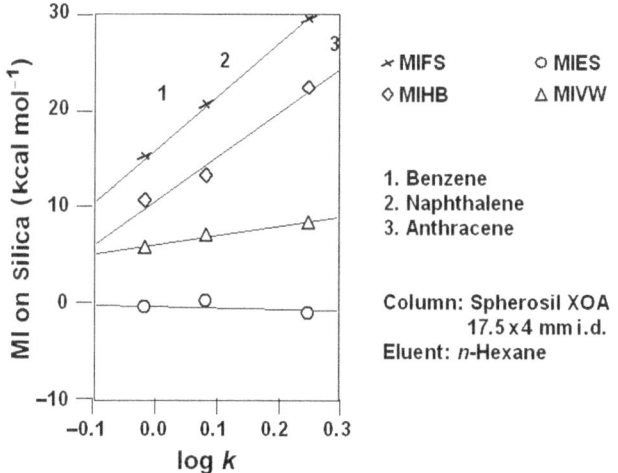

Figure 5.9 The relationship between log k and molecular interaction energy values on a Spherosil XOA column.

5.4 The Effect of Acidic and Basic Components in the Eluent

Usually, different types of solvent mixtures are used in normal-phase liquid chromatography, especially in thin-layer liquid chromatography. For example, sedative medications were separated by thin-layer liquid chromatography using different types of eluents (developing solvents).

These medications are usually in their molecular form in acidic solution, and in their ionized form in basic solution. The degree of ionization depends on their pK_a values and the eluent pH. The measurement of real pH values in organic solvent mixtures is difficult, therefore, the 100% molecular and ionized forms were used. The k values were converted from reference R_f values using $k = (1-R_f)/R_f$. The silica gel used was Silica Gel G.[13] The eluents were chloroform and acetone $(9:1)$, benzene and acetic acid $(9:1)$, and a dioxane, benzene and aqueous ammonia $(20:75:5)$ mixture. The chemical structures and pK_a values were obtained from ref. 15.

The molecular properties of the sedative medications were obtained from the optimized structures using the MM2 program. The fs, hb, es, and vw energy values are summarized in Table 5.6. The energy values of the complexes are also summarized in Table 5.6 as FS, HB, ES, and VW energy values.

Table 5.6 Molecular properties of the molecular forms of various sedative medications. log k_a represents the log k values measured in an acidic benzene and acetic acid $(9:1)$ eluent. log k_n represents the log k values measured in a neutral chloroform and acetone eluent $(9:1)$. Reproduced by permission of Taylor and Francis, ref. 17.

Medication	log k_a	fs	hb	es	vw
Allobarbital	0.122	−69.3903	−8.090	−80.052	3.733
Amobarbital	1.140	−58.4183	−7.984	−75.118	5.333
Barbital	0.213	−60.8260	−7.995	−75.123	4.358
Cyclobarbital	−0.087	−56.0490	−8.133	−68.414	6.040
Hexobarbital	0.017	−58.2625	−8.014	−68.372	5.176
Pentobarbital	0.176	−58.1968	−7.957	−75.095	5.995
Phenobarbital	0.250	−72.9552	−8.186	−71.094	5.648
Secobarbital	0.070	−62.9430	−8.312	−77.768	5.054
Thiopental	−0.432	−67.7351	−4.648	−80.264	4.814
Glutethimide	−0.070	−19.1407	−2.753	−37.118	4.086
Ethinamate	−0.269	−17.8171	−5.436	−27.310	4.671
Ethchlorvynol	0.327	−5.0240	−2.831	−5.919	1.782
Methylprylone	−0.631	−10.9322	−2.720	−31.976	4.487
	log k_n	FS	HB	ES	VW
Allobarbital	−0.087	−948.2816	−67.294	−786.472	−253.606
Amobarbital	−0.176	−934.4776	−64.359	−781.289	−251.590
Barbital	0.000	−940.6903	−68.974	−780.050	−251.964
Cyclobarbital	−0.328	−934.9919	−66.901	−773.484	−251.896
Hexobarbital	−0.525	−935.5317	−63.654	−772.910	−254.266
Pentobarbital	0.000	−937.8826	−66.416	−780.287	−252.078
Phenobarbital	−0.123	−952.8935	−67.179	−775.217	−253.355
Secobarbital	−0.250	−939.9469	−66.109	−781.364	−252.980
Thiopental	−1.195	−941.3991	−57.203	−784.773	−253.607
Glutethimide	−0.602	−899.6517	−57.714	−742.755	−255.640
Ethinamate	−0.631	−895.9510	−61.418	−730.119	−250.332
Ethchlorvynol	−1.284	−875.9328	−57.058	−707.056	−256.154
Methylprylone	0.000	−889.1583	−55.600	−737.374	−254.594

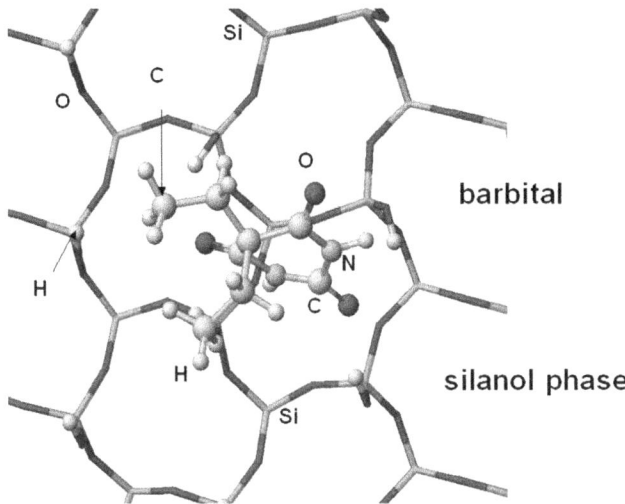

Figure 5.10 Adsorption of barbital on a silanol phase. White, light-gray, dark-gray, and black balls represent hydrogen, carbon, nitrogen, silicon, and oxygen, respectively. Light gray silicons are bonded to black oxygen. Reproduced by permission of Taylor and Francis, ref. 17.

Adsorption of barbital on the silanol phase is shown in Figure 5.10. The ionized form results are summarized in Table 5.7.

In the acidic eluent, a benzene and acetic acid (9:1) mixture, the following correlations were obtained between $\log k_a$ and the molecular interaction energy values:

$$\text{MIFS} = -0.015 \times (\log k_a) + 34.659, \ r = 0.002 \ (n = 13),$$
$$\text{MIHB} = 5.915 \times (\log k_a) + 22.058, \ r = 0.641 \ (n = 13),$$
$$\text{MIES} = -0.846 \times (\log k_a) + 4.919, \ r = 0.166 \ (n = 13),$$
$$\text{MIVW} = -0.527 \times (\log k_a) + 8.359, \ r = 0.113 \ (n = 13).$$

Ethchlorvynol is a neutral molecule, and is weakly retained in both neutral and basic eluents, but it retained strongly in acidic eluent. The reason for this is not clear; therefore, ethchlorvynol was eliminated from the correlation to give:

$$\text{MIFS} = 2.976 \times (\log k_a) + 35.343, \ r = 0.418 \ (n = 12),$$
$$\text{MIHB} = 7.905 \times (\log k_a) + 22.513, \ r = 0.833 \ (n = 12),$$
$$\text{MIES} = 0.561 \times (\log k_a) + 5.240, \ r = 0.149 \ (n = 12),$$
$$\text{MIVW} = -0.604 \times (\log k_a) + 8.342, \ r = 0.121 \ (n = 12).$$

The correlation coefficient was improved, and the contribution of the hydrogen-bonding energy is clear. When only barbital-related compounds were selected, the correlation coefficients were a little improved. The correlation coefficients for MIFS, MIHB, MIES and MIVW were 0.431, 0.838, 0.020 and 0.415 ($n = 11$), respectively.

Table 5.7 Molecular properties of the ionized forms of various sedative medications. log k_b represents the log k values measured in a basic dioxane, benzene and aqueous ammonia (20 : 75 : 5) eluent. Reproduced by permission of Taylor and Francis, ref. 17.

Medication	log k_b	fs	hb	es	vw
Allobarbital	0.035	−46.5193	−4.460	−53.738	3.763
Amobarbital	−0.035	−35.0332	−4.303	−48.209	5.492
Barbital	0.176	−37.5171	−4.307	−48.204	4.442
Cyclobarbital	0.213	−30.7984	−4.368	−40.720	6.350
Hexobarbital	−0.140	−32.7362	−4.331	−40.720	5.591
Pentobarbital	0.017	−34.3887	−4.202	−48.260	6.271
Phenobarbital	0.454	−47.9891	−4.405	−43.830	5.891
Secobarbital	−0.070	−40.2759	−4.406	−50.967	5.299
Thiopental	−0.368	−44.3983	−2.241	−58.241	5.014
Glutethimide	−1.996	6.3333	0.000	−12.266	5.420
Ethinamate	−0.087	−17.8171	−5.436	−27.310	4.671
Ethchlorvynol	−1.276	−5.0240	−2.831	−5.919	1.782
Methylprylone	0.087	−10.9322	−2.720	−31.976	4.487

	pK_a	FS	HB	ES	VW
Allobarbital	7.77	−919.9919	−59.156	−759.040	−250.696
Amobarbital	7.8	−911.0409	−57.181	−752.930	−255.930
Barbital	7.97	−911.5673	−58.920	−753.044	−252.290
Cyclobarbital	8.60	−908.8343	−63.525	−743.273	−252.699
Hexobarbital	8.2	−909.8929	−59.339	−745.939	−252.496
Pentobarbital	8.0	−908.5543	−58.448	−752.731	−250.514
Phenobarbital	7.4	−925.9632	−64.695	−747.063	−251.303
Secobarbital	7.90/12.60	−916.1206	−60.953	−755.808	−250.336
Thiopental	7.50	−917.0497	−53.620	−761.892	−254.307
Glutethimide	9.2	−863.1612	−44.706	−716.760	−255.538
Ethinamate	—	−897.7109	−63.981	−729.050	−251.009
Ethchlorvynol	—	−863.7277	−47.106	−706.995	−253.996
Methylprylone	12.0	−889.1583	−55.600	−737.374	−254.594

In the basic eluent, a dioxane, benzene and aqueous ammonia (20 : 75 : 5) mixture, the following correlations were obtained. Glutethimide was, however, eliminated from the calculation because its high R_f value (0.99) is close to the void volume in column liquid chromatography. The relationship between log k and the MI values is shown in Figure 5.11.

$$\text{MIFS} = 11.246 \times (\log k_b) + 32.650, \ r = 0.873 \ (n = 12),$$

$$\text{MIHB} = 8.589 \times (\log k_b) + 20.556, \ r = 0.870 \ (n = 12),$$

$$\text{MIES} = 1.672 \times (\log k_b) + 4.374, \ r = 0.494 \ (n = 12),$$

$$\text{MIVW} = 1.189 \times (\log k_b) + 7.904, \ r = 0.247 \ (n = 12).$$

In the neutral eluent, a chloroform and acetone (9 : 1) mixture, a solvent effect was apparent. Why methylprylone was retained strongly in the eluent is not clear, as the retention was weak in both acidic and basic eluents. The following correlations were obtained after elimination of methylprylone. The chromatographic behavior in neutral eluent is shown in Figure 5.12.

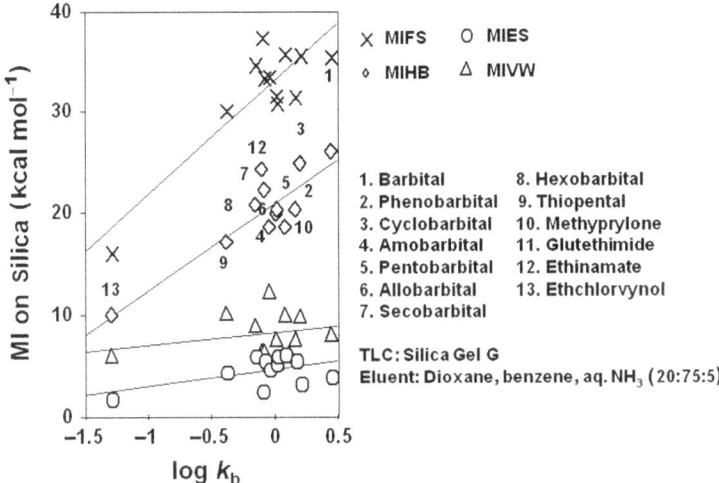

Figure 5.11 The relationship between log k and molecular interaction energy values on a Silica Gel G plate with a basic eluent.

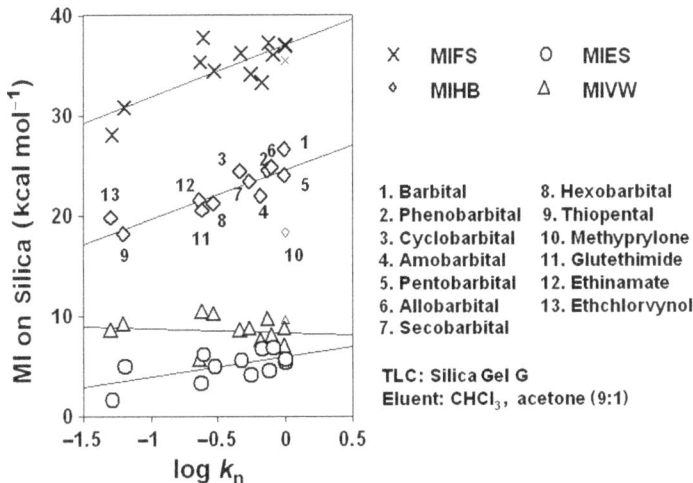

Figure 5.12 The relationship between log k and molecular interaction energy values on a Silica Gel G plate with a neutral eluent.

$$\text{MIFS} = 5.195 \times (\log k_n) + 36.861, \ r = 0.789 \ (n = 12),$$
$$\text{MIHB} = 4.957 \times (\log k_n) + 24.445, \ r = 0.884 \ (n = 12),$$
$$\text{MIES} = 2.088 \times (\log k_n) + 5.767, \ r = 0.617 \ (n = 12),$$
$$\text{MIVW} = -0.474 \times (\log k_n) + 8.066, \ r = 0.156 \ (n = 12).$$

When only barbitals were selected, the correlation coefficients were a little improved. The correlation coefficients for MIFS, MIHB, MIES and MIVW were 0.839, 0.895, 0.293 and 0.434 $(n = 9)$, respectively. The reproducibility was studied for another set of barbital R_f values measured on the same silica gel and eluent.[16] The correlation coefficients between log k and MIFS, MIHB, MIES and MIVW were 0.861, 0.935, 0.188 and 0.460, respectively.

The above results indicate that the hydrogen-bonding energy makes the major contribution to retention in normal-phase liquid chromatography on silica gels. The van der Waals energy effect is negligible for similar size compounds. These phenomena are different to the results obtained for reversed-phase liquid chromatography. Generally, the eluents in normal-phase liquid chromatography are relatively hydrophobic compared to those in reversed-phase liquid chromatography. Acidic and basic components are, however, usually added to the eluents for normal-phase liquid chromatography to improve the separation. Another motivation is improving the selectivity factor (α). Solubility depends on the similarity between an analyte and solvent. The comparison of the chromatographic behavior of different types of compounds is very difficult due to unpredictable solubilities in normal-phase liquid chromatography. Further study is required to predict the chromatographic behavior of partially ionized analytes in normal-phase liquid chromatography, like the pH effect in reversed-phase liquid chromatography.[17]

5.5   The Effect of Organic Solvents in the Eluent

A model silanol phase was constructed with 150 silicon atoms, 288 oxygen atoms, and 48 hydrogen atoms using the CAChe™ molecular modeling program, and the structure was optimized using the CAChe™ MM2 program. Thin-layer liquid chromatograms of triterpenes were analyzed to study the feasibility of the quantitative *in silico* analysis of retention in normal-phase liquid chromatography. The log k values were converted from R_f values measured for triterpenes separated in three eluents.[18] The capacity ratios were converted from R_f values and are summarized in Table 5.8. The properties of the solvents used in this experiment are summarized in Table 5.9.

The molecular properties of triterpenes and the model silanol phase calculated using the MM2 module of the CAChe™ program are summarized in Table 5.10, and the final structure, hydrogen-bonding, electrostatic, and van der Waals energy values of one-to-one complexes are given in Table 5.11. An example of a complex consisting of oleanolic acid adsorbed on the silanol phase is shown in Figure 5.13.

The molecular interaction energy values (MIFS, MIHB, MIES, and MIVW) were obtained by subtracting the complex energy values (Table 5.12) from the sum of the triterpene properties and the silanol phase energy values (Table 5.10). MIFS correlated well with MIHB (MIFS $= 1.021 \times$ MIHB-18.167, $r = 0.925$, $n = 20$), but did not correlate with MIES (MIFS $= 0.053 \times$ MIES $+ 0.224$, $r = 0.454$, $n = 20$) and MIVW (MIFS $= -0.191 \times$ MIVW $+ 22.686$, $r = 0.404$, $n = 20$). The results indicated that the hydrogen-bonding energy was the main contributor to retention on the silanol phase. The feasibility of this approach was analyzed by examining the correlation coefficients

Table 5.8 Converted log k values from the R_f values of triterpenes. The eluent for k_1 was a benzene and ethylacetate (7 : 3) mixture, the eluent for k_2 was an n-hexane and ethylacetate (7 : 3) mixture, and the eluent for k_3 was a diisopropylether and acetone (19 : 1) mixture. The adsorbent was Silica Gel G. Reproduced by permission of Institute for Chromatography, ref. 27.

No.	Analyte	log k_1	log k_2	log k_3
1	Allobetulin	−0.954	−0.550	−0.826
2	β-Amyrin	−1.061	−0.659	−1.061
3	Betulin	−0.659	−0.070	−0.602
4	Cerin	−0.630	−0.250	−0.659
5	Chorestanol	−0.477	−0.176	−0.550
6	Choresterol	−0.550	−0.176	−0.602
7	Cyclolaudenol	−1.061	−0.659	−1.061
8	3-Desoxyoleanolic acid	−1.061	−0.477	—
9	Dihydrobetulin	−0.689	−0.122	−0.630
10	*epi*-Friedelanol	−1.195	−0.689	—
11	Erythrodiol	−0.788	−0.105	−0.602
12	Friedelanol	−1.005	−0.575	−1.005
13	Hederagenin	0.602	1.195	0.308
14	Longispinogenin	0.017	0.410	−0.231
15	Oleanolic acid	−0.389	0.327	−0.550
16	Panaxadiol	0.176	0.525	0.000
17	Stigmastarol	−0.477	−0.176	−0.550
18	Stigmasterol	−0.550	−0.176	−0.602
19	Ursolic acid	−0.368	0.347	−0.525
20	Uvaol	−0.788	−0.105	−0.602

Table 5.9 Properties of various solvents. δ is the total solubility parameter and indicates solvent strength and polarity. A large δ means the solvent is non-polar. δ_d is the ability of a solvent to participate in dispersive interactions, which indicates its degree of solubility of aromatic compounds having halogen and sulfur substituents. Larger values of δ_d mean the solvent can participate in strong dispersive interactions. δ_o is the orientation interaction and indicates the degree of solubility of dipole compounds in the solvent. δ_a is the proton-donor property of the solvent and indicates the degree of solubility of alcohols, phenols, and carboxylic acids in the solvent. δ_h is the proton-acceptor property of the solvent and indicates the degree of solubility of basic compounds in the solvent. ε^o is the solvent strength in normal-phase liquid chromatography on alumina.[19,20] Reproduced by permission of Institute for Chromatography, ref. 27.

Solvent	δ	δ_d	δ_o	δ_a	δ_h	ε^o
Diisopropylether	7.0	6.9	0.50	0.5	0	0.28
n-Hexane	7.3	7.3	0.00	0.0	0	0.01
Ethylacetate	8.6	8.6	1.88	2.0	0	0.58
Benzene	9.2	9.2	0.00	0.5	0	0.32
Acetone	9.4	6.8	2.69	2.5	0	0.56

Table 5.10 Molecular properties of analytes on a model silanol phase. Reproduced by permission of Institute for Chromatography, ref. 27.

No.	Analyte	fs	hb	es	vw
1	Allobetulin	81.1546	−0.001	0.628	24.210
2	β-Amyrin	76.8250	−0.001	0.227	25.692
3	Betulin	79.1898	−1.604	−1.459	24.044
4	Cerin	11.1852	−3.377	8.607	33.528
5	Chorestanol	56.7039	0.000	0.000	17.418
6	Choresterol	53.4649	0.000	0.000	14.756
7	Cyclolaudenol	299.1196	0.000	−0.136	12.807
8	3-Desoxyoleanolic acid	57.5241	−3.491	−7.519	23.416
9	Dihydrobetulin	79.9680	−1.585	−1.452	23.585
10	*epi*-Friedelanol	90.1923	0.000	0.000	29.524
11	Erythrodiol	77.2215	−0.038	−0.034	25.875
12	Friedelanol	89.5079	0.000	0.000	29.217
13	Hederagenin	60.7910	−5.068	−8.656	25.732
14	Longispinogenin	73.5634	−1.690	−2.074	25.704
15	Oleanolic acid	60.6005	−3.494	−7.506	24.429
16	Panaxadiol	98.0599	−1.487	−0.551	29.406
17	Stigmastarol	65.4040	0.000	0.000	19.019
18	Stigmasterol	55.9746	0.000	0.000	13.910
19	Ursolic acid	60.7057	−3.504	−7.507	24.927
20	Uvaol	74.2697	−0.043	−0.047	24.748
	Silanol phase	−850.6494	−45.076	−599.891	−246.911

Table 5.11 Energy values of one-to-one complexes between triterpenes and a model silanol phase. Reproduced by permission of Institute for Chromatography, ref. 27.

No.	Analyte	FS	HB	ES	VW
1	Allobetulin	−793.7269	−47.991	−600.408	−243.022
2	β-Amyrin	−795.4569	−47.757	−600.339	−240.269
3	Betulin	−798.5065	−50.461	−603.325	−246.774
4	Cerin	−763.7993	−55.665	−593.770	−229.854
5	Chorestanol	−813.1291	−48.582	−602.138	−247.180
6	Choresterol	−818.4404	−48.618	−601.773	−252.682
7	Cyclolaudenol	−574.6130	−49.005	−600.998	−252.816
8	3-Desoxyoleanolic acid	−823.3986	−65.040	−609.325	−237.020
9	Dihydrobetulin	−793.3047	−49.448	−602.862	−244.707
10	*epi*-Friedelanol	−779.7433	−47.059	−600.569	−235.117
11	Erythrodiol	−792.4632	−48.758	−601.570	−236.602
12	Friedelanol	−782.1053	−48.047	−600.628	−235.325
13	Hederagenin	−825.3814	−67.268	−610.275	−238.368
14	Longispinogenin	−800.8490	−53.639	−603.020	−239.148
15	Oleanolic acid	−820.4357	−65.131	−609.390	−236.044
16	Panaxadiol	−773.6678	−51.716	−601.399	−236.486
17	Stigmastarol	−808.3902	−48.315	−601.423	−247.269
18	Stigmasterol	−818.7456	−48.367	−601.638	−255.684
19	Ursolic acid	−826.0251	−66.542	−609.997	−237.916
20	Uvaol	−795.9097	−48.751	−601.773	−239.433

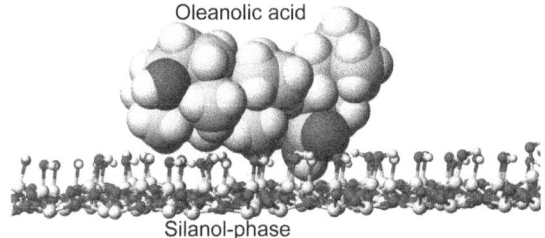

Figure 5.13 Adsorption of oleanolic acid on a silanol phase. White, light-gray, dark-gray and black balls represent hydrogen, carbon, silicon and oxygen, respectively. The atom size of the silanol phase is 20% of the original atom size, the atom distance is the original size.
Reproduced by permission of Institute for Chromatography, ref. 27.

Table 5.12 Solvent effects on the molecular interaction energies. Reproduced by permission of Institute for Chromatography, ref. 27.

Analyte	Ethylacetate		Acetone	
	$\Delta MIHB$	$\Delta MIES$	$\Delta MIHB$	$\Delta MIES$
Allobetulin	−3.465	−1.820	−3.447	−1.564
β-Amyrin	−3.475	−1.855	−3.445	−1.609
Betulin	−3.507	−1.537	−3.440	−1.695
Cerin	−6.717	−2.394	−6.555	−3.129
Chorestanol	−3.479	−1.537	−3.448	−1.613
Choresterol	−3.522	−1.762	−3.590	−0.303
Cyclolaudenol	−3.502	−1.528	−3.443	−1.598
3-Desoxyoleanolic acid	−7.149	−3.507	−6.618	−3.411
Dihydrobetulin	−3.503	−1.536	−3.447	−1.610
epi-Friedelanol	−3.494	−1.742	−3.440	−1.743
Erythrodiol	−3.496	−1.619	−3.444	−1.587
Friedelanol	−3.496	−1.645	−3.449	−1.771
Hederagenin	−7.465	−2.694	−6.756	−2.951
Longispinogenin	−3.501	−1.598	−3.445	−1.707
Oleanolic acid	−10.959	−4.294	−10.214	−4.441
Panaxadiol	−3.531	0.493	−3.475	−1.816
Stigmastarol	−3.463	−1.711	−3.524	−0.087
Stigmasterol	−3.502	−1.709	−3.572	−0.121
Ursolic acid	−10.610	−4.210	−10.163	−5.256
Uvaol	−3.477	−1.709	−3.453	−1.635
Ethylacetate[a]	0.000	−2.246	—	—
Acetone[a]	—	—	0.000	0.000

[a]The original hydrogen-bonding and electrostatic energy values.

between the log k values of triterpenes and the molecular interaction energy values:

$$MIHB = 5.335 \times (\log k_1) + 8.639, \ r = 0.435, \ n = 20,$$
$$MIHB = 6.785 \times (\log k_2) + 7.195, \ r = 0.579, \ n = 20,$$
$$MIHB = 7.862 \times (\log k_2) + 10.676, \ r = 0.509, \ n = 18.$$

Triterpenes with a carboxyl group demonstrated especially high MIHB values. Even if 3-desoxyoleanolic acid was eliminated as an exception, the correlation coefficient did not significantly improve. The correlation coefficients were 0.610 and 0.736 $(n = 19)$ for log k_1 and log k_2, respectively.

The results indicated that hydrogen bonding is the major contributor to the molecular interaction, which differs from reversed-phase[21] and ion-exchange[22] liquid chromatography, where van der Waals and electrostatic energy values are the major contributing factors, respectively. The retention of compounds with a carboxyl group is weaker than that expected from the calculated molecular interaction energy values. This is shown in Figure 5.14, where the hydrogen-bonding energies of 3-deoxyoleanolic acid, hederagenin, oleanolic acid and ursolic acid are very high. Why are the retention properties of carboxylic acids so exceptional? Steric hindrance may cause weak contact, as even the hydrogen-bonding energy values are quite high because of the poor surface area available for contact. The calculated energy values can indicate retention, but not desorption, and desorption should occur when the hydrogen-bonding site is replaced by solvent molecules in the eluent. Further studies were performed to determine the solvent effect.

The eluent strength of the three solvents was in the order: $k_1 > k_3 > k_2$. The ε° values were approximately 0.40, 0.29, and 0.18, respectively. The R_f values followed the solvent strength ε°. The eluent for log k_2 was a simple mixture, used to evaluate the hydrogen-bonding effect of ethylacetate. In an *n*-hexane and ethylacetate mixture, *n*-hexane is the dispersive solvent, but it is not responsible for hydrogen bonding. *n*-Hexane is also not critically involved in the desorption of triterpenes from the silica gel. The solvent effect of ethylacetate, therefore, was studied by forming complexes between ethylacetate and one of the polar groups of the triterpenes, such as a hydroxyl or carboxyl group.

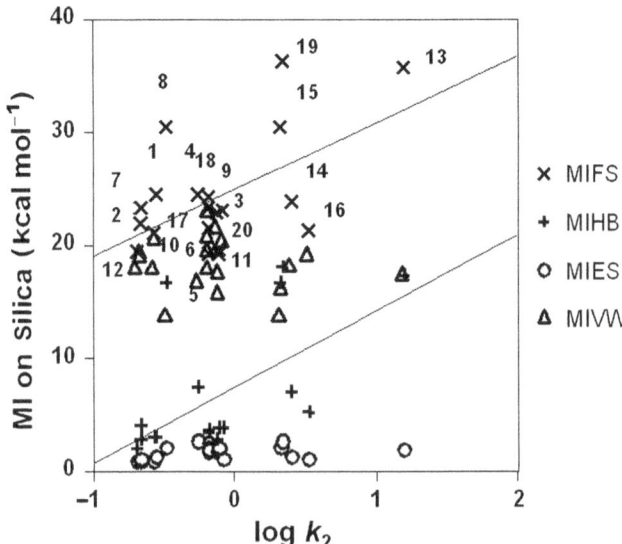

Figure 5.14 Contribution of hydrogen bonding energy to the retention.

First, the complex conformation for the solvent effect was studied using octanoic acid and octanol as model compounds to analyze the electronic strength of the carboxyl and hydroxyl groups based on the atomic partial charge (apc) of the terminal hydrogen. The carboxyl and hydroxyl groups of these compounds have more free space available to accept the solvent molecules than the triterpenes. A complex with a polar group was prepared with the solvent oxygen, and the complex conformation was optimized using MM2 calculations. The atomic partial charge was calculated using the MOPAC PM5 program. When the sum of the atomic partial charge of hydrogen and the adjacent oxygen is zero, the carboxyl and hydroxyl groups are neutralized, and no extra atomic partial charge exists to form an additional complex, like the hydrogens and carbons of an alkyl group. A large solvent molecule cannot form a multicomplex with one carboxyl and/or hydroxyl group due to steric hindrance. The actual solvated triterpene form is not known. A model complex was used as a simple solvation form. After forming a carboxyl group and one ethylacetate molecule complex, the sum of the atomic partial charge was -0.070 au compared to the original -0.091 au. The value decreased to -0.047 au for a complex of a carboxyl group with two ethylacetate molecules. A complex with three ethylacetate molecules could not be constructed due to steric hindrance. The atomic partial charge of a hydroxyl group with a single ethylacetate molecule was reduced to -0.162 from -0.169 au (atomic units). The atomic partial charge of the complex with two ethylacetate molecules was -0.171 au. The apc value for the hydroxyl group was essentially constant for complexes with either one or two ethylacetate molecules. Based on these results, two ethylacetate molecules were used to form complexes with the triterpene carboxyl group, and one ethylacetate molecule was used to form a complex with the

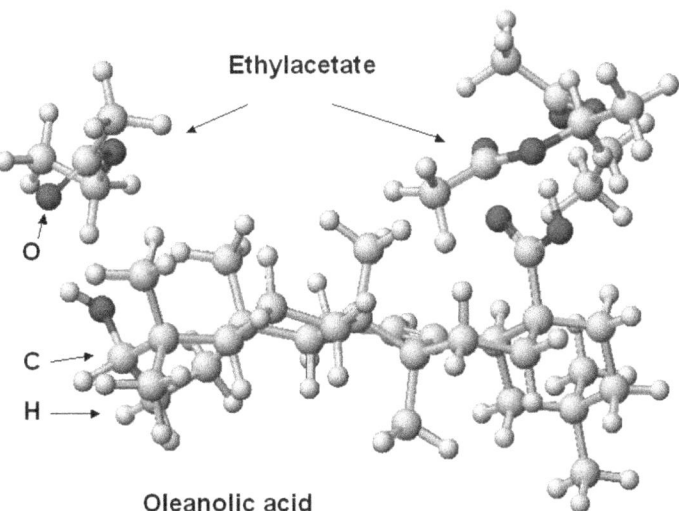

Figure 5.15 Conformation of the complex between oleanolic acid and ethylacetate. Small white, large light gray, and black balls represent hydrogen, carbon, and oxygen.

triterpene hydroxyl group. The model complex is shown in Figure 5.15. The calculated solvent effect is summarized in Table 5.12 as ΔMIHB and ΔMIES with those of acetone. The ΔMIHB values are the subtracted original energy values (hb in Table 5.10) from the complex hydrogen-bonding energy values. The ΔMIES values are obtained by subtracting the original energy values (ec in Table 5.10) from the complex electrostatic energy values. The adsorption side polar group ΔMIHB and ΔMIES values were used for further calculations because the other polar groups are not involved in a direct interaction (*via* hydrogen bonds) with the silanol phase. Hydroxyl groups with intramolecular hydrogen bonds were not included in the complex.

Which solvent preferentially forms hydrogen bonds with the triterpene polar group? In an *n*-hexane and ethylacetate (7 : 3) mixture, ethylacetate should selectively form hydrogen bonds with the triterpene hydroxyl and carboxyl groups. The hydrogen-bonding energy values obtained from the solvent and triterpene complexes (ΔMIHB) were used for additional correlation analyses with the log k_2 values. The following correlations were obtained:

$$\text{MIHB} + \Delta\text{MIHB} = 4.480 \times (\log k_2) + 2.181, \; r = 0.617, \; n = 20,$$

$$\text{MIHB} + \Delta\text{MIHB} = 5.348 \times (\log k_2) + 1.769, \; r = 0.847, \; n = 19.$$

Including ΔMIHB increased the correlation coefficients. When 3-desoxyoleanolic acid was eliminated as an exception, the correlation coefficient was 0.847 ($n = 19$). The electrostatic energy values of the ethylacetate and carboxyl group complex were very high. Therefore, the electrostatic energy values, ΔMIES, of the carboxyl and hydroxyl groups were added and the correlation coefficient recalculated as follows:

$$\text{MIHB} + \Delta\text{MIHB} + \Delta\text{MIES} = 4.889 \times (\log k_2) - 0.188, \; r = 0.921, \; n = 19.$$

The addition of a solvent effect, partly calculated from the complex, improved the precision of the predicted retention time. If the total solvation can be calculated, the precision should further improve.

In a benzene and ethylacetate (7 : 3) mixture, ethylacetate should be the predominant solvent over benzene, even if benzene can form hydrogen bonds with the polar groups of triterpene, and even if these solvents are classified similarly, because the ethylacetate δ_a (proton donor property) value (2.0) is higher than that of benzene (0.5). The correlation coefficient for log k_1 is less than that of log k_2 because benzene is also a weak proton donor solvent and contributes to desorption. It is difficult, however, to quantitatively analyze such competitive solvation. After adding the electrostatic energy values of the carboxyl group, the correlation coefficient increased:

$$\text{MIHB} + \Delta\text{MIHB} + \Delta\text{MIES} = 3.856 \times \log k_1 + 1.977, \; r = 0.611, \; n = 20,$$

$$\text{MIHB} + \Delta\text{MIHB} + \Delta\text{MIES} = 4.905 \times (\log k_1) + 2.179, \; r = 0.871, \; n = 19.$$

After eliminating 3-desoxyoleanolic acid as an exception, the correlation coefficient increased to 0.871 ($n = 19$).

The above results indicated that electrostatic energy values strengthened the correlation. Therefore, the sum of MIHB and MIES used to study the contribution to the retention time prediction is:

$$\text{MIHB} + \text{MIES} + \Delta\text{MIHB} + \Delta\text{MIES} = 5.318 \times (\log k_1) + 3.908, \ r = 0.869, \ n = 19,$$

$$\text{MIHB} + \text{MIES} + \Delta\text{MIHB} + \Delta\text{MIES} = 5.351 \times (\log k_2) + 1.346, \ r = 0.927, \ n = 19.$$

The relationships are shown in Figures 5.16 and 5.17. The above calculation was performed without the contribution of benzene due to the difficulty in

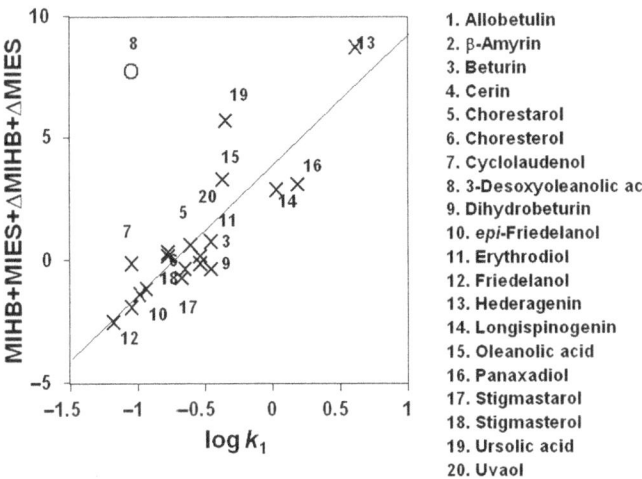

Figure 5.16 The relationship between log k_1 and molecular interaction energy values for a benzene and ethylacetate (7 : 3) eluent.

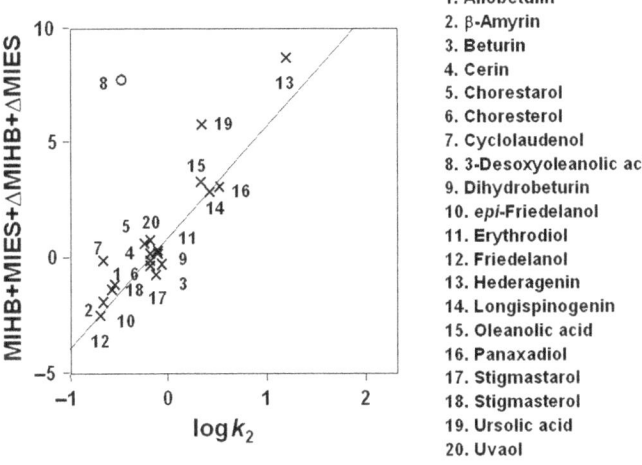

Figure 5.17 The relationship between log k_2 and molecular interaction energy values for an *n*-hexane and ethylacetate (7 : 3) eluent.

building the conformation with benzene molecules. The original electro-static energy values did not contribute to the retention, but the electrostatic energy induced by complex formation improved the precision.

In a diisopropylether and acetone (19:1) mixture, acetone should induce strong desorption. Therefore, the hydrogen-bonding effect was calculated for acetone with the triterpene polar group complex.

$$\text{MIHB} = 7.862 \times (\log k_3) + 10.676, \ r = 0.509, \ n = 18,$$

$$\text{MIHB} + \Delta\text{MIHB} = 6.245 \times (\log k_3) + 5.174, \ r = 0.652, \ n = 18.$$

Furthermore, adding the electrostatic energy effect only increased the cor-relation coefficient a little in this eluent:

$$\text{MIHB} + \Delta\text{MIHB} + \Delta\text{MIES} = 5.504 \times (\log k_3) + 2.833, \ r = 0.763, \ n = 18.$$

In addition, the molecular interaction energy values for triterpene–diisopropylether complexes were calculated to study the contribution of diisopropylether. The ΔMIHB and ΔMIES values were small, and did not demonstrate a solvent effect for the quantitative analysis of desorption. The hydrogen bonding in diisopropylether and acetone in different mixture ratios may control the desorption of triterpenes. Multisolvent mixtures form clus-ters, and not a homogeneous solvation form with an analyte. Part of the solvation process can be represented by complex formation, which improves retention time prediction. Further studies of solvation are necessary to im-prove the quantitative analysis of desorption from adsorbents in normal-phase liquid chromatography.[23] The unrestricted choice of solvents in nor-mal-phase liquid chromatography, especially planar liquid chromatography, is a suitable tool for studying solvation. The capacity ratio is a combination of adsorption and desorption. The quantitative analysis of desorption mech-anisms should permit the study of solvation and solubility.

References

1. B. Arwidi and O. Samuelson, Partition chromatography of sugars on ion-exchange resins, *Sven. Kem. Tidskr.*, 1965, 77, 84–90; L.-I. Larsson and O. Samuelson, An automated procedure for separation of monosaccharides on ion exchange resins, *Acta Chem. Scand.*, 1965, **19**, 1357–1364.
2. K. Koizumi, Y. Okada, M. Fukuda, K. Koizumi, Y. Okada and M. Fukuda, High-performance liquid chromatography of mono- and oligo-sacchar-ides on a graphitized carbon column, *Carbohydr. Res.*, 1991, **215**, 67–80.
3. R. Kaliszan, K. Osmialowski, B. J. Bassler and R. A. Hartwick, Mech-anism of retention in high-performance liquid chromatography on porous graphitic carbon as revealed by principal component analysis of structural descriptions of solutes, *J. Chromatogr.*, 1990, **499**, 333–344.
4. T. Hanai, Separation of polar compounds using carbon columns, *J. Chromatogr., A*, 2003, **989**, 183–196.

5. T. Hanai, Quantitative *in silico* analysis of ion-exchange from chromatography to protein, *J. Liq. Chromatogr. Relat. Technol.*, 2007, **30**, 1251–1275.
6. Steric hindrance for enantiomer separation is explained in detail in Chapter 8.
7. T. Hanai, Analysis of the mechanism of retention on graphitic carbon by a computational chemical method, *J. Chromatogr., A*, 2004, **1027**, 279–287.
8. J. H. Knox and M. T. Gilbert, Preparation of porous carbon, *Br. Pat.* 2035282, 1978.
9. T. Obayashi, H. Ozawa and T. Kawase, *Eur. Pat.*, 0458548A, 1990.
10. A. V. Kiselev and D. P. Poshkus, Chromatostructural analysis (chromatography) a new method of determination of molecular structure, *Faraday Symp. Chem. Soc.*, 1980, **15**, 13–24.
11. N. S. Kulikov and M. S. Bobyleva, Molecular modeling in chromatography as a new tool in the structure elucidation of novel isomers by GC/MS, *Struct. Chem.*, 2004, **15**, 51–64.
12. R. E. Kaiser, *Einfuhrung in die Hochleistungs-Dunnschicht-Chromatographie*, Institute fur Chromatographie, Bad Durkheim, 1976.
13. I. Suzuki, Y. Saitoh and M. Toyoda, *Practice of Thin-Layer Chromatography*, Hirokawa, Tokyo, 1990.
14. J. Vermont, M. Deleuil, A. J. De Vries and C. L. Guillemin, Modern liquid chromatography on Spherosil, *Anal. Chem.*, 1975, **47**, 1329–1337.
15. C. Hansch, P. G. Sammes and J. B. Taylor, *Comprehensive Medicinal Chemistry II*, Pergamon Press, Oxford, 1990.
16. K. Macek, *Pharmaceutical Application of Thin-Layer and Paper Chromatography*, Elsevier, Amsterdam, 1972, p. 221.
17. T. Hanai, Quantitative *in silico* analysis of retention in normal-phase liquid chromatography, *J. Liq. Chromatogr. Relat. Technol.*, 2010, **33**, 297–304.
18. S. Hara, O. Tanaka and S. Takitani, *Thin-Layer Chromatography*, Nankodo, Tokyo, 2nd edn, 1964, p. 126.
19. L. R. Snyder, *Principles of Adsorption Chromatography*, Marcel Dekker, New York, 1968, p. 174.
20. J. J. Kirkland, *Modern Practice of Liquid Chromatography*, Wiley, New York, 1971, p. 136.
21. T. Hanai, Chromatography *in silico*: basic concept in reversed-phase liquid chromatography, *Anal. Bioanal. Chem.*, 2005, **382**, 708–717.
22. T. Hanai, Y. Masuda and H. Homma, Chromatography *in silico*; retention of basic compounds on a carboxyl ion exchanger, *J. Liq. Chromatogr. Relat. Technol.*, 2005, **28**, 3087–3097.
23. T. Hanai, Quantitative *in silico* analysis of retention in normal-phase liquid chromatography II. http://www.internet-chromatography.com/html/toshihikbeitrage.html.
24. T. Hanai, Analysis of the mechanism of retention on graphitic carbon by a computational chemical method, *J. Chromatogr. A*, 2004, **1030**, 13–16.

25. T. Hanai and H. Homma, Quantitative *in silico* analysis of the selectivity of graphitic carbon synthesized by different methods, *Anal. Bioanal. Chem.*, 2008, **390**, 369–375.

26. T. Hanai, Quantitative *in silico* analysis of the specificity of graphitized (graphitic) carbons, *Adv. Chromatogr.*, 2011, **49**, 257–290.

27. T. Hanai, Quantitative *in silico* analysis of retention time on methyl-silicone and polyethyleneglycol phases in capillary gas chromatography . http://www.internet-chromatography.com/html/toshihikbeitrage.html.

Retention in Reversed-Phase Liquid Chromatography

6.1 Basic Concepts of Reversed-Phase Liquid Chromatography

The retention mechanism in reversed-phase liquid chromatography has been quantitatively analyzed using computational chemical methods based on solubility properties. The most important molecular interaction force in reversed-phase liquid chromatography is van der Waals force (hydrophobic interactions), which is a combination of van der Waals volume, repulsion, and London dispersion forces. Hydrogen bonding and electrostatic interactions between an analyte and the alkyl group brush of the bonded phase, which are very important in normal-phase and ion-exchange liquid chromatography, respectively,[1] are not be considered. Organic modifiers compete with an analyte for direct interaction with the bonded phase. Ions, as components of the eluent control the dissociation of the analyte. The predominant retention force between an analyte and the alkyl-chain bonded phase is hydrophobic. A system using the octanol–water partition coefficient (log P) can predict retention times in reversed-phase liquid chromatography. Log P is the sum of solubility properties and is not the best solution for studying the quantitative structure–retention relationship (QSRR) in reversed-phase liquid chromatography. However, the error is based on lack of log P values for ionized compounds. The solution can be obtained by controlling the solubility of the analyte under various conditions. If the steric effect, which is very important in affinity chromatography involving enantioseparations, is neglected, a one-to-one molecular interaction can be used to study the basic molecular interactions.

RSC Chromatography Monographs No. 19
Quantitative *In Silico* Chromatography: Computational Modelling of Molecular Interactions
By Toshihiko Hanai
© Toshihiko Hanai, 2014
Published by the Royal Society of Chemistry, www.rsc.org

One of the basic considerations in reversed-phase liquid chromatography using silica-based packing materials is the inertness of bonded-phase silica gels.[2-4] The phenomenon of pyridine adsorption was first analyzed using molecular mechanics calculations *via* the CAChe™ software, because inertness affects the selectivity of the chromatograms, the elution order, and the quantitative analysis of the chromatograms. The selection of alkyl-chain length is a very important factor in reversed-phase liquid chromatography.[5] Selectivity was analyzed using model compounds. An alkane is considered to be an alkyl brush, and the molecular interaction between an alkyl brush and the analyte was quantitatively analyzed using molecular mechanics calculations. Steric hindrance was studied by adsorption on a model graphitic carbon phase. Investigation of the QSRR in reversed-phase liquid chromatography was performed using the chromatographic data of phenolic compounds,[6] aromatic acids,[7-9] and acidic[10] and basic drugs.[11] Quantitative analysis was also achieved using molecular mechanics calculations.

6.1.1   Adsorption of Pyridine

A stable and inert bonded-phase silica gel is required to establish the quantitative analysis of retention time in chromatography. A quality control test used by column manufacturers is the *pyridine test*,[2-4] which measures the inertness of bonded-phase silica gels from the chromatographic behavior of pyridine. Pyridine is strongly retained on poorly treated bonded silica gels, and even aniline has a symmetrical peak shape. Phenol is used as the reference compound. The dissociation constants (pK_a) of pyridine and phenol are 5.25 and 10.02, respectively, therefore, these compounds are in their molecular form under neutral conditions. Because these compounds maintain their molecular form, the elution order depends on log P. The log P values of pyridine and phenol are 0.70 and 1.54, respectively. Pyridine should elute before phenol, but pyridine is retained longer than phenol in some cases. The abnormal chromatographic behavior of pyridine was analyzed by *in silico* chromatography. A model silica gel surface was constructed with 198 atoms, 189 bonds, and 1406 connectors. Twenty-seven silanol groups exist on one side of the surface. The top and side views of the model phase are shown in Figure 6.1. The silanol, siloxane, and ionized silica forms are shown in Figure 6.1(a)–(c), respectively. The molecular interaction of the silanol form with pyridine and phenol is shown in Figure 6.1(d) and (e), respectively. The interaction energy values of the final structure (MIFS) are summarized in Table 6.1. The interaction energy values (MI values) were obtained after subtracting the energy value of the complex from the sum of the individual energy values of the analyte and the model phase.

The results indicated that when silanol groups are ionized, pyridine is strongly retained on the surface. Generally, silica gels are stable at pH 6.5, and dissolve in higher pH solutions. This means that the silica gel might be partially ionized under neutral conditions and adsorb pyridine. Complete surface coverage is important to avoid unpredictable chromatographic behavior.

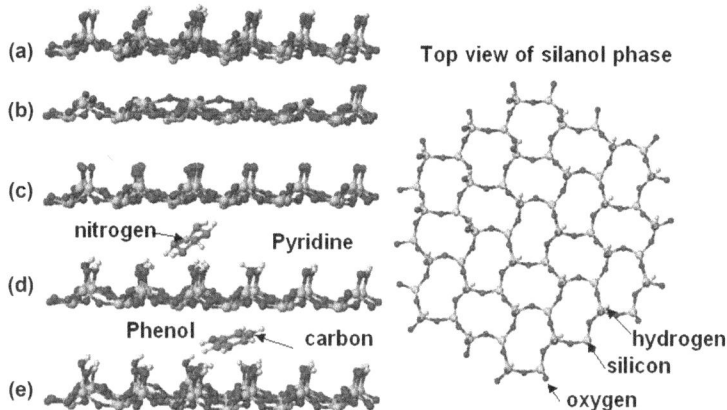

Figure 6.1 Retention of pyridine and phenol on a model silica surface. (a) The model silanol phase, (b) the siloxane phase, (c) the ionized silicon oxide phase, (d) adsorption of pyridine on a silanol phase, and (e) adsorption of phenol on a silanol phase. White, light-gray, gray, and black balls represent hydrogen, nitrogen, silicon, and oxygen, respectively. Reproduced by permission of Springer, ref. 22.

Table 6.1 Molecular properties of pyridine and phenol, and their molecular interaction energies (kcal mol^{-1}) with model silica gels.

Properties	Silica surface	Pyridine	Phenol
MIFS	SiOH	6.94	20.26
	Siloxane	6.01	12.91
	SiO$^-$	12.68	3.92
pK_a	—	5.25	10.02
log P	—	0.70	1.54

6.1.2 Alkyl Chain Length Effect of Bonded Phases

According to Berendsen and De Galan, a dodecylphase is suitable for practical applications in reversed-phase liquid chromatography.[5] The alkyl chain length effect was studied by computational chemical calculations. Dodecane was used as the brush for the model bonded phase, and the molecular interaction between the dodecane and a variety of analytes was quantitatively analyzed using molecular mechanics calculations. The analytes were alkanes, alkenes, and alkyl alcohols. Aromatic compounds were not included in this first analysis due to the requirements of steric effects. The molecular interactions of dodecane with various analytes are shown in Figure 6.2. The van der Waals energy was the predominant molecular interaction. The calculated energy values of individual compounds and the complexes are summarized in Table 6.2, with their properties. Alkyl chain length affected the retention of a variety of compounds. The retention of larger compounds

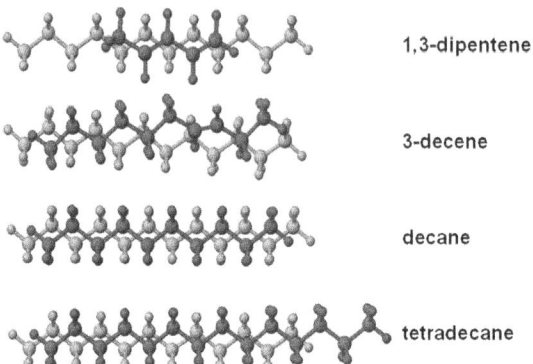

1,3-dipentene

3-decene

decane

tetradecane

Figure 6.2 Interaction of alkanes and alkenes on dodecane (a model alkyl chain of a bonded phase). Small and large balls represent hydrogen and carbon, respectively.
Reproduced by permission of Springer, ref. 22.

was constant when a shorter alkyl chain bonded phase was used.[5] A C30 bonded phase was developed for the separation of carotenoids, due to requirement for a longer alkyl chain compatible with the size of carotenoid molecules.[12,13] In *in silico* chromatography, the molecular interaction energy was constant for larger analytes. The results were the same for alkanes, alkenes, and alkyl alcohols.

Retention depended on the contact surface area of the molecules. The alkenes with multiple double bonds had less contact surface area and the molecular interaction energy was smaller than that expected from the carbon numbers (surface area). The electrons of double bonds and hydroxyl groups did not affect retention. The relationship between the molecular interaction energy values of alkyl alcohols, alkanes and alkenes and their carbon numbers is shown in Figure 6.3.

The molecular interaction energy values of alkenes were smaller than for their related alkanes. This result supports the idea that the hydrophobic interaction due to the van der Waals energy is the predominant molecular interaction in reversed-phase liquid chromatography. No dipole–dipole or π–π interactions influenced the direct interaction. The lack of dipole–dipole or π–π interactions can be studied from chromatographic behavior on a graphitized carbon phase.[14,15]

6.2   Prediction of Dissociation Constants

Dissociation constants, pK_a, can be calculated using Hammett's equation:

$$pK_a = A + B\Sigma\sigma \tag{1}$$

where A and B are constants for individual groups of compounds, and σ is Hammett's constant, measured by liquid chromatography or titration.[16] The equations obtained from ref. 16 are summarized in Table 6.3.

Table 6.2 Molecular properties of alkanes and alkenes in complexes with dodecane. fs, hb, es, and vw represent the energy of the final (optimized) structure, the hydrogen-bonding energy, the electrostatic energy, and the van der Waals energy of each analyte (kcal mol^{-1}), respectively. FS, VW, HB, and ES represent the same energy values for each complex. Reproduced by permission of Springer, ref. 22.

Analyte	fs	hb	es	vw	FS	HB	ES	VW
Methanol	0.0605	0	0.000	−0.077	5.1249	0	0.000	3.025
Ethanol	0.8019	0	0.000	0.583	4.9799	0	0.000	2.804
Propylalcohol	1.4662	0	0.000	1.086	4.7094	0	0.000	2.234
Butanol	2.1307	0	0.000	1.573	4.4579	0	0.000	1.820
Pentanol	2.7809	0	0.000	2.041	4.1053	0	0.000	1.319
Hexanol	3.4250	0	0.000	2.497	3.8288	0	0.000	0.866
Tetradecane	8.6083	0	0.000	6.312	4.6461	0	0.000	0.365
Tridecane	7.9663	0	0.000	5.850	4.0840	0	0.000	−0.006
Dodecane	7.3257	0	0.000	5.383	3.6762	0	0.000	−0.236
Undecane	6.6823	0	0.000	4.924	3.3824	0	0.000	−0.352
Decane	6.0348	0	0.000	4.462	3.3819	0	0.000	−0.171
Nonane	5.3978	0	0.000	3.996	3.5011	0	0.000	0.129
Octane	4.7556	0	0.000	3.535	3.6404	0	0.000	0.426
Heptane	4.1135	0	0.000	3.069	3.8965	0	0.000	0.873
Hexane	3.4743	0	0.000	2.611	4.2497	0	0.000	1.358
Pentane	2.8290	0	0.000	2.151	4.6560	0	0.000	1.950
Butane	2.1765	0	0.000	1.674	4.9004	0	0.000	2.405
Propane	1.5011	0	0.000	1.188	5.1439	0	0.000	2.857
Ethane	0.8167	0	0.000	0.682	5.4915	0	0.000	3.392
Methane	0.0000	0	0.000	0.000	5.3661	0	0.000	3.343
1-Ethene	0.4230	0	0.000	0.386	5.1134	0	0.000	3.158
1-Propene	2.1632	0	0.000	0.525	3.7718	0	0.000	2.385
1-Butene	2.4863	0	0.000	1.157	4.4793	0	0.000	2.543
2-Butene	3.9495	0	0.099	0.609	2.8872	0	0.098	1.957
1,3-Dibutene	−2.9639	0	0.000	1.132	−0.1779	0	0.000	1.958
1-Pentene	3.1544	0	0.000	1.632	4.2291	0	0.000	2.054
2-Pentene	4.2420	0	0.000	1.213	3.2818	0	0.098	1.671
1,3-Gipentene	−1.2575	0	0.000	1.230	−1.3309	0	0.000	1.385
1-Hexene	3.8062	0	0.000	2.102	4.1093	0	0.000	1.600
2-Hexene	4.9020	0	0.099	1.682	3.1446	0	0.098	1.216
3-Hexene	4.5252	0	0.099	1.812	3.5618	0	0.098	1.543
1,3-Dihexene	−0.9595	0	0.000	1.857	−0.3334	0	0.000	1.662
1,5-Dihexene	4.2232	0	0.100	1.579	4.3919	0	0.100	1.587
2,4-Dihexene	0.4513	0	0.010	1.304	−2.2327	0	0.010	0.962
1,3,5-Trihexene	−0.6347	0	0.000	1.872	−5.3345	0	0.000	0.959
1-Decene	6.3731	0	0.000	3.956	3.1719	0	0.000	0.181
2-Decene	7.4747	0	0.000	3.539	2.7221	0	0.098	−0.039
3-Decene	7.1127	0	0.000	3.665	2.7233	0	0.098	−0.039
4-Decene	7.1256	0	0.000	3.671	2.7172	0	0.098	0.009
5-Decene	7.1276	0	0.000	3.679	2.8538	0	0.098	0.122

However, several pK_a values could not be calculated, due to the lack of Hammett equations. The simplification of Hammett's equations has been studied. A new approach was based on the correlation between the reference

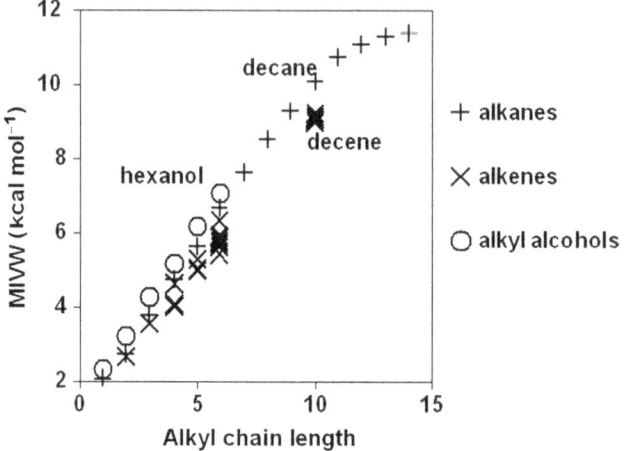

Figure 6.3 The relationship between alkyl chain length (number of carbons) and van der Waals energy (MIVW).
Reproduced by permission of Springer, ref. 22.

Table 6.3 Hammett's equations for pK_a calculation.

Analyte	A	$+ B\Sigma\sigma$
Benzoic acid	4.20	$-1.00 \, \Sigma\sigma$
2-chloro benzoic acid	3.69	$-0.86 \, \Sigma\sigma$
2-hydroxy benzoic acid	4.00	$-1.10 \, \Sigma\sigma$
2-methyl benzoic acid	3.90	$-1.22 \, \Sigma\sigma$
2-nitro benzoic acid	2.21	$-0.91 \, \Sigma\sigma$
Phenylacetic acids (1)	3.18	$-0.23 \, \Sigma\sigma$
Phenylacetic acids (2)	3.17	$-0.30 \, \Sigma\sigma$
Phenolic compounds	9.92	$-2.23 \, \Sigma\sigma$
Anilines	4.58	$-2.88 \, \Sigma\sigma$
Pyridines (1)	5.25	$-5.90 \, \Sigma\sigma$
Pyridines (2)	5.39	$-5.70 \, \Sigma\sigma$
Quinoline (2-substituted)	5.12	$-9.04 \, \Sigma\sigma$
Quinoline (8-substituted)	4.64	$-3.11 \, \Sigma\sigma$
α-Naphthylamines	3.85	$-2.81 \, \Sigma\sigma$
β-Naphthylamines	4.29	$-2.81 \, \Sigma\sigma$
Benzylamines (ring-substituted)	9.39	$-1.05 \, \Sigma\sigma$
N,N-Dimethylanilines	5.06	$-3.46 \, \Sigma\sigma$
1-Aminoanthracenes	substituted aniline	$+0.17$
Quinolines	substituted pyridine	-0.06

pK_a values and the atomic partial charge calculated using the MOPAC program.[17] The following equation was proposed based on Hammett's method and the CAChe program for the prediction of pK_a values using atomic partial charge.[7]

$$pK_a(\text{predicted}) = pK_a(\text{base compounds}) + \Delta pK_a(\text{substituent effect})$$

where pK_a (base compounds) was derived from the atomic partial charge of the basic compounds, benzoic acid, phenylacetic acid, 3-phenylpropionic acid, mandelic acid, *trans*-cinnamic acid, indole-3-acetic acid, indole-3-butylic acid and phenol. The ΔpK_a (substituent effect) was derived from the difference in the atomic partial charge between derivatized and base compounds. These equations were further modified for the investigation of QSRR of acidic compounds including aromatic acids and phenolic compounds in reversed-phase liquid chromatography.

First, the apc values of the hydroxyl group hydrogens and oxygens of phenolic compounds were correlated with measured pK_a values and used to study the QSRR in the reversed-phase liquid chromatography of phenolic compounds.[18] Later, two equations were derived from the atomic partial charge of the carboxyl group hydrogens on aromatic acids for the prediction of pK_a values instead of using Hammett's equations. The predicted pK_a values were successfully used to study the QSRR of aromatic acids in reversed-phase liquid chromatography at various eluents pH values.[7]

The AM1 and PM3 calculation methods in the MOPAC program have been used for a variety of applications. Therefore, the atomic partial charge was calculated using AM1 and PM3 in the CAChe program, and a suitable calculation method was evaluated for the prediction of pK_a values. The calculated atomic partial charges using PM3 are a little larger, except the partial charge of oxygen. Such a difference is due to the precision of the atomic heat of formation, according to J. Stewart[19] who is developing the MOPAC program. The precision of the calculation of atomic partial charges by AM1 was better than the predicted pK_a values of phenolic compounds and aromatic acids.[20] Therefore, these atomic partial charges were correlated with measured pK_a values[21] in Table 6.4.

$$pK_a = -209.611 \times (\text{H by AM1}) + 55.568, \ r = 0.996, \ n = 8, \qquad (2)$$

$$pK_a = 53.365 \times (\text{O by AM1}) + 21.412, \ r = 0.984, \ n = 8. \qquad (3)$$

The atomic partial charge of hydrogen was shown to be suitable for predicting the pK_a values of aromatic acids,[7] and that of oxygen was better than that of hydrogen for phenolic compounds. Which value derived from the

Table 6.4 pK_a and atomic partial charges of standard compounds.

| Chemicals | Atomic partial charge | | Predicted pK_a | | |
	O	H	pK_a[7]	O (AM1)	H (AM1)
Benzoic acid	−0.3171	0.2455	4.200	4.490	4.108
Phenylacetic acid	−0.3211	0.2447	4.334	4.276	4.276
Mandelic acid	−0.3262	0.2476	3.420	4.004	3.668
trans-Cinnamic acid	−0.3145	0.2436	4.376	4.629	4.507
Phenylpropionic acid	−0.3209	0.2440	4.691	4.287	4.423
Indoleacetic acid	−0.3180	0.2424	4.590	4.442	4.758
Indolebutylic acid	−0.3215	0.2430	4.781	4.255	4.632
Phenol	−0.2134	0.2172	10.020	10.024	10.040

atomic partial charges of hydrogen or oxygen is better for acidic compounds overall? The partial charges of hydrogen and oxygen for 52 phenolic compounds were calculated using AM1, and are listed in Table 4 of the Appendix (p. 292) with the measured pK_a values from ref. 21.

The substituent effect (Δapc) on atomic partial charge was balanced between the atomic partial charge of the substituted phenol and phenol using: Δapc = apc (substituted phenol) − apc (phenol). The substituent effects of the atomic partial charge were correlated with the substituent effects of the pK_a values. The correlations between ΔpK_a (pK_a of the substituted phenol $-pK_a$ of phenol) and Δapc were as follows, where the pK_a values were predicted using individual equations, and the atomic partial charges were calculated.

$$\Delta pK_a = -153.659 \times (\Delta \text{apc O by AM1}) - 0.355,\ r = 0.953,\ n = 38, \quad (4)$$

$$\Delta pK_a = -202.919 \times (\Delta \text{apc H by AM1}) + 0.180,\ r = 0.951,\ n = 38. \quad (5)$$

The results suggested that the atomic partial charge of either the hydrogen or the oxygen of a hydroxy group can be used for the prediction of pK_a. The recalculated pK_a values of 12 benzoic acid derivatives are listed in Table 6.5. The correlations between the ΔpK_a and Δapc of benzoic acid derivatives are:

$$\Delta pK_a = -189.852 \times (\Delta \text{apc O by AM1}) + 0.042,\ r = 0.970,\ n = 12, \quad (6)$$

$$\Delta pK_a = -160.438 \times (\Delta \text{apc H by AM1}) + 0.042,\ r = 0.982,\ n = 12. \quad (7)$$

where the atomic partial charges of benzoic acid, 3-methylbenzoic acid, 4-methylbenzoic acid, 3,4-dimethylbenzoic acid, 3,5-dimethylbenzoic acid, 4-ethylbenzoic acid, 3-chlorobenzoic acid, 4-chlorobenzoic acid, 3,4-dichlorobenzoic acid, 3,5-dichlorobenzoic acid, 3-bromobenzoic acid, and 4-bromobenzoic acid were used for the calculation. However, proton partial atomic charge was selected when the approach was applied to other compounds.

Then, the term $B\Sigma\sigma$ of eqn (1) was derived by AM1 from the relationship between the ΔpK_a and Δapc H of phenolic compounds and benzoic acid derivatives, found in eqns (5) and (7):

$$B\Sigma\sigma = \Delta pK_a = (-7.299 \times \Delta \text{apc} + 0.024) \times pK_a + (-129.782 \times \Delta \text{apc} - 0.057) \quad (8)$$

Table 6.5 pK_a values of benzoic acid derivatives.[7]

Molecule	pK_a	Molecule	pK_a
Benzoic acid	4.20	3-Chlorobenzoic acid	3.83
3-Methylbenzoic acid	4.26	4-Chlorobenzoic acid	3.96
4-Methylbenzoic acid	4.34	3,4-Dichlorobenzoic acid	3.50
3,4-Dimethylbenzoic acid	4.40	3,5-Dichlorobenzoic acid	3.46
3,5-Dimethylbenzoic acid	4.32	3-Bromobenzoic acid	3.86
4-Ethylbenzoic acid	4.35	4-Bromobenzoic acid	3.98

Finally, eqns (2) and (8) were used to predict the pK_a values of phenolic compounds. The predicted pK_a values are listed in Table 4 of the Appendix (p. 292). The correlation between the predicted and measured pK_a values is:

$$pK_a(\text{reference}) = 1.000 \times (pK_a \text{ predicted}) - 0.004, \, r = 0.951, \, n = 38.$$

Furthermore, the pK_a values predicted using the atomic partial charge of hydrogen were re-examined according to the *ortho*-effect (intra-molecular hydrogen bonding) and the phenol pK_a values. The *ortho*-effect for pK_a values was 0.659. The predicted phenol pK_a value was 0.184 higher than the reference value. The addition of the *ortho*-effect and the phenol value improved the correlation coefficient between measured and predicted pK_a values to 0.987 from 0.951 ($n = 38$), and the pK_a values are summarized in Table 4 of the Appendix (p. 292). The relationship is shown in Figure 6.4. The same approach was applied to aromatic acids[7] and nitrogen-containing compounds.[18] These pK_a values are summarized in Tables 5–8 of the Appendix (pp. 294–296).

The *ortho*-effect of halogen-substituted benzoic acids was very high, as shown in Figure 6.5 where error values are plotted against the predicted pK_a values calculated using Hammett's equation. Further investigation of the *ortho*-effect will require additional experimental pK_a values, for other acids.

The prediction of pK_a values for nitrogen-containing compounds was not simple. The atomic partial charge of primary amino group hydrogens can be used for the prediction of pK_a values, as shown in Figure 6.6, but not for quarternary amino groups. Nitrogen atomic partial charges were also used for this purpose. The r value for the correlation between the predicted and reference pK_a values was 0.951 ($n = 23$). Benzylamine, 1-aminoindan,

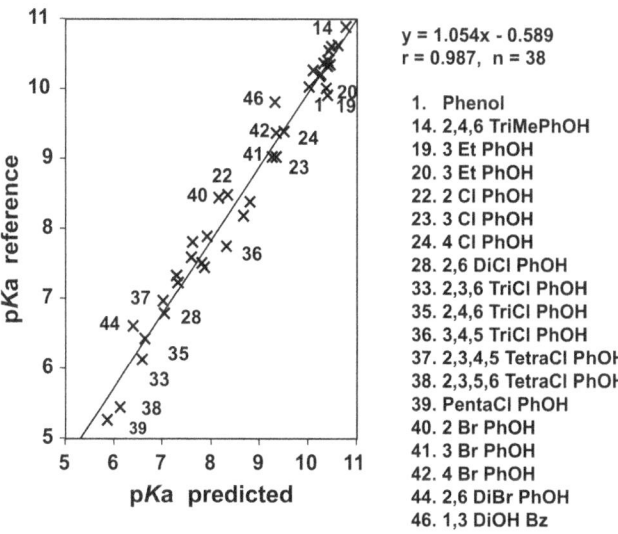

y = 1.054x - 0.589
r = 0.987, n = 38

1. Phenol
14. 2,4,6 TriMePhOH
19. 3 Et PhOH
20. 3 Et PhOH
22. 2 Cl PhOH
23. 3 Cl PhOH
24. 4 Cl PhOH
28. 2,6 DiCl PhOH
33. 2,3,6 TriCl PhOH
35. 2,4,6 TriCl PhOH
36. 3,4,5 TriCl PhOH
37. 2,3,4,5 TetraCl PhOH
38. 2,3,5,6 TetraCl PhOH
39. PentaCl PhOH
40. 2 Br PhOH
41. 3 Br PhOH
42. 4 Br PhOH
44. 2,6 DiBr PhOH
46. 1,3 DiOH Bz

Figure 6.4 The relationship between the predicted and reference pK_a values of phenolic compounds.

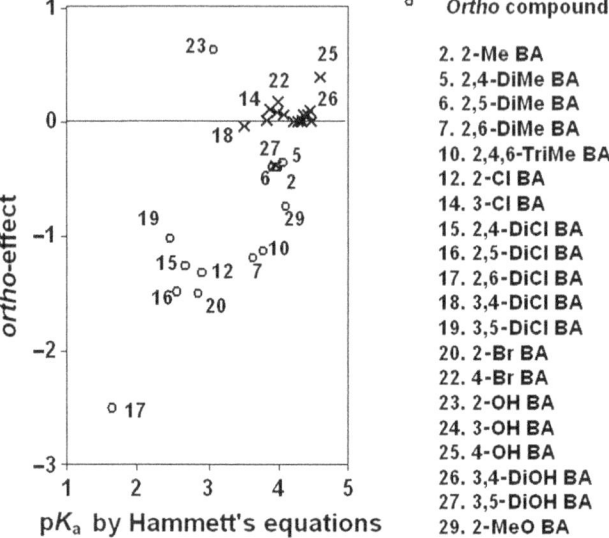

Figure 6.5 The relationship between the *ortho*-effect of benzoic acid derivatives and pK_a.

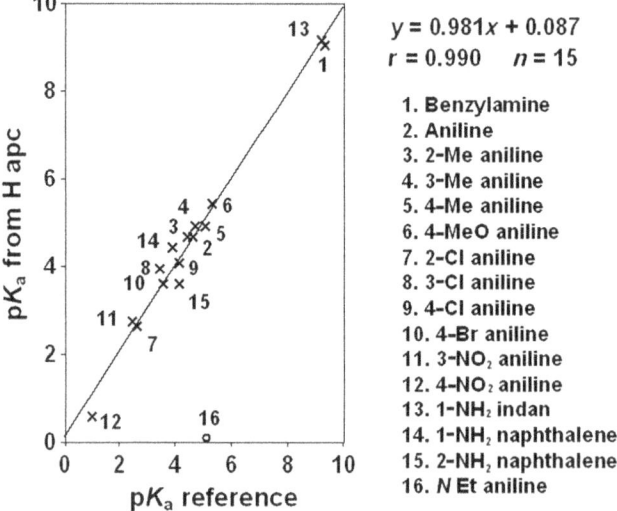

Figure 6.6 Predicted aniline derivative pK_a values.

quaternary amines, 2- and 4-amino pyridine, and 8-hydroxyquinoline were eliminated from the calculations.

The prediction of pK_a values using a modified Hammett's equation with atomic partial charge can be performed for simple compounds. Improving the computational chemical calculation method, and increasing the number of reference pK_a values can improve the precision of predicted pK_a values.

6.3 Chromatographic Behavior on Graphitized Carbon

Graphitized carbon is a special phase that adsorbs entire compounds. These materials are used, mainly for sample collection, in several analytical fields. Small, porous graphitized carbon particles are used as a packing material for gas analysis in gas chromatography, and for saccharides and anions in liquid chromatography. Only cations are not adsorbed.[14,15] The computational chemical analysis of a model graphitized carbon phase was performed using a 196-aromatic-ring phase and alkanes and alkenes. The model phase is shown in Figure 6.7. The relationship between the molecular interaction energy values and the number of decene double bonds is summarized in Figure 6.8.

The results indicate that increasing the number of double bonds decreases the molecular interaction energy values, especially those of *cis*-form compounds. The predominant interaction force is the van der Waals force. The electrostatic energy did not change after formation of the complex. No π–π interactions influenced the direct interaction. The calculated results supported the chromatographic behavior of fatty acids in reversed-phase liquid chromatography.

The chromatographic behavior of the analytes listed in Table 6.3, with their properties, was analyzed to determine whether or not the retention

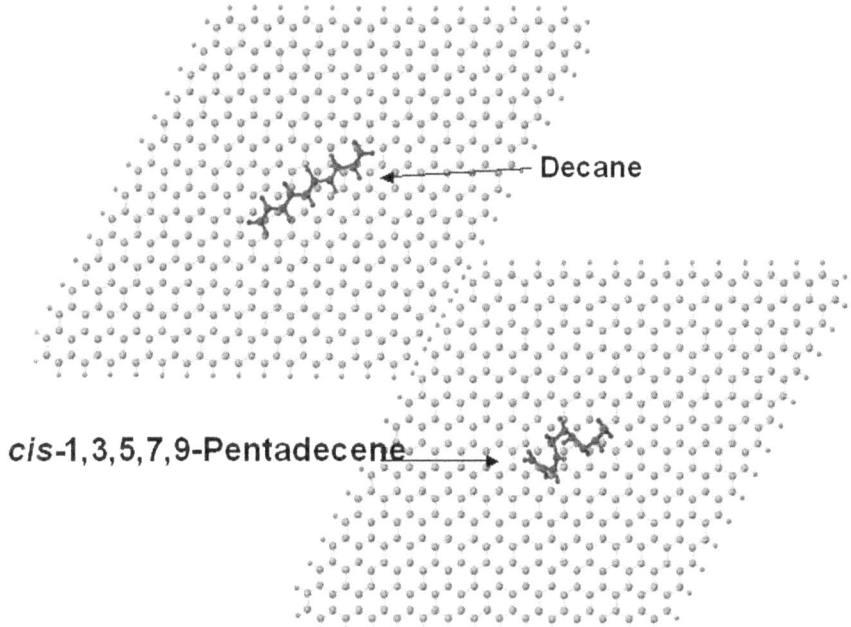

Figure 6.7   Adsorption of decane and *cis*-1,3,5,7,9-pentadecene on a model graphitized carbon phase. Small and large balls represent hydrogen and carbon, respectively.

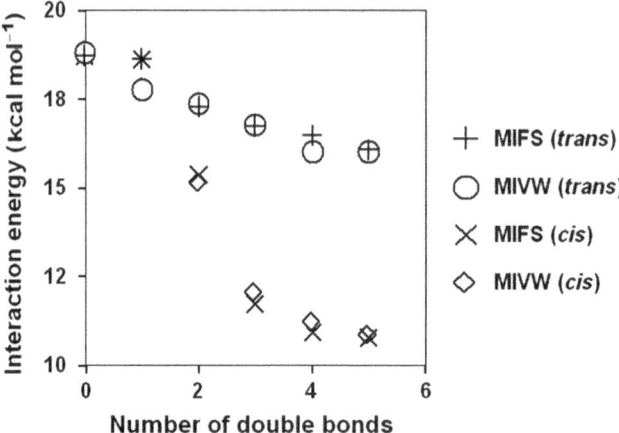

Figure 6.8   The relationship between the molecular interaction energy and the number of decene double bonds.

mechanism was based on reversed-phase liquid chromatography. These compounds were classified into two groups according to their acidity and basicity. The log P values of the analytes were related to the log k values measured at pH 2 and 10. There were only six compounds. The correlation was better for basic compounds ($r=0.825$, $n=6$). Silanol groups existed in their sodium salt form and did not contribute as a hydrogen-bonding partner at high pH, but did affect the retention of the acidic compounds at low pH. The correlation coefficient was 0.352 ($n=6$) for acidic compounds at pH 2.

The MI energy values were calculated using the model carbon phase shown in Figure 6.9, with *p*-anisidine on the model phase. The calculated energy values are summarized in Table 6.6.

The correlation coefficient between the calculated individual MI and the logarithmic capacity ratios indicated the contribution of each individual factor to the retention. MIVW was the main contributor to the retention in reversed-phase liquid chromatography,[22] and MIES was the main contributor to the retention in ion-exchange liquid chromatography.[23] Steric hindrance affected the molecular interactions in enantioseparation.[24] The MI energy values were correlated with reference log k values, and the results are given in the following equations, and demonstrated in Figure 6.10(a) and (b):

At pH 10:

$\quad$ MIFS $= 2.854 \times (\log k) + 12.303$, $r=0.902$, $n=6$,

$\quad$ MIHB $= 0.048 \times (\log k) + 2.769$, $r=0.061$, $n=6$,

$\quad$ MIES $=$ No correlation was calculated due to identical MIES values,

$\quad$ MIVW $= 2.796 \times (\log k) + 9.703$, $r=0.822$, $n=6$.

Carbon phase

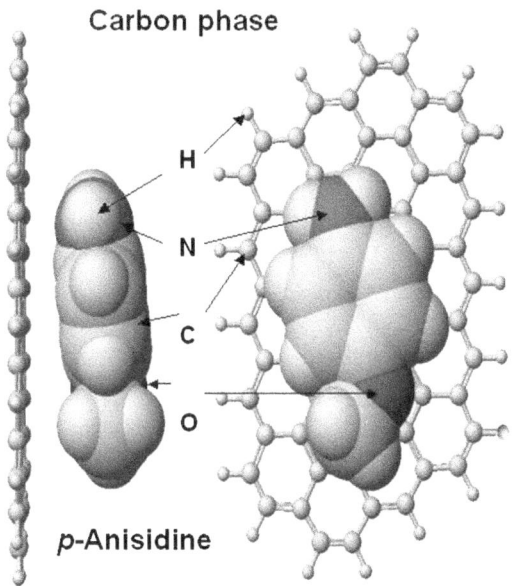

p-Anisidine

Figure 6.9 Adsorption of *p*-anisidine on a model graphitized carbon phase. White, light-gray, dark-gray, and black balls represent hydrogen, carbon, nitrogen, and oxygen, respectively.
Reproduced by permission of Taylor and Francis, ref. 50.

At pH 2:

$$MIFS = 0.328 \times (\log k) + 16.078, \, r = 0.140, \, n = 6,$$

$$MIHB = -0.969 \times (\log k) + 7.595, \, r = 0.844, \, n = 6,$$

$$MIES = 0.018 \times (\log k) - 0.104, \, r = 0.372, \, n = 6,$$

$$MIVW = 1.588 \times (\log k) + 9.346, \, r = 0.534, \, n = 6.$$

Why were these compounds retained so strongly on the graphitized carbon column? According to one study,[25] the graphitized carbon was synthesized by washing the silica from graphitized carbon using potassium hydroxide. Using this method, a high-porosity HPLC silica gel was impregnated with a phenol–formaldehyde resin. The resin was carbonized at 2000–2800 °C in nitrogen or argon, and the silica particles dissolved out with alkali. This process results in the possibility of trace amounts of silica and metals remaining in the graphitized carbon. This is likely to have occurred, given the relatively high correlation coefficient for the hydrogen-bonding energy values with log k.

The probability of silanol groups affecting the graphitized carbon was studied using a model silanol phase that was used to study retention of silica gels.[26] The MI energy values were calculated using the model silanol phase

Table 6.6 The molecular properties and energies of analytes on two different phases.

Analyte	log P	fs	hb	es	vw	FS^a
o-Aminobenzoic acid	0.979	−18.0208	−5.740	−11.260	6.452	−881.8233
m-Aminobenzoic acid	1.053	−15.8272	−5.526	−7.258	5.435	−887.3857
p-Aminobenzoic acid	0.995	−15.8508	−5.514	−7.265	5.384	−884.0685
o-Anisic acid	1.621	−15.2144	−4.250	−7.200	6.069	−880.2007
m-Anisic acid	1.708	−16.2962	−4.186	−7.375	5.303	−886.5956
p-Anisic acid	1.670	−16.3114	−4.182	−7.309	5.207	−889.3311
o-Anisidine	1.275	−4.3759	−1.783	0.580	3.729	−866.7141
m-Anisidine	1.391	−4.5510	−1.340	−0.010	3.718	−865.0760
p-Anisidine	1.351	−4.5652	−1.330	−0.008	3.695	−864.4962
o-Phenetidine	1.764	−3.8094	−1.787	0.580	4.229	−869.4632
m-Phenetidine	1.892	−3.9153	−1.340	−0.010	4.289	−868.0431
p-Phenetidine	1.864	−3.9271	−1.330	−0.080	4.270	−865.4588
o-Toluic acid	0.979	−14.6095	−4.184	−6.088	6.476	−883.2097
m-Toluic acid	1.990	−18.4149	−4.180	−7.371	5.029	−889.9217
p-Toluic acid	2.020	−18.4331	−4.179	−7.351	4.971	−889.6820
Silanol phase	—	−842.9634	−34.699	−699.650	−249.575	—
Carbon phase	—	38.9017	0.000	0.000	65.672	—

Analyte	HB^a	ES^a	VW^a	FS^b	HB^b	ES^b	VW^b
o-Aminobenzoic acid	−57.448	−712.530	−247.004	3.9764	−16.312	−11.135	64.200
m-Aminobenzoic acid	−61.144	−710.219	−249.389	6.7896	−13.697	−7.185	61.433
p-Aminobenzoic acid	−63.337	−709.647	−245.441	7.0117	−12.840	−7.190	60.945
o-Anisic acid	−56.102	−708.568	−248.902	6.9083	−10.888	−7.119	60.539
m-Anisic acid	−60.135	−709.679	−251.425	6.0108	−10.550	−7.297	59.597
p-Anisic acid	−60.788	−710.279	−250.815	5.7458	−10.465	−7.230	59.121
o-Anisidine	−49.563	−698.803	−252.456	21.0089	−4.823	0.580	58.778
m-Anisidine	−48.068	−699.812	−251.824	20.7514	−4.255	−0.010	58.536
p-Anisidine	−46.271	−700.104	−252.685	20.8385	−3.914	−0.008	58.289
o-Phenetidine	−51.005	−699.663	−252.966	20.0854	−4.432	0.580	57.397
m-Phenetidine	−49.567	−699.975	−252.754	19.8381	−4.339	−0.010	57.642
p-Phenetidine	−46.325	−700.131	−253.641	19.9858	−3.961	−0.008	57.405
o-Toluic acid	−57.534	−706.953	−248.830	8.4151	−10.880	−5.990	61.791
m-Toluic acid	−59.044	−710.164	−250.663	4.1319	−10.814	−7.286	59.673
p-Toluic acid	−58.758	−710.165	−250.938	4.2232	−10.543	−7.268	59.497

[a]Values on a silanol phase.
[b]Values on a model graphitized carbon phase.

shown in Figure 6.11, with p-anisidine adsorbed. The correlation coefficients are given in the following equations, and demonstrated in Figure 6.12(a) and (b).

At pH 10:

$$MIFS = 7.695 \times (\log k) + 14.023, \ r = 0.922, \ n = 6,$$

$$MIHB = 4.718 \times (\log k) + 8.993, \ r = 0.706, \ n = 6,$$

$$MIES = 0.482 \times (\log k) - 0.051, \ r = 0.408, \ n = 6,$$

$$MIVW = 1.909 \times (\log k) + 5.803, \ r = 0.595, \ n = 6.$$

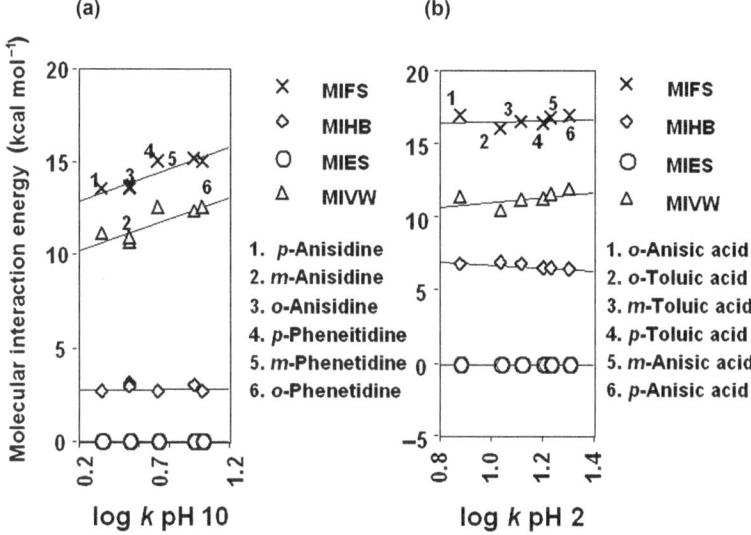

Figure 6.10 Relationship between molecular interaction energy and log *k* on a model graphitized carbon phase at (a) pH 10 and (b) pH 2.

Silanol phase

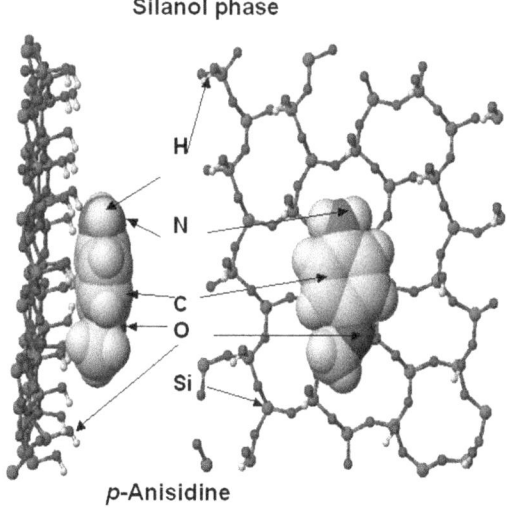

p-Anisidine

Figure 6.11 Adsorption of *p*-anisidine on a model silanol phase. White, light-gray, large white, dark-gray, and black balls represent hydrogen, carbon, silicon, nitrogen, and oxygen, respectively.
Reproduced by permission of Taylor and Francis, ref. 50.

At pH 2:

$$MIFS = 16.926 \times (\log k) + 7.799, \; r = 0.929, \; n = 6,$$
$$MIHB = 10.791 \times (\log k) + 7.606, \; r = 0.969, \; n = 6,$$
$$MIES = 4.310 \times (\log k) - 2.348, \; r = 0.768, \; n = 6,$$
$$MIVW = 3.319 \times (\log k) + 2.435, \; r = 0.846, \; n = 6.$$

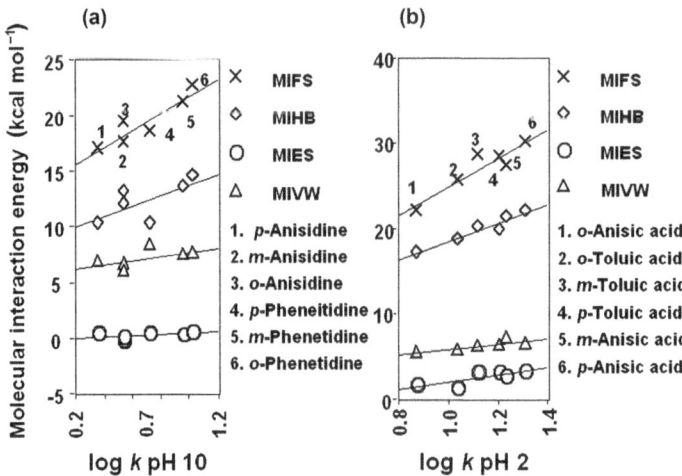

Figure 6.12 The relationship between molecular interaction energy and log k on a model silanol phase at (a) pH 10 and (b) pH 2.

There was a strong correlation between log k and the hydrogen-bonding energy values. This result is typical of normal-phase liquid chromatography using silica gels, even when the log k values are measured using a graphitized carbon column.

These results support the speculation regarding the silanol activity of one of the graphitized carbon columns. It seems that the matrix silica gel was not completely washed, and some silanol activity persisted.

$$MIFS = 0.183 \times (\log P) + 15.305, \ r = 0.057, \ n = 15,$$

$$MIHB = -2.656 \times (\log P) + 9.448, \ r = 0.407, \ n = 15,$$

$$MIES = 0.032 \times (\log P) - 0.099, \ r = 0.265, \ n = 15,$$

$$MIVW = 2.387 \times (\log P) + 7.380, \ r = 0.765, \ n = 15.$$

Log P values do not relate to MI energy values. They relate weakly with van der Waals energy values, because molecular size is a very important factor for estimating log P values.

A semi-empirical molecular statistical theory of adsorption based on an atom–atom approximation for the potential function of the intermolecular adsorbate–adsorbent interaction was studied to obtain Henry's constant based on gas chromatography data using thermally graphitized carbon, even though porous adsorbents are inhomogeneously prepared. A simple quantitative correlation of the thermodynamic characteristics of adsorption was applied in liquid chromatography.[27] Based on several experimental measurements of the isotherms for one of the test solutes followed by non-linear model fitting, the graphitized carbon surface was considered to be homogeneous and to have only one type of adsorption site.[28] Hence, an MM2 calculation was performed to obtain Henry's constant using a flat model.[28] The latter method is straightforward for analyzing a variety of chromatographic data, but it is not straightforward for synthesizing model phases other than a graphitized carbon phase. Chromatographic phases are synthesized homogeneously but their steric structures are inhomogeneous. The original computer software is used for the conformational analysis of proteins, therefore such an approach can be applied to the analysis of chromatographic retention if a prospective model is designed. The MI energy values calculated using the MM2 program support a set of data measured on a graphitized carbon column synthesized using 100% organic materials, but not the retention data measured on a graphitized carbon column synthesized using a silica matrix. The chromatographic behavior of these graphitized carbon columns differs, as clearly demonstrated by the above computational chemical approach.

6.4   Model Phase Selectivity for Phenolic Compounds

A model butyl-bonded phase was constructed with highly dense butyl-groups, but without silanol groups based on the chromatographic performance of pentyl-bonded silica gels.[29] The butyl-bonded phase was a modified carbon phase, and consisted of 628 carbons, 216 hydrogens, 1197 bonds and 6768 connectors. The molecular weight was 7752. The optimized energy value change was less than 0.00001 kcal mol^{-1}.[30,31] The molecular size and alkyl chain length were decided by the calculation capacity of the computer used and the alkyl chain length effect on the hydrophobicity.[32] The adsorption form of 2,4,6-trimethylphenol on the butyl phase is shown in Figure 6.13.

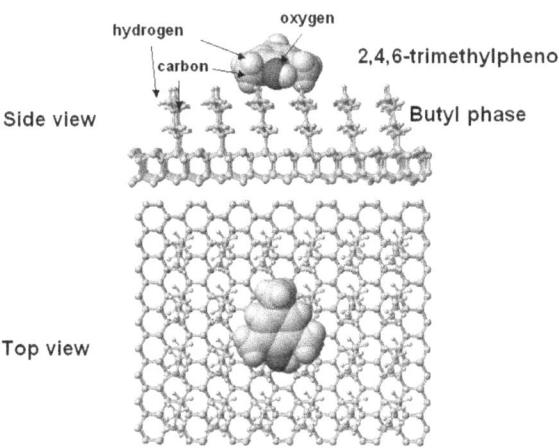

Figure 6.13   Adsorption of 2,4,6-trimethylphenol on a butyl-bonded phase. White, light-gray, and black balls represent hydrogen, carbon, and oxygen, respectively. The atom size of the butyl phase is 20%. Reproduced by permission of Springer, ref. 22.

2,4,6-Trimethylphenol

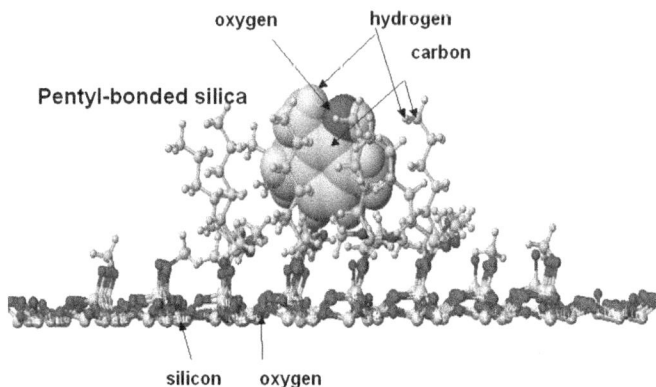

Figure 6.14   Adsorption of 2,4,6-trimethylphenol on a model pentyl-bonded silica phase. White, light-gray, gray, and black balls represent hydrogen, carbon, silicon, and oxygen, respectively. The atom size of the pentyl phase is 20% of the original size for easy demonstration.

A silica-gel-based pentyl-bonded phase was also constructed. It consisted of 686 atoms, 746 bonds, and 5130 connectors, containing 158 silicons, 304 oxygens, 64 carbons and 160 hydrogens. The monolayer shown in Figure 3.11 was locked to avoid deformation of the structure by further optimization, because silica gel does not change its atomic distances under liquid chromatographic conditions. The structure of the model phase consisted of eight pentyl groups and many oxygens that were kept free to reduce the number of atoms. A side view of an optimized structure with an adsorbed 2,4,6-trimethylphenol is shown in Figure 6.14.

The calculated energy values of the individual compounds, these two phases and their complexes calculated using the MM2 program are listed in Table 9 of the Appendix (p. 298) along with the properties, log P and pK_a, of the phenolic compounds used. The calculated energy values hb and es are not listed due to their lack of meaningful contribution to molecular interaction. The details are explained later. The energy values of individual complexes between a model butyl phase and a phenolic compound are also listed in Table 9 of the Appendix (p. 298). FS1 and VW1 are the final (optimized) structure and the van der Waals energy values of a complex between the butyl-bonded phase and a phenolic compound, and FS2 and VW2 are the final (optimized) structure and van der Waals energy values of a complex between the pentyl-bonded phase and a phenolic compound, respectively.

The interaction energy values between molecular-form phenolic compounds and the model butyl phase were calculated using the MM2 program to qualitatively analyze the retention of the molecular form of analytes. The r values between MIFS1 or MIVW1 calculated using the model butyl phase and the measured log k values of molecular-form phenolic compounds, listed as pH 4.01_{mes} in Table 10 of the Appendix (p. 303), were 0.916 ($n = 38$). The retention times were measured by reversed-phase liquid chromatography

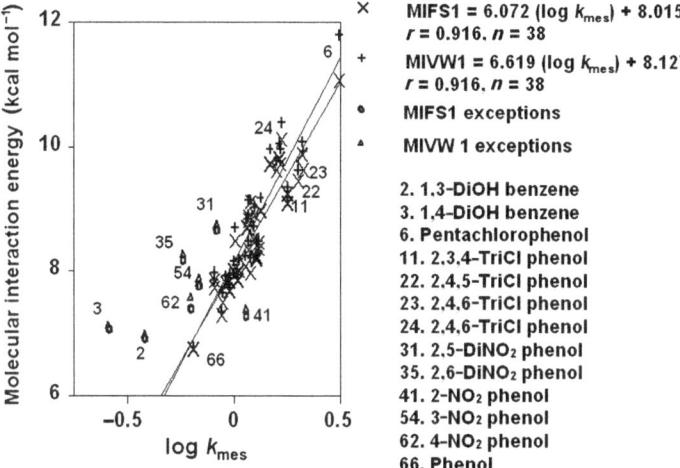

Figure with axes: y-axis "Molecular interaction energy (kcal mol⁻¹)" from 6 to 12, x-axis "log k_{mes}" from −0.5 to 0.5.

Legend:
MIFS1 = 6.072 (log k_{mes}) + 8.015
$r = 0.916$, $n = 38$

MIVW1 = 6.619 (log k_{mes}) + 8.127
$r = 0.916$, $n = 38$

● MIFS1 exceptions
▲ MIVW 1 exceptions

2. 1,3-DiOH benzene
3. 1,4-DiOH benzene
6. Pentachlorophenol
11. 2,3,4-TriCl phenol
22. 2,4,5-TriCl phenol
23. 2,4,6-TriCl phenol
24. 2,4,6-TriCl phenol
31. 2,5-DiNO₂ phenol
35. 2,6-DiNO₂ phenol
41. 2-NO₂ phenol
54. 3-NO₂ phenol
62. 4-NO₂ phenol
66. Phenol

Figure 6.15 The relationship between molecular interaction energy and log k_{mes} on a model butyl phase.

using an octadecyl-bonded silica gel column at 40 °C in 70% aqueous acetonitrile containing 20 mM sodium phosphate solution, at pH 2–10.[18] The results are shown in Figure 6.15.

The outliers were nitro-substituted phenols, dihydroxybenzenes and *tert*-butylphenols and ethylphenols. The capacity ratios of dihydroxybenzenes were too short. In this model system, one side of the analyte contacted with this model phase, and the steric effect cannot be neglected for *tert*-butyl-phenol and ethylphenols.

An improvement in the correlation was expected if a low-density phase was used as a model phase because an analyte should be buried in the alkyl chains. The interaction energy values between a phenolic compound and the silica gel-based-pentyl phase were calculated. The *r* value between MIFS2 calculated using the model pentyl phase and the measured log *k* values (k_{mes}) of molecular-form phenolic compounds, listed as pH 4.01$_{mes}$ in Table 10 of the Appendix (p. 303), improved to 0.956 ($n = 42$), and all alkyl-substituted phenols were included in the calculation. The *r* value was 0.983 ($n = 42$) from the MIVW2 energy values in Table 12 of the Appendix (p. 306), as shown in Figure 6.16. The outliers were nitro-substituted phenols and dihydrox-ybenzenes. The poor results for nitro-substituted phenols may be due to the difficulty of the computational chemical calculation of the nitro group that was observed for the log *P* prediction.[18,34]

The contribution of the HB2 and ES2 values listed in Table 6.15 can be neglected due to their very poor *r* values at present. The HB2 energy value of this model phase is zero. The *r* values for MIHB2 and MIES2 were 0.183 and 0.558, respectively. The contribution of the MIVW energy indicated that hydrophobic interactions were the predominant molecular interaction in the retention of these phenolic compounds on an alkyl-bonded phase in reversed-phase liquid chromatography.

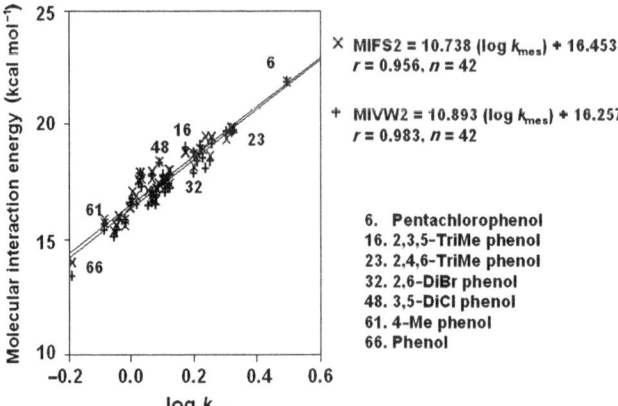

Figure 6.16   The relationship between molecular interaction energy and log k_{mes} on a model pentyl phase.

The capacity ratios of the partially ionized phenolic compounds were calculated using eqn (9)[33] and pK_a values predicted from the partial charge of the hydrogen of the phenolic hydroxy group.[34] The reversed-phase liquid chromatographic results of several compounds such as benzoic acid, phenol, 4-chlorophenol, 2,4-dichlorophenol, 2,4,6-trichlorophenol and 2,3,4,6-tetrachlorophenol, using a pentyl-bonded silica gel were analyzed using molecular mechanics. The correlation coefficient between their interaction energy values and logarithmic capacity ratios was very high, even when these compounds were completely ionized, as shown in Figure 6.17. The precision of the predicted capacity ratios of partly ionized compounds using eqn (9) was very satisfactory. The correlation coefficients between the predicted and measured retention factors were greater than 0.9 for eluents at pH 3–9.

$$k = \frac{k_m + k_i(K_a/[H^+])}{1 + (K_a/[H^+])} \tag{9}$$

where k_m and k_i are the capacity ratios of the molecular and ionized analytes, respectively, K_a is the dissociation constant of the analytes, and $[H^+]$ represents the hydrogen ion concentration in the eluent.

The correlation between those capacity ratios that were measured, those predicted by the former method using NlogP,[34] and those predicted by this new method using MIFS2, at pH 8.49 are summarized in Figure 6.18. The measured and predicted capacity ratios of phenolic compounds are given in Table 10 of the Appendix (p. 303).

The above results indicate that the retention time of phenolic compounds can be predicted using both energy value changes in the optimized structure calculated using the MM2 program, and log P values calculated using the MOPAC program. The addition of pK_a values predicted from the atomic partial charge calculated using the MOPAC program enables the capacity ratios in a given pH eluent to be predicted.

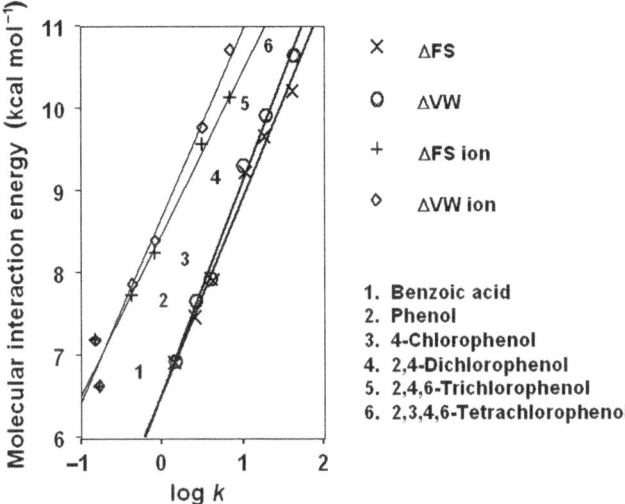

Figure 6.17 The relationship between molecular interaction energy and log k for both the molecular and ionized forms of phenolic compounds on a model pentyl phase.

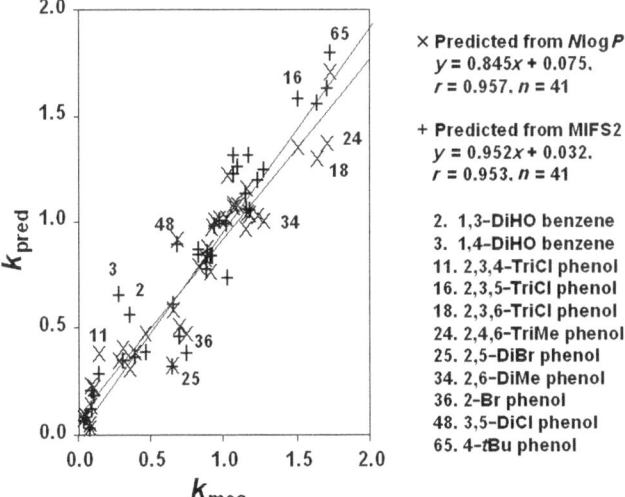

Figure 6.18 The relationship between predicted (k_{pred}) and measured (k_{mes}) capacity ratios at pH 8.49.

The retention time of phenolic compounds in reversed-phase liquid chromatography was predicted *via* molecular interaction energy values calculated using the MM2 program. The precision of the capacity ratios predicted by this new method was equivalent to a former method in which the retention time was predicted by log P calculated using the MOPAC program. Furthermore, the prediction of capacity ratios of phenolic compounds in reversed-phase

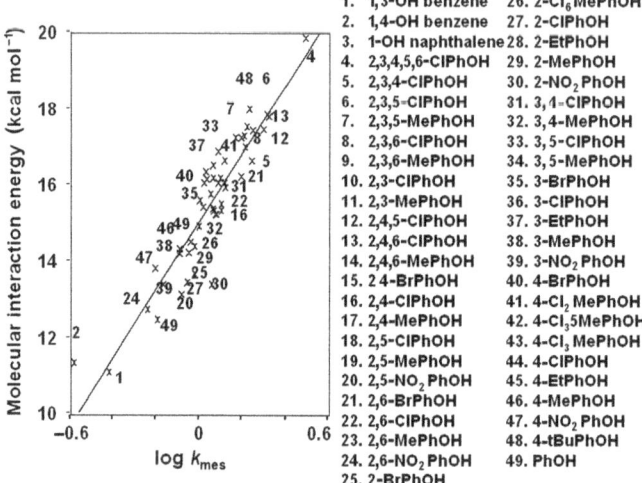

1. 1,3-OH benzene
2. 1,4-OH benzene
3. 1-OH naphthalene
4. 2,3,4,5,6-ClPhOH
5. 2,3,4-ClPhOH
6. 2,3,5-ClPhOH
7. 2,3,5-MePhOH
8. 2,3,6-ClPhOH
9. 2,3,6-MePhOH
10. 2,3-ClPhOH
11. 2,3-MePhOH
12. 2,4,5-ClPhOH
13. 2,4,6-ClPhOH
14. 2,4,6-MePhOH
15. 2 4-BrPhOH
16. 2,4-ClPhOH
17. 2,4-MePhOH
18. 2,5-ClPhOH
19. 2,5-MePhOH
20. 2,5-NO$_2$PhOH
21. 2,6-BrPhOH
22. 2,6-ClPhOH
23. 2,6-MePhOH
24. 2,6-NO$_2$PhOH
25. 2-BrPhOH
26. 2-Cl$_6$MePhOH
27. 2-ClPhOH
28. 2-EtPhOH
29. 2-MePhOH
30. 2-NO$_2$PhOH
31. 3,4-ClPhOH
32. 3,4-MePhOH
33. 3,5-ClPhOH
34. 3,5-MePhOH
35. 3-BrPhOH
36. 3-ClPhOH
37. 3-EtPhOH
38. 3-MePhOH
39. 3-NO$_2$PhOH
40. 4-BrPhOH
41. 4-Cl$_2$MePhOH
42. 4-Cl$_3$5MePhOH
43. 4-Cl$_3$MePhOH
44. 4-ClPhOH
45. 4-EtPhOH
46. 4-MePhOH
47. 4-NO$_2$PhOH
48. 4-tBuPhOH
49. PhOH

Figure 6.19   The relationship between molecular interaction energy and log k_{mes} for molecular-form phenolic compounds.

liquid chromatography in a given pH eluent was performed using pK_a values from the atomic partial charge calculated using the MOPAC program.

Further study was performed, and the precision was improved. Both nitro- and hydroxyl-substituted phenolic compounds were included in the calculation. The nitro-substituted phenolic compound recognized the existence of hydrogen bonding for adsorption, however the hydrogen-bonding energy values of the complexes were identical to the hydrogen-bonding energies of the phenolic compounds themselves. Those of *ortho*-substituted compounds were very high, due to intramolecular hydrogen bonding. Hydrogen bonding was not considered to contribute to the retention. The subtraction of the hydrogen-bonding energy values of phenolic compounds from the interaction energy values improved the correlation coefficient.

$$\text{MIFS} = 7.721 \times (\log k_{mes}) + 16.834, \; r = 0.879, \; n = 49,$$

$$(\text{MIFS} - \text{hb}) = 8.919 \times (\log k_{mes}) + 14.996, \; r = 0.926, \; n = 49.$$

Catechols and nitrophenols were not outliers (Figure 6.19). Addition of MIES improved r to 0.930. The reason for the outliers was mainly intramolecular hydrogen bonding. The hydrogen-bonding energy values were very high compared to the other compounds. The results demonstrated that hydrogen bonding did not contribute to their molecular interactions.

6.5   Model Phase Selectivity for Benzoic Acid Derivatives

The simplest model phase in reversed-phase liquid chromatography is a graphitized carbon phase that is a polycyclic aromatic hydrocarbon (PAH).

A large phase consisting of 49 aromatic rings was first constructed to calculate molecular interaction energies.

After subtraction of the complex energy value from the sum of the individual energy values of an analyte and the model phase (MI value = energy value of the analyte + energy value of the model phase – energy value of the complex), the capacity ratios obtained by liquid chromatography were related to their energy values calculated using the MM2 program and are listed in Table 11 of the Appendix (p. 304). The correlation coefficients between the model phase and the log k values of benzoic acid derivatives (log k_1) listed in Table 6.7 were calculated as follows:

$$\text{MIFS1} = 3.239 \times (\log k_1) + 13.396, \ r = 0.724, \ n = 22,$$

$$\text{MIVW1} = 3.765 \times (\log k_1) + 7.492, \ r = 0.776, \ n = 22.$$

where MIFS and MIVW are the interaction energy value of the final structure and the van der Waals energy, respectively. The log k_1 values were measured in reversed-phase liquid chromatography using an octadecyl-bonded silica gel column in a pH 2 eluent.[34]

The correlation coefficient indicated that the predominant interaction was hydrophobicity, with the van der Waals energy values demonstrating the highest correlation coefficient. The hydrogen-bonding and electrostatic energy values did not show this correlation. The correlation between the MI energy values and log k measured on an octadecyl-bonded polyvinylalcohol gel[36] was poor. The r values for MIFS and MIVW were 0.405 and 0.505, respectively ($n = 22$). The low r values indicated that the retention mechanisms on the polymer gel were not the same as those on the octadecyl-bonded silica gel. Some selectivity should be considered when a chemically stable octadecyl-bonded polyvinylalcohol gel is used for reversed-phase liquid chromatography. No correlation existed between these MI energy values and log k values measured on polystyrene gels in reversed-phase liquid chromatography.[37] The slope indicated that the retention of benzoic acid derivatives did not correlate with van der Waals energy values, even though the log k values correlated with the log P values.[38] These results indicated that the retention mechanism on a polystyrene gels is different to that on an alkyl-chain-bonded silica gel. The model carbon phase did not explain the retention mechanism even when the polystyrene gel was an aromatic hydrocarbon.

The predominant molecular interaction was hydrophobic, and van der Waals energy values contributed to the retention when reversed-phase liquid chromatography was performed using an alkyl-chain-bonded silica gel for phenolic compounds.[39] The model butyl-bonded phase shown in Figure 6.13 was used to study the alkyl group effects on the molecular interactions of benzoic acid derivatives in reversed-phase liquid chromatography. After subtraction of the complex energy from the sum of the individual energies of the analytes and the butyl phase, the capacity ratios obtained by liquid

Table 6.7 Molecular properties of benzoic acid derivatives on a model carbon phase. k values are capacity ratios where k_1 values were measured on an octadecyl-bonded silica gel phase,[35] k_2 the molecular form, and k_{2i} the ionized form.[38] k_3 values were measured on an octadecyl-bonded polyvinylalcohol phase,[34] k_4 and k_5 were measured on a polystyrene gel (Hitachi 3013 and 3011),[37] and k_6 values were measured on an ODS silica gel (41OODS).[38]

Analyte	$\log k_1$	$\log k_2$	$\log k_{2i}$	$\log k_3$	$\log k_4$	$\log k_5$	$\log k_6$
2,4,6-Trimethylbenzoic acid	1.116	—	—	0.944	—	—	−0.495
2,4-Dichlorobenzoic acid	1.178	—	—	1.672	—	—	—
2,4-Dimethylbenzoic acid	1.117	—	−0.856	1.133	—	—	—
2,5-Dichlorobenzoic acid	1.110	—	—	0.600	—	—	—
2,5-Dimethylbenzoic acid	1.117	—	—	1.118	—	—	—
2,6-Dichlorobenzoic acid	0.789	—	—	0.652	—	—	—
2,6-Dimethylbenzoic acid	0.815	—	—	0.880	—	—	—
2-Bromobenzoic acid	0.774	—	—	0.875	—	—	—
2-Chlorobenzoic acid	0.710	—	—	0.771	—	—	—
2-Methylbenzoic acid	0.824	—	—	0.867	—	—	—
2-Hydroxybenzoic acid	—	(0.912)	(−0.645)	—	0.819	0.940	0.912
3,4,5-Trihydroxybenzoic acid	—	−0.495	−1.987	—	−0.332	−0.022	−0.495
3,4-Dichlorobenzoic acid	1.371	—	—	1.673	—	—	—
3,4-Dimethylbenzoic acid	1.082	—	—	1.109	—	—	—
3,4-Dihydroxybenzoic acid	—	−0.181	−1.510	—	−0.123	0.120	−0.181
3,5-Dichlorobenzoic acid	1.442	—	—	1.314	—	—	—
3,5-Dimethylbenzoic acid	1.160	—	—	1.171	—	—	—
3,5-Dihydroxybenzoic acid	—	−0.242	−1.287	—	−0.161	0.084	−0.242
3-Bromobenzoic acid	1.073	—	—	1.307	—	—	—
3-Chlorobenzoic acid	0.993	—	—	1.169	—	—	—
3-Methylbenzoic acid	0.867	—	—	0.883	—	—	—
3-Methoxybenzoic acid	—	0.883	−0.369	1.185	0.669	0.910	0.883
3-Hydroxybenzoic acid	—	0.246	−0.908	—	0.180	0.370	0.246
4-Bromobenzoic acid	1.096	1.429	−0.374	1.189	—	—	1.429
4-Chlorobenzoic acid	1.010	1.325	0.163	1.189	—	—	1.325
4-Ethylbenzoic acid	1.157	1.515	0.212	1.164	—	—	1.515
4-Methylbenzoic acid	0.847	1.128	−0.167	0.868	—	—	1.128
4-Methoxybenzoic acid	—	—	−0.187	—	—	—	—
4-Hydroxy-3-methoxybenzoic acid	—	0.148	−1.084	—	0.133	0.390	0.148
4-Hydroxybenzoic acid	—	0.262	−1.142	—	0.104	0.310	0.262
Benzoic acid	0.574	0.765	−0.517	0.595	0.631	0.800	0.765

chromatography were related to the energy values calculated using the MM2 program and are listed in Table 12 of the Appendix (p. 306).

The energy values of flat molecules correlated well with log k values, but the energy values of compounds having steric hindrance were less than those expected from log k values for phenolic compounds. The correlation coefficients between the molecular interaction energy values and the log k values of aromatic acids were not good, due to the steric hindrance of some compounds.[8]

$$MIFS2 = 2.068 \times (\log k_1) + 7.107, \, r = 0.687, \, n = 22,$$

$$MIVW2 = 1.587 \times (\log k_1) + 7.886, \, r = 0.521, \, n = 22.$$

Models of several alkyl-chain-bonded silica gels were constructed to eliminate steric effects. A model dimethylpentylsilane-bonded silica phase was useful for the quantitative analysis of the retention of phenolic compounds.[6] The molecular interaction energy of benzoic acid derivatives was calculated using the model dimethylpentylsilane bonded phase shown Figure 6.14. The energy values of the complexes are listed in Table 12 of the Appendix (p. 306) as MIFS3, MIHB3, MIES3 and MIVW3.[8] The correlation coefficients are:

$$MIFS3 = 5.082 \times (\log k_1) + 12.880, \, r = 0.831, \, n = 22,$$

$$MIVW3 = 5.413 \times (\log k_1) + 13.052, \, r = 0.768, \, n = 22.$$

These correlation coefficients were not satisfactory enough for the application of this computational chemical method for the optimization of liquid chromatographic conditions. Notably 3,4-dichloro- and 3,5-dichlorobenzoic acids were outliers. The origins of the error were analyzed by making a comparison with the data for phenolic compounds analyzed previously. The comparison with MIFS values of benzoic acid derivatives and phenolic compounds measured previously[6] indicated steric hindrance for 2,6-dimethyl-, 2,6-dichloro- and 2-ethylbenzoic acids, but not for 3,4-dichloro- and 3,5-dichlorobenzoic acids. This means that the theoretically calculated energy values should be acceptable, but the retention times of these dichlorobenzoic acids were longer than expected. The reason for this may be the bonded phase in old-type bonded silica gels, the inertness of which was not guaranteed. In addition, these dichlorobenzoic acids were not outliers in the chromatography using the octadecyl-bonded polyvinylalcohol gel that had no silanol effects.[32] The correlation was recalculated without these dicholobenzoic acids, and the correlation coefficients improved. The results are shown in Figure 6.20.

The correlation coefficients between log k_1 and log P values calculated with NlogP[7] or VlogP^{TM} (the trade name of the log P calculation program)[7] were 0.638 and 0.680 $(n = 20)$, respectively. The correlation coefficient of this new system was far higher than that of the old system using log P. The above system seemed to work well for the quantitative analysis of the reversed-phase liquid chromatography of benzoic acid derivatives. This

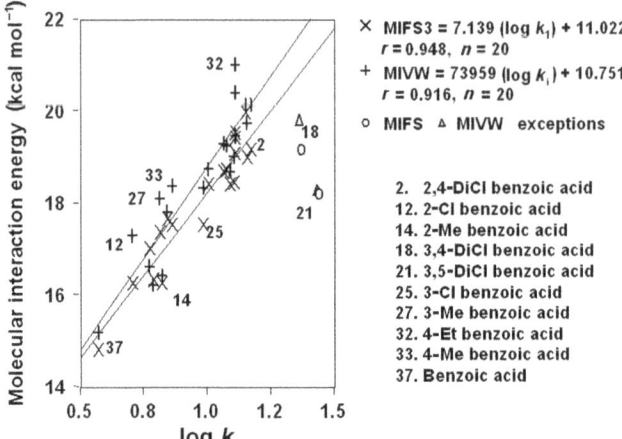

Figure 6.20 The relationship between molecular interaction energy and log k for benzoic acid derivatives on a pentyl-bonded silica gel.

approach was applied to another set of data for aromatic acids used to identify aromatic acids in urine.[40] The capacity ratios were measured using various pH-controlled eluents. The correlation coefficient for MIFS of the benzoic acid derivatives was 0.947 ($n = 12$), where this MIFS3 was fixed by subtraction of individual hydrogen-bonding energy values. The hydrogen-bonding energy value is zero for the model bonded phase, but the hydrogen-bonding energy values of hydroxyl-substituted benzoic acid derivatives were very high. The MIHB3 values were identical to the hydrogen-bonding energy values of the original molecules, and were not related to the molecular interaction. The adsorption method can calculate the molecular interaction of ionized compounds that the log P system cannot handle. The energy values of ionized benzoic acid derivatives and their complexes with Phase 3 are listed in Table 13 of the Appendix (p. 308).

The correlation coefficient between the log k values of ionized benzoic acid derivatives and MIFS3i was 0.942 (MIFS3i is the molecular interaction energy of the ionized molecules calculated from FS3i and fsi. FS3i is the molecular interaction energy between the ionized analyte and phase 3 given Appendix Table 13). These relations are summarized in Figure 6.21. However, the value for *ortho*-hydroxyl benzoic acid was excluded as an outlier. The value of the molecular form was included, and the correlation coefficient was slightly improved, but the log k value of the ionized form was relatively high, as shown in Figure 6.21.

The capacity ratios of partially ionized compounds can be predicted using eqn (9). For evaluation by the above approach, k_m and k_i were replaced with energy values calculated using eqns (4) and (5), respectively. The predicted pK_a values from the atom partial charge related to the measured pK_a values in liquid chromatography[7] were used for the calculation. The results are summarized in Figure 6.22.

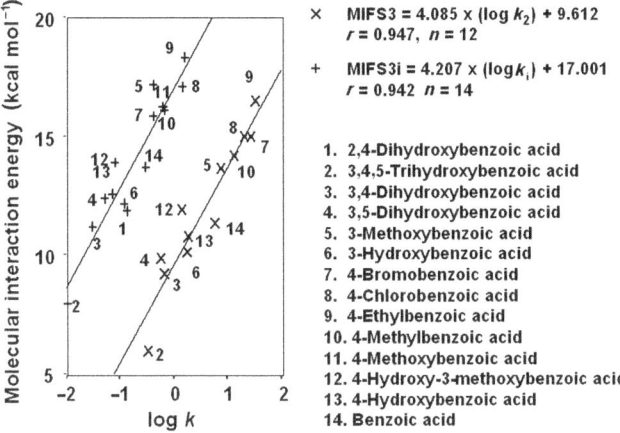

Figure 6.21 The relationship between molecular interaction energy and log k for molecular and ionized benzoic acid derivatives.
Reproduced by permission of Oxford University Press, ref. 9.

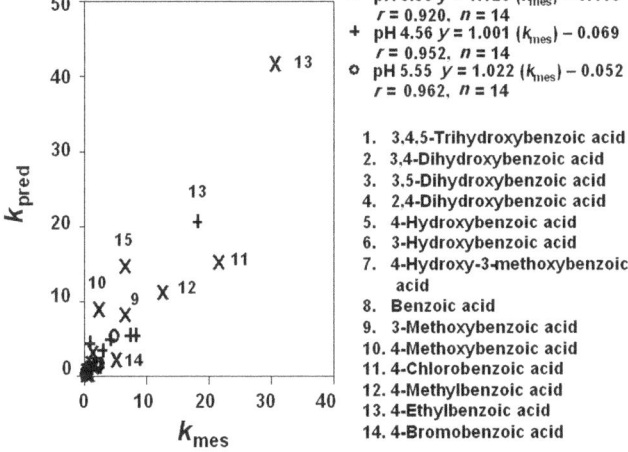

Figure 6.22 The relationship between predicted (k_{pred}) and measured (k_{mes}) capacity ratios for benzoic acid derivatives in eluents of various pH.

The predicted pK_a values from the atom partial charge related to the pK_a values derived from Hammett's equation[7] were used for further evaluation:

$$k_{pred} = 1.083 \times (k_{mes}) + 0.339, \ r = 0.918, \ n = 14, \ \text{at pH } 3.55,$$

$$k_{pred} = 0.988 \times (k_{mes}) + 0.024, \ r = 0.932, \ n = 14, \ \text{at pH } 4.05,$$

$$k_{pred} = 0.893 \times (k_{mes}) - 0.109, \ r = 0.949, \ n = 14, \ \text{at pH } 4.56,$$

$$k_{pred} = 0.772 \times (k_{mes}) - 0.059, \ r = 0.960, \ n = 14, \ \text{at pH } 5.05,$$

$$k_{pred} = 0.931 \times (k_{mes}) - 0.046, \ r = 0.962, \ n = 14, \ \text{at pH } 5.55.$$

The measured retention times at low pH were shorter than those expected from eqn (9) due to an ion-exclusion effect.[7] Those at high pH were too small to obtain precise values. Several approaches have been proposed for the prediction of pK_a values, but it is still difficult to obtain precise values for a variety of compounds at present. However, the above approach using a direct calculation of molecular interaction energy values and pK_a values derived from atom partial charge is better than the old model using log P and Hammett's pK_a values. This is because the correlation coefficient is equivalent and the slope is close to 1. Further development of a model phase suitable for a variety of compounds and a pK_a prediction system based on atom partial charge is necessary to establish a standard method for determining a QSRR for a variety of compounds in liquid chromatography.

6.6   Chromatography of Aromatic Acids

The simplest model phase in reversed-phase liquid chromatography (see Section 6.5) was not satisfactory for the quantitative analysis of the retention behavior of benzoic acid derivatives *in silico*.[8] Therefore, several model phases were constructed and the energy values of their optimized structures were calculated using the MM2 program. The chromatographic retention data from ref. [37] are listed in Table 6.8.

A practical model phase for calculating the molecular interaction energy of benzoic acid derivatives was an alkylsilane-bonded polysiloxane, shown in

Table 6.8   Molecular properties of aromatic acids. pK_{a1} values were predicted from the atomic partial charge, pK_{a2} values were calculated using Hammett's equation, k_m values are for the molecular form and k_i values are for the ionized form. Reproduced by permission of Springer, ref. 9.

No.	Analyte	pK_{a1}	pK_{a2}	k_m	k_i
1	2,5-Dihydroxyphenylacetic acid	4.648	—	0.5567	0.1237
2	2-Hydroxyphenylacetic acid	3.919	—	2.4120	0.5617
3	3,4-Dihydroxycinnamic acid	4.444	4.55	1.1650	0.0894
4	3,4-dihydroxyphenylacetic acid	4.353	4.39	0.6976	0.0516
5	3-Methoxymandelic acid	3.577	3.33	2.0720	0.2938
6	3-Methoxyphenylacetic acid	4.446	4.26	6.4230	0.4536
7	3-Hydroxymandelic acid	3.532	—	—	—
8	4-Chlorophenylacetic acid	4.275	4.22	18.6000	1.2940
9	4-Methoxyphenylacetic acid	4.663	4.40	6.4230	0.4536
10	4-Hydroxy-3-methoxycinnamic acid	4.523	4.56	2.7780	0.1959
11	4-Hydroxy-3-methoxyphenylacetic acid	4.462	4.39	1.5050	0.1031
12	4-Hydroxycinnamic acid	4.586	4.63	2.2580	0.1546
13	4-Hydroxyphenylacetic acid	4.570	4.43	1.2940	0.0928
14	Cinnamic acid	4.523	4.38	14.6300	0.9742
15	Mandelic acid	3.681	3.38	1.7060	0.2062
16	Phenylacetic acid	4.291	4.30	6.1270	0.3866
17	3-Phenylpropionic acid	4.460	4.59	12.8100	0.8763
18	3-Methoxycinnamic acid	4.444	4.31	17.9200	1.2160
19	4-Phenylbutyric acid	4.544	4.72	26.1600	1.7470

Figure 6.14, with eight dimethoxypentylsilanes and eight methylsilanes bonded to the polysiloxane phase.

The molecular interaction energy values were calculated using this pentyl-bonded phase for aromatic acids whose molecular sizes were usually greater than those of benzoic acid derivatives. The correlation coefficients were poor:

$$\text{MIFS} = 0.474 \times (\log k_m) + 18.765, \ r = 0.133, \ n = 15,$$
$$\text{MIVW} = 0.869 \times (\log k_m) + 19.246, \ r = 0.241, \ n = 15,$$
$$\text{MIFS} + \text{hb} = 5.910 \times (\log k_m) + 9.516, \ r = 0.759, \ n = 15.$$

The individual hydrogen-bonding energy (hb) did not contribute to the molecular interaction, but it affected the MIFS value.

It seemed that the contact surface area of the model phase was rather small for such aromatic acids. Therefore, the pentyl groups were replaced with octyl groups. However, the correlation coefficients were not improved:

$$\text{MIFS} = 0.455 \times (\log k_m) + 27.321, \ r = 0.108, \ n = 15,$$
$$\text{MIVW} = 1.343 \times (\log k_m) + 34.191, \ r = 0.174, \ n = 15,$$
$$\text{MIFS} + \text{hb} = 5.890 \times (\log k_m) + 18.071, \ r = 0.752, \ n = 15.$$

Therefore, a multi-octyl-bonded phase was constructed to increase the contact surface area, as shown in Figure 6.23. The model phase consisted of 1237 atoms, 1296 bonds and 8471 connectors including 20 dimethoxy-octylsilanes and 6 trimethylsilanes. The correlation coefficient was subsequently improved:

$$\text{MIFS} + \text{hb} = 6.653 \times (\log k_m) + 9.799, \ r = 0.894, \ n = 15.$$

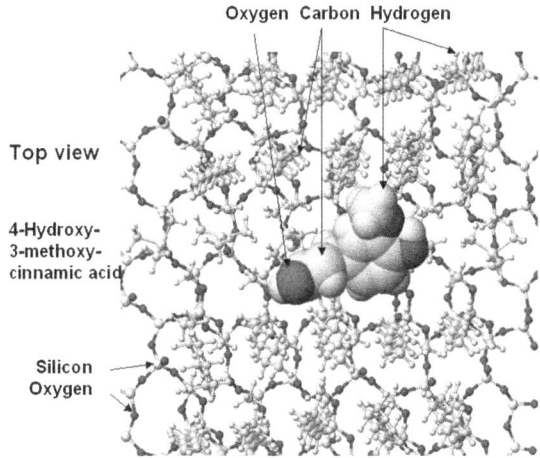

Figure 6.23 Docking of 4-hydroxy-3-methoxy-cinnamic acid on a modified octyl-bonded phase. White, light-gray, gray, and black balls represent hydrogen, carbon, silicon, and oxygen, respectively. The atom size is 20% of the original size for the phase for clear demonstration.

Further improvement was attempted using longer alkyl chains *i.e.*, decyl and dodecyl groups instead of octyl groups. However, the longer the alkyl chain, the narrower the pocket entrance where an analyte could dock, and the correlation coefficient decreased.

For the decyl-bonded phase:

$$\text{MIFS} + \text{hb} = 5.193 \times (\log k_m) + 12.990, \, r = 0.708, \, n = 15.$$

For the dodecyl-bonded phase:

$$\text{MIFS} + \text{hb} = 3.201 \times (\log k_m) + 16.894, \, r = 0.445, \, n = 15.$$

The direct hydrogen bonding between these model phases and the analytes was negligible, therefore an alkyl-bonded phase having a large flowerpot-type pocket was constructed using a carbon phase based on the structure shown in Figure 3.13. The carbon phase is like a double layer of PAHs.[7] The pocket was redesigned using shorter alkyl groups. The first pocket consisted of 1 methyl, 6 pentyl and 30 dodecyl groups. One methyl group was located at the center, 6 pentyl groups surrounded the methyl group and 30 dodecyl groups stood like double circles. This model phase consisted of 1572 atoms, 1810 bonds and 26 943 connectors. The pocket was not deep enough, therefore 3 ethyl groups replaced 3 pentyl groups, and octyl groups replaced 12 of the dodecyl groups in the first circle. The structure is shown in Figure 6.24.

The model-phase consisted of 1536 atoms, 1774 bonds and 26 727 connectors. The correlation coefficient from the first model was 0.756 $(n = 15)$, and that from the final model was 0.954. The energy values of molecular-form aromatic acid derivatives and their interaction energy values

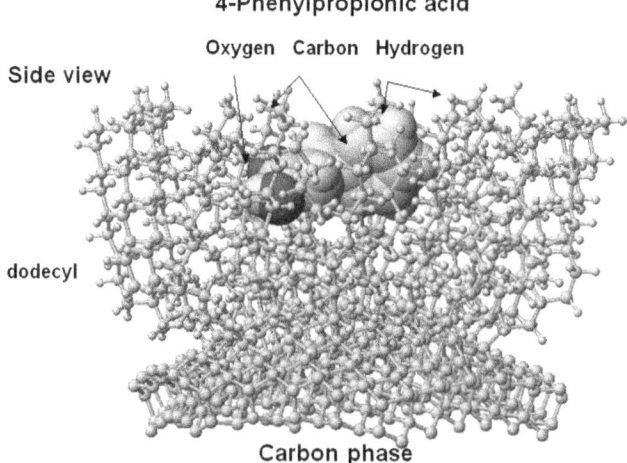

4-Phenylpropionic acid

Figure 6.24   Docking of 4-phenylpropionic acid on a model dodecyl-bonded phase. Black, dark-gray, gray, and white balls represent oxygen, carbon, silicon, and hydrogen, respectively. The atom size is 20% of the original size for the phase for clear demonstration.

with the final phase are listed in Table 6.8. The above system seemed to work well for developing a quantitative analysis of the reversed-phase liquid chromatography of aromatic acids. This adsorption method can calculate the molecular interaction energy values of ionized compounds that the log P system cannot handle. The energy values of ionized aromatic acid derivatives and their molecular interaction energies with the final phase are listed in

Table 6.9 Molecular properties of molecular-form and ionized-form aromatic acids. Reproduced by permission of Springer, ref. 9.

Analyte (molecular form)	fs	es	hb	vw	MIFS	MIVW
2,5-Dihydroxyphenylacetic acid	−26.7311	−10.589	−6.627	4.225	18.1935	12.978
2-Hydroxyphenylacetic acid	−24.9922	−8.839	−6.572	4.277	17.8561	12.733
3,4-Dihydroxycinnamic acid	−30.3850	−12.384	−7.672	2.567	19.7512	19.099
3,4-Dihydroxyphenylacetic acid	−26.0470	−10.228	−5.306	3.229	18.0391	17.212
3-Methoxymandelic acid	−10.0315	−8.738	4.408	5.612	19.2834	18.168
3-Methoxyphenylacetic acid	−16.0667	−3.708	−5.043	4.611	16.5710	16.336
3-Hydroxymandelic acid	−15.4549	−10.563	4.399	4.232	15.4113	14.856
4-Chlorophenylacetic acid	−19.4455	−3.689	−4.926	3.439	18.7079	17.650
4-Methoxyphenylacetic acid	−16.0715	−3.700	−5.025	4.639	18.4817	18.156
4-Hydroxy-3-methoxycinnamic acid	−22.9642	−7.975	−8.070	3.875	18.0576	18.479
4-Hydroxy-3-methoxyphenylacetic acid	−20.3167	−8.267	−5.188	4.608	17.5654	17.277
4-Hydroxycinnamic acid	−24.1484	−5.138	−7.996	2.492	16.1451	17.840
4-Hydroxyphenylacetic acid	−21.4308	−5.421	−5.038	3.213	14.4359	14.544
Cinnamic acid	−22.5637	−3.422	−8.008	2.543	17.9216	17.714
Mandelic acid	−12.6247	−7.262	4.664	3.574	16.0063	13.873
Phenylacetic acid	−19.7695	−3.701	−5.006	3.280	16.2275	16.160
3-Phenylpropionic acid	−20.4447	−3.437	−8.035	4.483	19.9719	18.417
3-Methoxycinnamic acid	−15.2138	−3.425	−8.030	3.099	21.5041	20.422
4-Phenylbutyric acid	−18.4651	−3.419	−6.718	4.911	21.9969	20.046

Analyte (ionized form)	fsi	esi	hbi	vwi	MIFSi	MIVWi
2,5-Dihydroxyphenylacetic acid	−9.6851	−3.280	1.858	3.414	17.8459	12.490
2-Hydroxyphenylacetic acid	−8.0653	−1.545	1.833	3.451	17.2921	12.181
3,4-Dihydroxycinnamic acid	−19.2825	−8.954	−2.602	2.928	18.8781	18.343
3,4-Dihydroxyphenylacetic acid	−11.9728	−6.499	2.868	3.153	17.6432	16.913
3-Methoxymandelic acid	18.0931	−3.007	25.157	4.552	18.9182	18.281
3-Methoxyphenylacetic acid	−2.0279	0.000	3.134	4.531	15.7044	15.572
3-Hydroxymandelic acid	12.6663	−4.759	25.073	3.162	14.7299	14.721
4-Chlorophenylacetic acid	−5.6219	0.000	3.059	3.356	18.2888	17.136
4-Methoxyphenylacetic acid	−2.0410	0.000	3.177	4.546	18.1018	17.740
4-Hydroxy-3-methoxycinnamic acid	−11.9530	−4.548	−3.086	4.249	17.3643	17.883
4-Hydroxy-3-methoxyphenylacetic acid	−6.2064	−4.549	3.036	4.533	17.8708	17.413
4-Hydroxycinnamic acid	−13.1029	−1.715	−2.967	2.845	16.0735	17.966
4-Hydroxyphenylacetic acid	−7.3955	−1.713	3.157	3.119	14.3173	14.288
Cinnamic acid	−11.5115	0.000	−2.980	2.902	17.3816	17.258
Mandelic acid	14.3203	−2.907	24.971	3.387	19.3249	14.181
Phenylacetic acid	−5.7880	0.000	3.161	3.174	15.8504	15.819
3-Phenylpropionic acid	−9.8148	0.000	−2.951	4.453	18.9959	17.564
3-Methoxycinnamic acid	−4.9862	0.000	−0.017	6.692	19.8273	23.116
4-Phenylbutyric acid	−6.2695	0.000	−0.048	4.858	20.9734	19.045

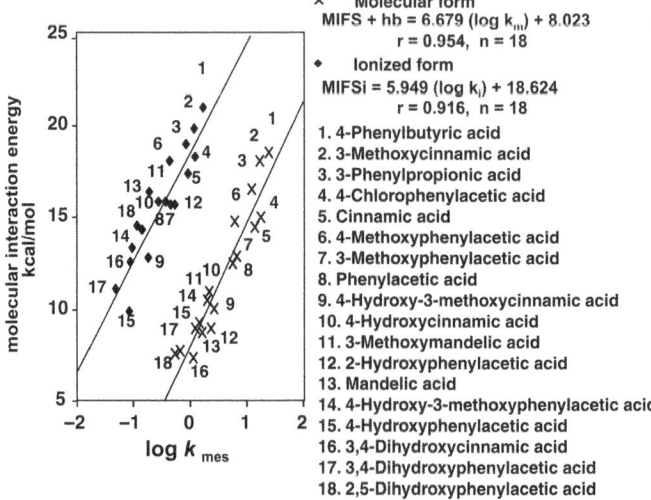

×   Molecular form
MIFS + hb = 6.679 (log k_m) + 8.023
r = 0.954, n = 18

♦   Ionized form
MIFSi = 5.949 (log k_i) + 18.624
r = 0.916, n = 18

1. 4-Phenylbutyric acid
2. 3-Methoxycinnamic acid
3. 3-Phenylpropionic acid
4. 4-Chlorophenylacetic acid
5. Cinnamic acid
6. 4-Methoxyphenylacetic acid
7. 3-Methoxyphenylacetic acid
8. Phenylacetic acid
9. 4-Hydroxy-3-methoxycinnamic acid
10. 4-Hydroxycinnamic acid
11. 3-Methoxymandelic acid
12. 2-Hydroxyphenylacetic acid
13. Mandelic acid
14. 4-Hydroxy-3-methoxyphenylacetic acid
15. 4-Hydroxyphenylacetic acid
16. 3,4-Dihydroxycinnamic acid
17. 3,4-Dihydroxyphenylacetic acid
18. 2,5-Dihydroxyphenylacetic acid

Figure 6.25   The relationship between molecular interaction energy and the measured (k_{mes}) capacity ratios for molecular- and ionized-form aromatic acids.

Table 6.9. The correlation coefficient between the log k values of ionized aromatic acid derivatives and MIFSi was 0.916. The relationships between the molecular interaction energy values and the log k values of the molecular- and ionized-form aromatic acid derivatives are shown in Figure 6.25.

    The capacity ratio of partially ionized compounds can be predicted using eqn (9). For evaluation by the above approach, k_m and k_i were replaced with molecular interaction energy values. The pK_a values predicted from the atomic partial charge, related to the measured pK_a values in liquid chromatography,[7] were used for the calculation, and the concentration effect of an organic modifier (0.022 pH units per % methanol from ref. 36) was added for the calculation. The predicted pK_a values derived from Hammett's equation[7] were used for further evaluation. The results are shown in Figure 6.26.

$$k_{pred} = 1.066 \times (k_{mes}) - 0.294, \ r = 0.916, \ n = 16, \ \text{at pH 3.55},$$

$$k_{pred} = 1.104 \times (k_{mes}) - 0.352, \ r = 0.930, \ n = 16, \ \text{at pH 4.05},$$

$$k_{pred} = 1.200 \times (k_{mes}) - 0.327, \ r = 0.950, \ n = 16, \ \text{at pH 4.56},$$

$$k_{pred} = 1.178 \times (k_{mes}) - 0.279, \ r = 0.966, \ n = 16, \ \text{at pH 5.05},$$

$$k_{pred} = 1.179 \times (k_{mes}) - 0.248, \ r = 0.965, \ n = 16, \ \text{at pH 5.55}.$$

The measured retention times at low and high pH were shorter than those expected from eqn (9) due to an ion-exclusion effect.[7] However, the above approach, using a direct calculation of molecular interaction energy values, was better than the old model using log P. This is because the correlation coefficient is equivalent or better, and the slope is close to 1. This new method, using pK_a values derived from the atomic partial charge, was simpler than older methods for acidic compounds, including phenolic

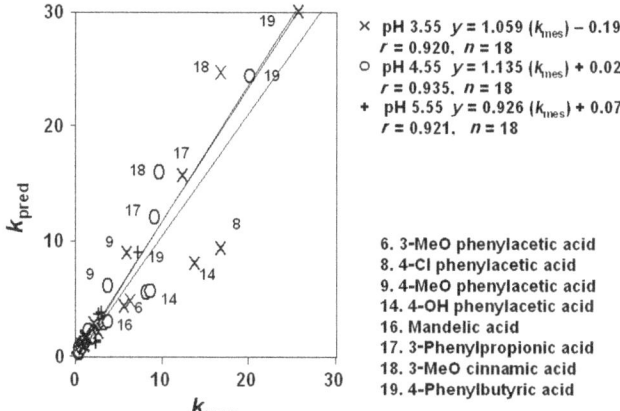

Figure 6.26 The relationship between predicted (k_{pred}) and measured (k_{mes}) capacity ratios for aromatic acids in eluents of various pH.

compounds. Several approaches have been proposed for the prediction of pK_a values, but it is still difficult to obtain precise values for a variety of compounds at present.

6.7 Chromatography of Acidic Drugs

A model butyl-bonded phase that was used for the development of a new optimization system in silico[38] was applied to develop a common optimization system for the analysis of chromatographic behavior measured using a pentyl-bonded silica gel column. The docking between an acidic drug and the butyl phase was simple. The lowest energy value of a complex was easily obtained. An example of the optimized complex formed between 2,4,6-trimethylphenol and the butyl phase is shown in Figure 6.13, where sticks and balls represent the structure of the optimized complex between the model phase and 2,4,6-trimethylphenol. The butyl groups of the model butyl-bonded phase are highly dense and not pushed down by the analyte, which lies on top of the butyl-group brush.

The energy values of individual compounds calculated using the MM2 prorgam are listed in Table 6.10 along with the properties, log P and pKa, of the acidic drugs used.[16] The energy values of the individual complexes for the model butyl phase and each acidic drug are listed in Table 14 of the Appendix (p. 310) as FS1, HB1, ES1 and VW1.

The interaction energy values between a molecular-form compound and the model butyl phase were calculated using the MM2 program to analyze the retention of molecular-form analytes qualitatively. The interaction energy value (the MI value) = energy value of the individual molecule + energy value of the model phase – energy value of the complex.

The r value between MIFS1 or MIVW1 calculated using the model butyl phase and the measured log k values of molecular-form acidic drugs, listed as log k_2 in Table 6.10, was 0.596 ($n = 19$). In this model system, one side of the analyte was in contact with the model phase, and the steric effect was

Table 6.10 Molecular properties and capacity ratios of acidic drugs. k_2 values were measured at pH 2, $k_{4.5}$ at pH 4.50, k_6 at pH 6, and $k_{7.4}$ at pH 7.4. Reproduced by permission of Oxford University Press, ref. 10.

No.	Acidic drug	VlogP	pK_a	$log\ k_2$	$log\ k_{4.5}$	$log\ k_6$
1	p-Aminohippuric acid	0.232	3.83	−1.155	−1.854	−1.886
2	Amoxicillin	−2.502	9.60	−1.444	—	−1.796
3	Barbituric acid	0.822	—	−1.131	−1.699	−2.097
4	Benzoic acid	1.485	4.20	−0.021	−0.489	−0.759
5	Furosemide	1.901	3.90	−0.136	−0.479	−0.511
6	p-Hydroxybenzoic	1.002	9.46	−0.775	−0.963	−1.538
7	Ibuprofen	3.550	5.20	1.204	0.910	0.634
8	Indomethacin	3.426	4.50	1.054	0.696	0.594
9	Iopanoic acid	3.873	—	1.346	1.087	0.692
10	Mefenamic acid	4.971	4.20	1.352	0.935	0.652
11	Nalidixic acid	0.966	6.00	0.054	0.008	−0.189
12	Naproxen	3.047	4.20	0.586	0.262	−0.015
13	Nicotinic acid	0.477	4.95	−0.796	−1.161	−1.237
14	Phenylbutazone	3.251	4.40	0.964	0.522	0.346
15	Probenecid	2.652	—	0.610	0.088	0.048
16	Salicylic acid	1.060	3.00	0.007	−0.666	−0.688
17	Sulfamethoxazole	0.791	5.81	−0.623	−0.717	−0.971
18	Tolazamide	1.448	5.70	0.407	0.343	0.139
19	Tolbutamide	2.266	5.30	0.372	0.284	0.086
20	Warfarin	2.866	5.10	0.733	0.383	−0.081

No	Acidic drug	$log\ k_{7.4}$	fs	hb	es	vw
1	p-Aminohippuric acid	−2.097	−18.7895	−10.241	−9.047	7.603
2	Amoxicillin	−1.310	39.4660	−8.568	0.402	5.705
3	Barbituric acid	−1.921	−59.1784	−8.299	−75.242	−4.501
4	Benzoic acid	−0.785	−13.9182	−3.458	−6.671	4.877
5	Furosemide	−0.511	9.9626	−5.541	−1.038	6.003
6	p-Hydroxybenzoic	−1.678	−16.0982	−4.931	−6.668	4.790
7	Ibuprofen	0.596	−16.9561	−3.737	−5.043	4.654
8	Indomethacin	0.581	−24.0717	−5.284	−12.458	5.883
9	Iopanoic acid	0.607	−8.2455	−5.634	−4.501	7.048
10	Mefenamic acid	0.577	12.5077	−3.951	−11.362	18.894
11	Nalidixic acid	−0.455	−37.4073	−4.051	−40.545	11.771
12	Naproxen	−0.048	−27.7018	−3.755	−5.025	6.778
13	Nicotinic acid	−1.174	−18.5217	−4.047	−10.511	3.675
14	Phenylbutazone	0.325	18.1704	0.000	−11.325	19.458
15	Probenecid	0.041	8.8530	−3.455	−3.682	8.863
16	Salicylic acid	−0.706	−15.3507	−5.355	−6.437	5.438
17	Sulfamethoxazole	−1.301	7.0614	−2.202	2.679	3.090
18	Tolazamide	0.078	−3.1534	−2.847	−12.721	8.547
19	Tolbutamide	0.032	−29.9856	−2.920	−25.539	4.886
20	Warfarin	−0.162	−17.5045	−2.808	−5.999	7.411

neglected. The difference between MIFS1 and MIVW1 was large for barbituric acid, probenecid and mefenamic acid.

The new silica-gel-based pentyl-bonded phase was constructed based on silica-gel-based pentyl-bonded phase as shown Figure 6.14. The new phase contains 753 atoms, 828 bonds, and 6056 connectors. 15

methylpentyldichlorosilanes were bonded on the surface. The calculated FS, HB, ES and VW energy values of complexes between the pentyl-bonded phase and each acidic drug are listed in Table 14 of the Appendix (p. 310) as FS2, HB2, ES2 and VW2. An improvement in the correlation would expected if a low-density phase is used as a model phase, because the analyte should be buried in the alkyl chains. The interaction energy values between an acidic drug and the silica-gel-based pentyl-phase were calculated. The r between MIFS2 and the measured $\log k$ values of molecular-form acidic drugs, listed as $\log k_2$ in Table 6.10, improved to 0.773 $(n = 19)$. The correlation (r) was 0.700 $(n = 19)$ from MIVW2. The contribution of HB2 and ES2 values was very poor. The HB2 energy value of these model phases is zero. The r for MIHB2 and MIES2 was 0.034 and 0.124, respectively. The contribution of MIVW energy indicated that hydrophobic interaction is the predominant molecular interaction in the retention of these acidic drugs on an alkyl-bonded phase in reversed-phase liquid chromatography. The difference between MIFS2 and MIVW2 was large for barbituric acid, probenecid and mefenamic acid, even when the silica-gel-based phase diminished the steric effect. The correlation coefficient was still very poor, therefore, further improvement of the model phase was studied.

During the synthesis of the alkyl-bonded silica gel, two chloro groups of the alkylchlorosilane may bind with two silanol groups on the poly-silicondioxide phase. Therefore, a monomethylpentyl-bonded phase was constructed as a model phase on which there is no free silanol group at the adsorption site. It consisted of 753 atoms, 828 bonds and 6056 connectors containing 165 silicons, 304 oxygens, 90 carbons and 210 hydrogens. Fifteen monomethylpentylsilicons bind with two oxygens from the polysilicondi-oxide phase within a 900 Å2 area.

In this phase, there is not enough space to stick a molecule between brushes. Only one side of molecule made contact with the model phase. This means that this type of model phase is not ideal, even when longer alkyl chains are used to construct a model phase, as these require longer calculation times. The FS, ES, HB and VW energy values of a complex between this monomethylpentyl-bonded phase and each acidic drug are listed in Table 14 of the Appendix (p. 310) as FS3, ES3, HB3 and VW3. The r between MIFS and $\log k_2$ in Table 6.10 was 0.486 $(n = 19)$. The r was 0.549 for MIVW3 and $\log k_2$. The correlation coefficients were very poor. This type of bonding may not be realistic for an alkyl-bonded silica gel for acidic drugs. The difference between MIFS3 and MIVW3 was large for barbituric acid, probenecid and mefenamic acid even for this bonded phase.

Therefore, a new phase was constructed based on dimethylpentylsilane. It consisted of 991 atoms, 1051 bonds and 15 193 connectors, containing 171 silicons, 328 oxygens, 143 carbons and 349 hydrogens. Twenty dimethyl-pentylsilanes and one trimethylsilane were bonded within 900 Å2 on the polysilicondioxide phase. The trimethylsilane was considered an end-capped molecule. The optimized structure of a complex of this model phase and iopanoic acid is shown in Figure 6.27. The upper view of the space-filled structure indicates how a molecule fits in the pocket. The trimethylsilane is the center of the pocket. The atomic size is 1 instead of 0.2 for the stick and

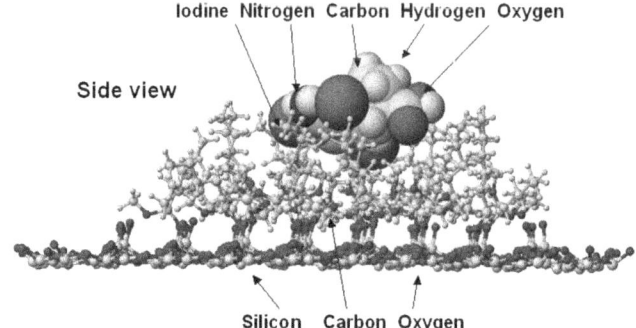

Figure 6.27 Adsorption of iopanoic acid on a dimethylpentylsilane-bonded phase. White, light-gray, gray, dark-gray, and black balls represent hydrogen, carbon, silicon, iodine, and oxygen, respectively. The atom size is 20% of the original size for the phase for clear demonstration. The oxygens and silicons of the dimethylpentylsilane-bonded phase are hidden behind the alkyl groups. The black ball representing the nitrogen atom of the iopanoic acid is indicated with an arrow.

ball model. Dimethylpentyl groups stand straight at beginning then the pentyl groups become closer together; some of them lay in free space after the optimization.

On this new bonded phase, dimethylpentyl groups surrounded one trimethyl group. Silanol groups around the trimethylsilane group are completely covered by alkyl groups. Therefore the silanol groups may not have contributed. The first circle of dimethylpentyl groups may not be pushed down in the presence of an analyte. The second circle of dimethylpentyl groups should support the first. The interaction energy values between each acidic drug and the new model phase were calculated and are listed as FS4, HB4, ES4 and VW4 in Table 14 of the Appendix (p. 310). The correlation between MIFS and $\log k_2$ was improved.

$$\text{MIFS4} = 6.483 \times (\log k_2) + 23.145, \; r = 0.878, \; n = 19,$$

$$\text{MIVW4} = 6.071 \times (\log k_2) + 19.864, \; r = 0.833, \; n = 19,$$

$$\log P = 1.514 \times (\log k_2) + 1.788, \; r = 0.925, \; n = 19.$$

It was not expected that hydrogen bonding would have any direct effect, and the hydrogen-bonding energy values were identical after the complex formation. The hydrogen-bonding energy values were subtracted from the MIFS energy values. The correlation coefficient with $\log k_m$ values, however, did not improve as was observed for phenolic compounds.

$$(\text{MIFS} - \text{hb}) = 7.824 \times (\log k_m) + 18.516, \; r = 0.888, \; n = 19.$$

The hydrogen-bonding energy did not affect retention or the correlation coefficient because no intramolecular hydrogen bonding exists for these acidic drugs.

These results are better than the results for the previous three models, but the difference between MIFS4 and MIVW4 was still large for barbituric acid, probenecid and mefenamic acid, at more than 10 kcal mol^{-1}. The r between log P and log k_2 was 0.925 ($n = 19$). This r value is not significantly high compared to the results for phenolic compounds. Therefore, log k_2 values measured by liquid chromatography may not be the maximum capacity ratios. Further development is, therefore, necessary for the simulation of the chromatography of drugs. The masses of drugs are quite large and their structures are complicated compared to those of phenolic compounds.

The capacity ratios of partially ionized compounds were calculated using eqn (9). For evaluation by the above approach, km and ki were replaced with MI energy values calculated from FS4i, HB4i, ES4i, VW4i, FSi, HBi, ESi and VWi in Table 14 of the Appendix (p. 310). The correlation between the retention factors measured and predicted with this new method using MIFS4, was obtained, and the results are summarized in Figure 6.28.

$$\text{MIFS4} = 7.395 \times (\log k_2) + 22.328, \ r = 0.891, \ n = 15, \text{ at pH} = 2.00,$$

$$\text{MIFS4} = 7.603 \times (\log k_{4.5}) + 24.172, \ r = 0.936, \ n = 15, \text{ at pH} = 4.50,$$

$$\text{MIFS4} = 6.954 \times (\log k_6) + 25.512, \ r = 0.851, \ n = 15, \text{ at pH} = 6.00,$$

$$\text{MIFS4} = 6.185 \times (\log k_{7.4}) + 25.766, \ r = 0.783, \ n = 15, \text{ at pH} = 7.40.$$

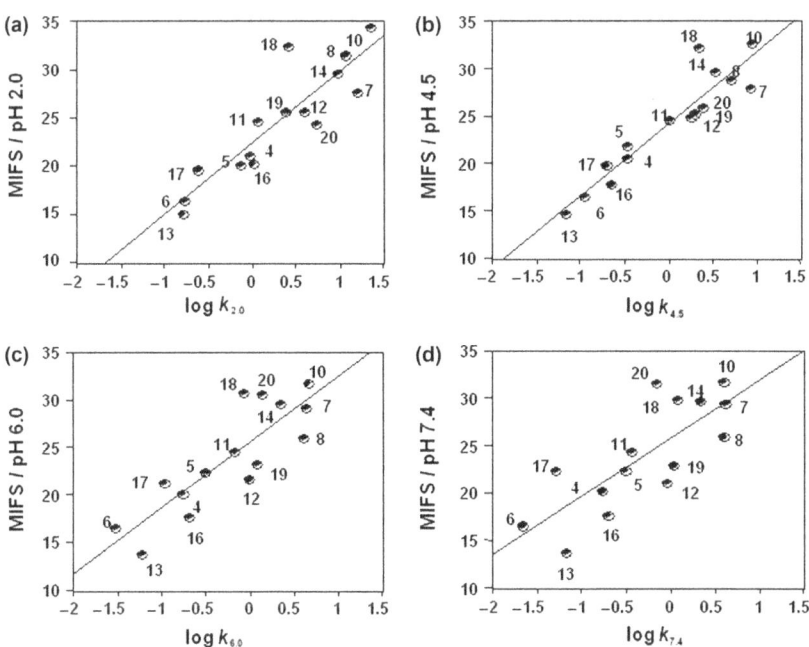

Figure 6.28 The relationship between molecular interaction energy and log k for acid drugs in eluents of various pH. Chemicals: see Table 14 of the Appendix.

The above results indicated that the retention time of acidic drugs can be predicted using both energy values in the optimized structure calculated using the MM2 program. The addition of pKa values predicted from the atomic partial charge calculated using the MOPAC program enables the retention factors in a given pH eluent to be predicted.

An octyl-bonded phase was constructed similar to the first pentyl phase without end-capping, and its molecular interactions with various acidic drugs were examined. However, the longer alkyl chains did not improve the correlation coefficient between the MI energies and log k_2. Addition of one water molecule beside an analyte polar group changed its MI energy value. However, this MM2 calculation method cannot handle multisolvent molecules.

6.8 Chromatography of Basic Drugs

The pentyl-bonded silica gel used in this experiment did not show silanol group activity.[41] Therefore, a model butyl-bonded phase previously used for phenolic compounds[6] was used for analyzing the chromatographic behavior of basic drugs. The surface is flat, and docking on the surface is simple.

The molecular properties of some nitrogen-containing compounds are listed in Table 6.11. The calculated energy values of individual complexes with the model butyl phase and each nitrogen-containing compound are listed in Table 15 of the Appendix (p. 312) as FS1 and VW1 where the hydrogen-bonding and electrostatic energy values are not given, because these energy values did not show any meaningful relationship with retention, even when the retention time was measured using a pentyl-bonded silica gel column. Hydrogen-bonding and electrostatic energy values are important in ion-exchange liquid chromatography.[42]

The interaction energy between a molecular-form compound and the model butyl phase was calculated using the MM2 program, where the interaction energy values (MI values) = energy value of the individual molecule + energy value of the model phase – energy value of the complex. The relationships between MIFS1 or MIVW1 calculated using the model butyl phase and the measured log k_{am} values of molecular-form nitrogen containing-compounds (log k_m) were poor:

$$\text{MIFS1} = 4.792 \times (\log k_{max}) + 11.101, \ r = 0.610, \ n = 13,$$

$$\text{MIVW1} = 2.910 \times (\log k_{max}) + 11.797, \ r = 0.609, \ n = 13.$$

As observed in previous analyses, this model phase worked well for simple phenolic compounds. The correlation, r, between the measured and predicted molecular interaction energy values was more than 0.92 ($n = 6$) at pH 3–9.[39] However, the molecular interaction energy values of larger compounds were smaller than expected. The reason for this is likely to be poor contact.

Table 6.11 Molecular properties of basic drugs and nitrogen-containing compounds. VlogP represents log P measured using the VlogP program, ClogP represents predicted log P values,[43] and MlogP represents log P values measured *via* the octanol–water partition coefficient.[43] pK_{a1} values are from ref. 43, and pK_{a2} values were measured by reversed-phase liquid chromatography. Reproduced by permission of Taylor and Francis, ref. 51.

Analyte	VlogP	ClogP	MlogP	k_{max}	k_{min}	pK_{a1}	pK_{a2}
Ajmaline	1.593	1.26	—	—	—	8.2	—
Allopurinol	−2.927	−0.92	−0.55	—	—	9.4	—
Amoxicillin	—	0.33	—	0.08	—	2.4/9.6	—
Aniline	—	—	—	0.31	—	4.63	—
Atropine	—	1.32	1.83	2.81	0.08	9.8	7.40
Benzylamine	—	—	—	—	—	9.33	—
Caffeine	−0.716	0.07	−0.07	0.17	0.17	0.6/14.0	—
Carbamazepine	2.524	1.98	2.45	1.71	1.50	1.98	6.47
Chloramphenicol	—	—	1.14	—	—	—	—
Dextromethorphan	2.801	3.99	—	19.57	0.76	8.3	8.54
Diazepam	—	3.18	2.80	5.66	4.78	3.3	—
Ethambutol	−1.993	0.12	—	—	—	6.3/9.5	—
Homatropine	1.213	1.45	—	—	—	9.9	—
Imipramine	4.654	4.41	4.80	—	—	9.5	—
Isoproterenol	—	0.08	—	0.17	0.00	8.6/10.1/12.0	—
Lidocaine	2.558	1.98	2.26	7.55	0.18	7.9	6.79
p-Methoxyaniline	—	—	—	—	—	5.34	—
Prazosin	1.315	2.16	—	0.90	0.09	6.5	5.65
Procaine	1.397	2.24	1.87	1.73	0.00	8.11/8.80	6.63
Pyridine	—	—	—	0.44	0.22	5.19	—
Quinine	−0.214	3.2	3.44	7.21	0.25	4.1/8.5	—
Rifampicin	5.961	2.99	—	—	—	1.7/7.9	—
Phenethylamine	—	—	—	—	—	9.84	—
Scopolamine	2.280	−0.20	1.2	0.90	0.03	7.75	—
Terbutaline	—	0.48	—	0.26	0.00	8.8/10.1/11.2	—
Tetracycline	—	−2.56	—	—	—	3.3/7.7/9.7	—
Theobromine	−1.175	−1.01	−0.78	—	—	0.12/10.05	—
Theophylline	−0.218	−0.25	−0.02	0.12	0.12	3.5/8.6	—
Triamterene	—	1.99	1.11	6.2	—	—	—

A model phase was constructed to increase the contact surface area. The model support consisted of 365 carbons, 248 hydrogens, 848 bonds and 3684 connectors. The molecular weight was 4579. The 7 center hydrogens were replaced by methyl groups, and 12 hydrogens of the 2nd and 18 hydrogens of the 3rd circles were replaced with octyl groups. The retention of lidocaine on the octyl phase is shown in Figure 6.29.

The relationships between MIFS2 or MIVW2 values calculated using the model phase and the measured log k values of the molecular-form nitrogen-containing compounds were:

$$\text{MIFS2} = 5.526 \times (\log k_{max}) + 22.232, \; r = 0.743, \; n = 14,$$

$$\text{MIVW2} = 4.980 \times (\log k_{max}) + 24.126, \; r = 0.586, \; n = 14.$$

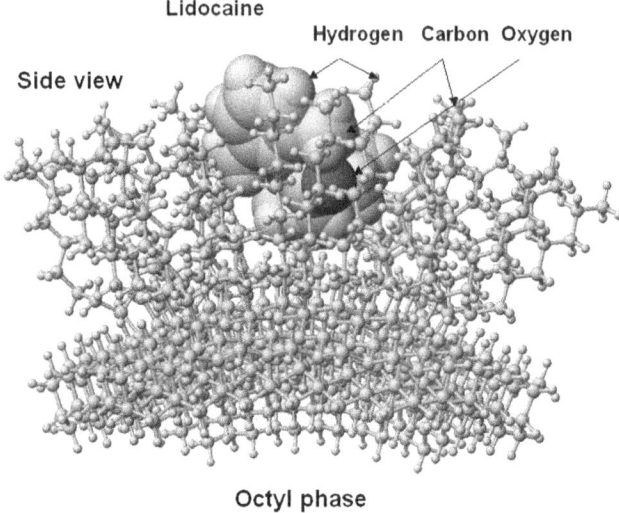

Figure 6.29 Docking of lidocaine on an octyl phase. Small white, large white, dark-gray, and black balls represent hydrogen, carbon, nitrogen, and oxygen, respectively. The atomic size of lidocaine is five times that of the model phase.
Reproduced by permission of Taylor and Francis, ref. 51.

The correlation coefficient did not demonstrate any improvement, even when analytes were buried in the octyl brushes.

Therefore, these octyl groups were replaced by dodecyl groups to increase the contact surface area. This new model phase was useful for aromatic acids. As an example, homatropine was buried in the dodecyl groups as shown in Figure 6.30.

$$\text{MIFS3} = 6.032 \times (\log k_{mam}) + 30.206, \; r = 0.562, \; n = 14,$$

$$\text{MIVW3} = 5.380 \times (\log k_{max}) + 31.128, \; r = 0.466, \; n = 14.$$

There was no meaningful correlation between the molecular interaction energy values and their capacity ratios. These results indicate that the above two models did not reflect the chromatographic behavior of nitrogen-containing compounds.

Therefore, a silica-gel-based bonded phase was constructed. The new phase was constructed based on the dimethoxypentylsilane-bonded poly-silicondioxide phase shown in Figure 3.13, and consisted of 991 atoms, 1051 bonds and 15 193 connectors, containing 171 silicons, 328 oxygens, 143 carbons and 349 hydrogens (Phase 4). Twenty dimethoxypentylsilanes and one trimethylsilane were bonded within an 900 Å^2 area on the poly-silicondioxide phase. The trimethylsilane was considered an end-capped molecule. A pocket caused by a small molecule, trimethylsilane, was designed to follow the V-shape model of a porous silica gel. The optimized structure of a complex formed between this model phase and quinine is

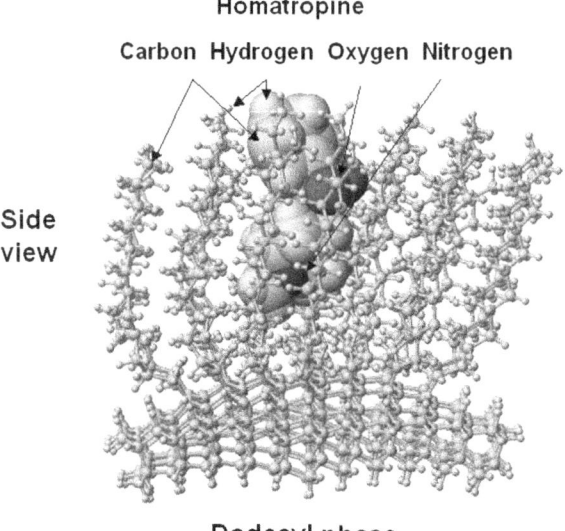

Homatropine

Carbon Hydrogen Oxygen Nitrogen

Side
view

Dodecyl phase

Figure 6.30 Docking of homatropine on a dodecyl phase. Small white, large white, dark-gray, and black balls represent hydrogen, carbon, nitrogen, and oxygen, respectively. The atomic size of lidocaine is five times that of the model phase.
Reproduced by permission of Taylor and Francis, ref. 51.

Carbon Hydrogen Oxygen Nitrogen

Quinine

Side view

Silicon   Oxygen

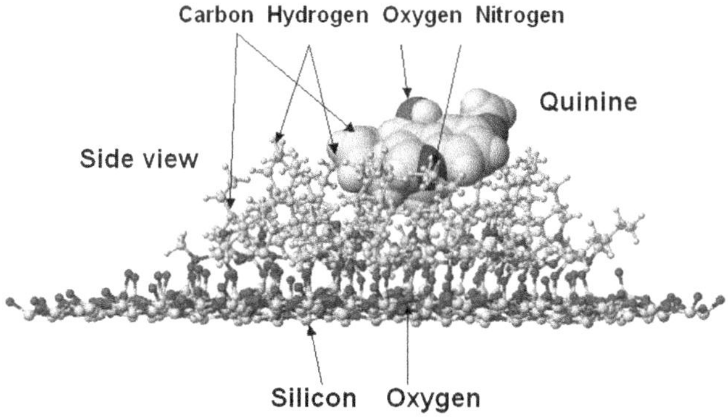

Figure 6.31 Docking of quinine on a dimethoxypentyl-bonded silica phase. Small white, large white, dark-gray, gray, and black balls represent hydrogen, carbon, silicon, nitrogen, and oxygen, respectively. The atomic size of quinine is five times that of the model phase.
Reproduced by permission of Taylor and Francis, ref. 51.

shown in Figure 6.31. Dimethoxypentyl groups stand close together; some of them lay in free space after the optimization.

On this new bonded-phase, dimethoxypentyl groups surrounded one trimethyl group. Dimethoxypentyl groups of the 2nd circle should support the dimethoxypentyl groups of the 1st circle. The interaction energies between this new model phase and each nitrogen-containing compound were calculated and the values are listed as FS4m and VW4m in Table 15 of the Appendix. The r between MIFSm and log k_{max} was improved.

$$MIFS4m = 7.619 \times (\log k_{max}) + 20.924, \ r = 0.941, \ n = 17,$$

$$MIVW4m = 6.700 \times (\log k_{max}) + 18.600, \ r = 0.919, \ n = 17.$$

The subtraction of the hydrogen-bonding energy values did not significantly improve the correlation coefficient, as given by the following equation:

$$(MIFS - hb) = 8.788 \times (\log k_{max}) + 17.800, \ r = 0.949, \ n = 17.$$

The effect of the hydrogen-bonding energy was only important for compounds with intramolecular hydrogen bonding, *ortho*-effect. Furthermore, the above results were examined using measured (log P_m) and calculated (log P_c) log P values from ref. 42.

$$\log P_m = 1.395 \times (\log k_{max}) + 0.491, \ r = 0.897, \ n = 14,$$

$$\log P_c = 1.275 \times (\log k_{max}) + 0.541, \ r = 0.842, \ n = 16.$$

These results are better than those of the three previous models. This r value is not significantly high compared to the results for phenolic compounds.[6] The r value between the log P and log k_m values was smaller than that obtained using MIFS and MIVW. This comparison indicated that this new approach using molecular interaction energy values should work for the quantitative analysis of retention in chromatography, but further development is necessary for the development of the simulation of chromatography for drugs.

Phase 4, the dimethoxypentyl-bonded silica gel, was further modified using dimethoxyoctyl groups. It consisted of 2021 atoms, 2081 bonds and 21 450 connectors, containing 171 silicons, 328 oxygens, 143 carbons and 349 hydrogens (Phase 5). Forty-seven dimethoxyoctylsilanes and one trimethylsilane were bonded to it. The retention of triamterene on the dimethoxyoctyl-bonded silica gel is shown in Figure 6.32.

$$MIFS5 = 3.034 \times (\log k_{max}) + 18.054, \ r = 0.514, \ n = 15$$

$$MIVW5 = 4.278 \times (\log k_{max}) + 18.744, \ r = 0.602, \ n = 15.$$

However, the longer alkyl chain did not improve the correlation coefficient. Therefore, the correlation was studied for ionized nitrogen-containing compounds using Phase 4. The final (optimized) structure and van der Waals energy values are listed as FS4i and VW4i in Table 15 of the Appendix (p. 312). The relationships are:

$$MIFS4i = 4.325 \times (\log k_{min}) + 29.751, \ r = 0.799, \ n = 15,$$

$$MIVW4i = 4.414 \times (\log k_{min}) + 26.636, \ r = 0.781, \ n = 15.$$

The retention times of the ionized compounds were very short, and should include the experimental error, especially the size- and ion-exclusion effects.

Triamterene

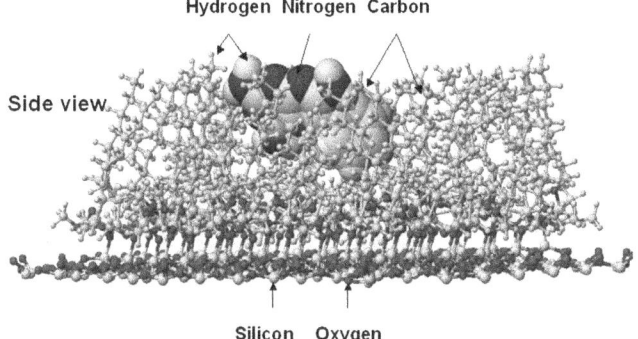

Figure 6.32 Docking of triamterene on a dimethoxyoctyl-bonded silica phase. Small white, large white, dark-gray, gray, and black balls represent hydrogen, carbon, silicon, nitrogen, and oxygen, respectively. The atomic size of triamterene is five times that of the model phase. Reproduced by permission of Taylor and Francis, ref. 51.

When log k values of less than -1 were eliminated, the correlation coefficient improved from 0.799 ($n = 15$) to 0.807 ($n = 11$) and from 0.781 ($n = 15$) to 0.804 ($n = 11$). This new system, using molecular interaction energy values, demonstrated it was possible to predict the capacity ratios of ionized compounds, but an old system using log P values cannot do this. Furthermore, the capacity ratios of partially ionized compounds were predicted using eqn (9).[33] The pK_a values measured in the above experiment were compared with reference pK_a values to predict the capacity ratios of partly ionized compounds:

$$pK_a(\text{reference}) = 1.010 \times (pK_a \text{ measured}) + 0.770, \ r = 0.919, \ n = 18. \quad (10)$$

The correlation coefficient was improved to 0.952 when the values for theobromine were excluded. The difference in the pK_a value, 0.770, may be due to solvent effects. The pH of the eluents was measured before mixing with methanol. However, the difference was similar to that reported previously. The pH values of the buffer solutions were influenced by the addition of methanol or acetonitrile as an organic modifier.[18] How to obtain an absolute pH value in eluent containing organic modifier is still unclear for practical purposes.

The correlation between the capacity ratios measured and predicted with this new method using the molecular interaction energy values, MIFS4m, MIVW4m, MIFS4i and MIVW4i, was obtained from previous equations. From MIFS4$_m$, MIFS4$_i$ and experimental pK_a values (calculated using eqn (10)):

$$k_{pred}(\text{pH } 6.00) = 0.526 \times (k_{mes}) + 0.428, \ r = 0.863, \ n = 16,$$

$$k_{pred}(\text{pH } 7.00) = 0.711 \times (k_{mes}) + 0.444, \ r = 0.933, \ n = 16,$$

$$k_{pred}(\text{pH } 8.00) = 0.726 \times (k_{mes}) + 1.487, \ r = 0.858, \ n = 16.$$

From MIVW4$_m$, MIVW4$_i$ and experimental pK_a (calculated using eqn (10)):

$$k_{pred}(pH\ 6.00) = 0.487 \times (k_{mes}) + 0.741,\ r = 0.645,\ n = 16,$$

$$k_{pred}(pH\ 7.00) = 0.735 \times (k_{mes}) + 0.866,\ r = 0.799,\ n = 16,$$

$$k_{pred}(pH\ 8.00) = 0.645 \times (k_{mes}) + 2.561,\ r = 0.871,\ n = 16.$$

Furthermore, the pH effect was calculated using reference pK_a values. The pK_a values used for the calculations were added the organic modifier concentration effect, $\Delta pK_a = 0.022 \times$ (% of methanol). The constant 0.022 was experimentally obtained.[18] The predicted k values were calculated from MIFS4$_m$, MIFS4$_i$ and the modified reference pK_a values (calculated using eqn (10)).

The results shown in Figure 6.33 indicated that the retention times of basic drugs can be predicted from molecular interaction energy values calculated using the MM2 program. However, the molecular interaction energy of dextromethorphan was smaller than the expected value from the long retention time, which indicates strong hydrophobicity. The interaction energy was small even when values for the longer alkyl chain phases like Phases 2 and 3, shown in Table 15 of the Appendix (p. 312) were used. Further development of a model phase is required to analyze a variety of compounds. With the addition of pK_a values one can predict the retention time of partially ionized compounds. At present, pK_a values can be predicted without Hammett's equation from the atomic partial charge calculated using the MOPAC program for phenolic compounds and aromatic acids. However, no such simple calculation method has been established for nitrogen-containing compounds, because of a lack of standard pK_a values.

The retention times of nitrogen-containing compounds including basic drugs in reversed-phase liquid chromatography were quantitatively analyzed

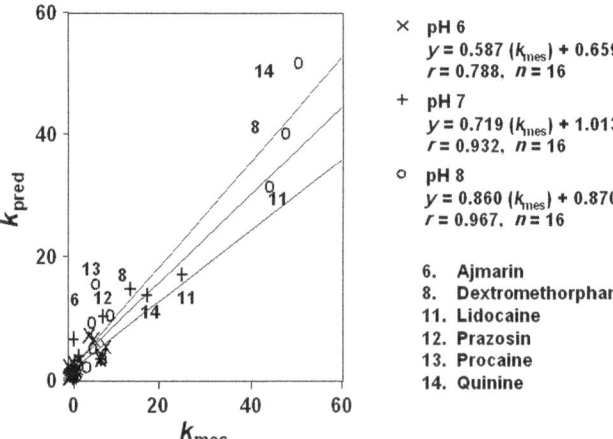

Figure 6.33 The relationship between predicted (k_{pred}) and measured (k_{mes}) capacity ratios for nitrogen-containing compounds in eluents of various pH.

from molecular interaction energy values calculated using the MM2 program. The precision of the capacity ratios predicted with this new method was better than that for a former method in which the retention time was predicted from log *P*. Furthermore, the prediction of the capacity ratios of these compounds in reversed-phase liquid chromatography in a given pH eluent was performed using their pK_a values. Computational chemical calculations demonstrated the possibility of *in silico* chromatography of the retention of basic drugs on a pentyl phase. The addition of solvent effects and the construction of a better model phase should improve the precision of qualitative analysis of capacity ratios in liquid chromatography.

6.9 The Organic Modifier Effect

6.9.1 Solvents in Liquid Chromatography

Consistent with the concept of "like dissolves like", the retention mechanisms of normal- and reversed-phase liquid chromatographies are the same. The retention mechanism depends on the relationship between the analyte and the packing material surface. Predicting the retention time in reversed-phase liquid chromatography is relatively simple due to the limited selection of organic modifiers. A direct *in silico* calculation of molecular interaction energies using model phases was developed instead of calculating enthalpies, due to computer capabilities. Measured enthalpies did not linearly related to the log *k* values of a variety of compounds. Enthalpies were only related to similar compounds' log *k* values.[1] In addition, a new pK_a prediction method was developed in which Hammett σ constants were replaced with atomic partial charges. By combining these methods, retention in both reversed-phase[22] and ion-exchange[44] liquid chromatographies could be predicted when similar compounds were analyzed because of their similar solvation mechanisms. In normal-phase liquid chromatography, the adsorption strength can be quantitatively analyzed *in silico*, but the desorption strength cannot be quantitatively analyzed. Although a variety of organic solvents can be used as eluent components, the solvent strength does not directly relate to the elution order. The total solubility parameter does not relate to solvent strength in normal-phase liquid chromatography,[45,46] and at present, no solubility-prediction method exists. The addition of acids and bases to the eluent modifies the silica gel surface and affects the ionization of the analyte. Therefore, quantitative analysis of retention in normal-phase liquid chromatography, other than for enantioseparation, remains problematic. However, the retention of triterpenes in normal-phase liquid chromatography using a silica gel was not simply related to the direct MI energy values and was affected by the desorption power of the organic solvent. The correlation between the log *k* and MI energy values was improved by including the solvent effects obtained from the analyte–solvent complexes.[47]

The correlation coefficient between the individual MI energy values and the log *k* values indicates the contribution of the main molecular interaction factor to retention in chromatography. MIHB, MIES, and MIVW should represent the main interaction factors in normal-phase, ion-exchange, and reversed-phase liquid chromatographies, respectively. Steric hindrance affects the molecular interactions in enantioseparation.[2,3] However, solubility has to be quantitatively involved, especially in liquid chromatography, even if solubility cannot be predicted at the present stage.

6.9.2   Basic Study of the Organic Modifier Effect

In reversed-phase liquid chromatography, the retention times (log k values) were related to the MI energy values of similar compounds because of the lack of a solubility-prediction system. However, the precision can be improved when solvent effects are included.[48] A similar approach to that used in gas chromatography was therefore studied by using a variety of simple analytes in reversed-phase liquid chromatography, even though alkanes and alkyl alcohols are not generally used for column calibration because of their very poor detectability and solubility compared to common analytes. The retention times of the 18 compounds listed in Tables 6.12 and 6.13 were measured using a variety of hydrophobic columns in aqueous acetonitrile to study the feasibility of this approach. Acetonitrile is a common organic modifier in reversed-phase liquid chromatography, especially when theoretically discussing the relationship between retention time and log P values.[1] The log k values were related to the MI energy values using a model hydrophobic phase, and the solvent effect was studied using a model solvent phase.

The analytes listed in Tables 6.12 and 6.13 are flat, so steric hindrance can be neglected when studying their molecular interactions, therefore, a simple and homogeneous model was used instead of a brush-type bonded phase silica gel model whose blush types and density were designed to reduce the steric hindrance effects of the analytes.[7] A model butyl phase with adsorbed naphthalene is shown in Figure 6.34.

The calculated MI energy values (summarized in Table 6.12) were correlated with the log k values of these compounds. For example, MIFS $= a \times \log k + b$, where the slope a and constant b of all groups should be the same in an ideal system for predicting retention time. The relationships are summarized in Figure 6.35 for log k values measured on the octadecyl-bonded silica gel column in 80% aqueous acetonitrile at 40 °C.

The coefficients of the individual groups were close to 1. The order of the MIFS values was alkyl alcohols > alkyl benzenes $\geq$ PAHs > alkanes. If the relationship between alkane log k values and their MIFS values is considered as the standard for *in silico* analysis and the retention is based on a hydrophobic interaction, the retention of other types of compounds should follow the relationship of the alkanes. The retention strength was as follows: alkanes > PAHs $\geq$ alkyl benzenes > alkanes, based on the MI energy values. The van der Waals energy values support the strength of hydrophobic retention. If their solvation mechanisms are the same, all compounds should

Table 6.12 Molecular properties of some common analytes. Log k was measured on an ODS silica gel in 80% aqueous acetonitrile at 40 °C. ac represents molecular properties calculated using a model acetonitrile solvent phase. Reproduced by permission of Oxford University Press, ref. 48.

No.	Analyte	*log* k	*fs*	*hb*	*es*	*vw*
1	Benzene	−0.038	−8.0770	0	0	3.006
2	Naphthalene	0.189	−8.6880	0	0	5.767
3	Phenanthrene	0.434	−4.6360	0	0	9.988
4	Chrycene	0.734	−9.9860	0	0	14.628
5	Toluene	0.104	−8.6003	0	0	3.036
6	Ethylbenzene	0.224	−6.7841	0	0	4.054
7	Butylbenzene	0.514	−5.4670	0	0	4.995
8	Hexylbenzene	0.814	−4.1871	0	0	5.916
9	Heptylbenzene	0.970	−3.5456	0	0	6.382
10	Butanol	−0.529	2.1290	0	0	1.566
11	Hexanol	−0.259	3.4226	0	0	2.502
12	Octanol	0.047	4.7082	0	0	3.425
13	Decanol	0.126	5.9900	0	0	4.355
14	Pentane	0.498	2.8289	0	0	2.149
15	Hexane	0.656	3.4720	0	0	2.614
16	Heptane	0.816	4.1135	0	0	3.070
17	Octane	0.978	4.7556	0	0	5.537
18	Decane	1.302	6.0410	0	0	4.461
Alkyl phase		—	3373.0355	0	0	419.967
Acetonitrile phase		—	44.7183	0	47.055	−8.952

No.	FS	HB	ES	VW	acFS	acHB	acES	acVW
1	3358.6760	0	0	416.659	32.2908	0	47.054	−10.308
2	3345.5813	0	0	416.684	19.5243	0	47.054	− 9.700
3	3337.2844	0	0	418.612	12.1285	0	47.054	−6.918
4	3330.4387	0	0	421.323	5.5540	0	47.039	−4.034
5	3356.7158	0	0	415.127	30.1996	0	47.086	−11.743
6	3355.7514	0	0	413.835	31.5723	0	47.075	−11.290
7	3355.6050	0	0	412.577	32.6733	0	47.208	−11.246
8	3354.1305	0	0	411.082	32.3498	0	46.642	−11.482
9	3353.4853	0	0	409.954	32.3409	0	46.672	−11.755
10	3367.5463	0	0	413.740	39.8065	0	44.209	−11.710
11	3365.9773	0	0	411.814	40.6882	0	44.373	−12.211
12	3364.1596	0	0	409.337	41.1229	0	44.373	−12.211
13	3363.1108	0	0	408.187	41.4577	0	44.992	−12.692
14	3366.9124	0	0	412.978	43.0103	0	47.046	−11.605
15	3366.3628	0	0	412.277	42.5589	0	47.049	−12.039
16	3365.4766	0	0	411.003	43.2786	0	47.056	−11.700
17	3365.0100	0	0	410.387	43.6990	0	47.050	−11.806
18	3363.5446	0	0	408.650	43.5787	0	47.045	−12.245

demonstrate the same linear relationship between their MI energy values and log k values measured in liquid chromatography.

The MIFS differences (ΔMIFS) between alky alcohols, alkyl benzenes, PAHs and alkanes have to be due to difference in their solubility. The prediction of

Table 6.13 Log k values of some common analytes. ODP represents an octadecyl-bonded polyvinylalcohol copolymer gel. Reproduced by permission of Oxford University Press, ref. 48.

Column	Octyl-bonded silica				Propyl-bonded silica			
Acetonitrile/%	90	80	70	60	80	70	60	50
No.								
Analyte								
1 Benzene	−0.561	−0.271	−0.006	0.249	−0.416	−0.165	0.060	0.280
2 Naphthalene	−0.441	−0.122	0.169	0.469	−0.321	−0.041	0.225	0.507
3 Phenanthrene	−0.319	0.018	0.332	0.660	−0.238	0.065	0.365	0.700
4 Chrycene	−0.208	−0.159	0.499	0.877	−0.149	0.181	0.516	0.903
5 Toluene	−0.465	−0.170	0.100	0.376	−0.366	−0.090	0.157	0.406
6 Ethylbenzene	−0.400	−0.081	0.205	0.505	−0.300	−0.008	0.258	0.539
7 Butylbenzene	−0.233	0.121	0.441	0.791	−0.159	0.168	0.480	0.827
8 Hexylbenzene	−0.081	0.319	0.677	1.077	−0.017	0.346	0.701	1.115
9 Heptylbenzene	−0.002	0.420	0.797	1.222	0.055	0.436	0.813	1.259
10 Butanol	−0.791	−0.578	−0.404	−0.267	−0.710	−0.497	−0.336	−0.235
11 Hexanol	−0.613	−0.388	−0.185	0.014	−0.557	−0.318	−0.117	0.055
12 Octanol	−0.438	−0.221	0.050	0.297	−0.409	−0.145	0.096	0.335
13 Decanol	−0.274	−0.010	0.281	0.578	−0.267	0.025	0.308	0.614
14 Pentane	−0.234	0.077	0.379	0.672	−0.196	0.106	0.380	0.663
15 Hexane	−0.150	0.181	0.499	0.820	−0.123	0.196	0.493	0.810
16 Heptane	−0.069	0.288	0.624	0.971	−0.046	0.289	0.610	0.959
17 Octane	0.015	0.393	0.748	1.127	0.029	0.382	0.727	1.117
18 Decane	0.177	0.608	1.004	—	0.185	0.579	0.971	—

Column	ODS silica		ODP			Capcellpack[a]		
Acetonitrile/%	90	70	90	80	70	80	70	60
No. Analyte								
1 Benzene	−0.283	0.199	−0.309	−0.075	0.135	−0.453	−0.204	0.030
2 Naphthalene	−0.068	0.452	−0.062	0.202	0.450	−0.265	0.007	0.272
3 Phenanthrene	0.145	0.728	0.208	0.503	0.786	−0.063	0.229	0.525
4 Chrycene	0.408	1.062	0.510	0.843	1.161	0.165	0.485	0.816
5 Toluene	−0.136	0.351	−0.240	0.013	0.242	−0.318	−0.063	0.191
6 Ethylbenzene	−0.039	0.492	−0.195	0.082	0.333	−0.210	0.062	0.334
7 Butylbenzene	0.203	0.825	−0.023	0.300	0.594	0.045	0.362	0.676
8 Hexylbenzene	0.453	1.168	0.174	0.544	0.882	0.303	0.669	1.026
9 Heptylbenzene	0.584	1.346	0.284	0.677	1.037	0.437	0.827	1.207
10 Butanol	−0.660	−0.360	−0.711	−0.626	−0.493	−0.799	−0.703	−0.561
11 Hexanol	−0.393	−0.064	−0.540	−0.399	−0.241	−0.565	−0.430	−0.280
12 Octanol	−0.141	0.264	−0.344	−0.164	0.034	−0.337	−0.131	0.052
13 Decanol	0.126	0.612	−0.134	0.089	0.326	−0.061	0.171	0.397
14 Pentane	0.226	0.769	−0.039	0.239	0.497	0.118	0.420	0.673
15 Hexane	0.359	0.948	0.068	0.370	0.652	0.254	0.582	0.859
16 Heptane	0.493	1.131	0.179	0.506	0.809	0.392	0.746	1.045
17 Octane	0.629	1.306	0.293	0.646	0.972	0.531	0.910	1.232
18 Decane	0.901	—	0.530	0.934	—	0.808	1.239	—

[a]Capcellpack™: ODS silica gel column from Shiseido.

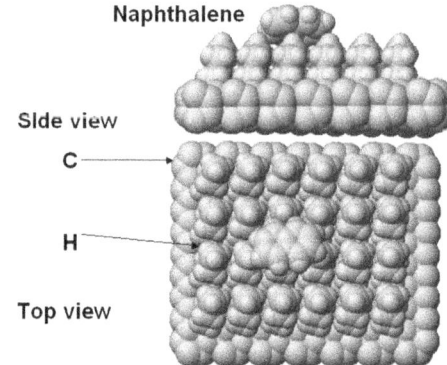

Figure 6.34 Docking of naphthalene on a model butyl phase. White and gray balls represent hydrogen and carbon, respectively.

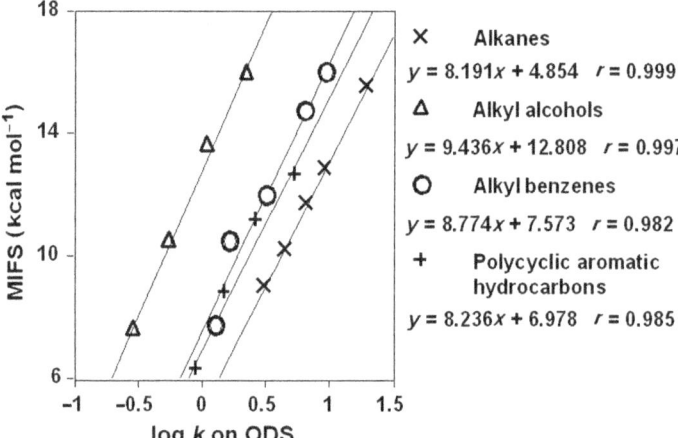

Figure 6.35 The relationship between molecular interaction energy and log k on an octadecyl-bonded silica gel column in 80% aqueous acetonitrile at 40 °C for some common analytes.

solubility is impossible at this stage. Acetonitrile molecules form complexes between their cyano groups but not their methyl groups, as shown in Figure 6.36. The complex formed in Figure 6.36(b) demonstrated a very low electrostatic energy drop. Methyl groups were repulsed in Figure 6.36(c) and did not form a complex. The MI energy values of the analyte and solvent complexes were calculated using a model solvent layer, shown in Figure 6.37, where benzene interacted with the model phase.

The molecular interaction energy values are summarized in Table 6.12. The ΔMIFS values were related to the acMIES values between the analytes and the model solvent phase. The relationship is given by the following equation:

$$\Delta\text{MIFS} = 1.498x^2 - 6.598x - 0.954, \; r = 0.951, \; \text{where } x = \text{acMIES}.$$

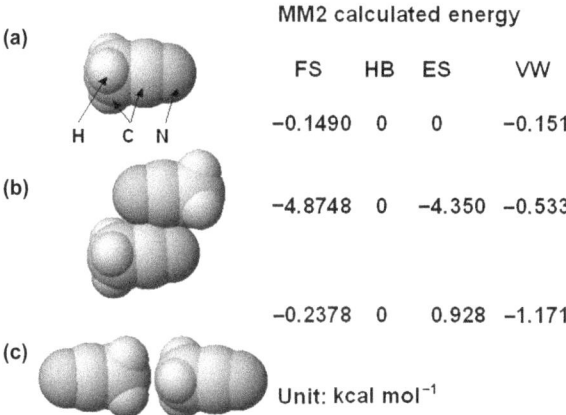

MM2 calculated energy

	FS	HB	ES	VW
(a)	−0.1490	0	0	−0.151
(b)	−4.8748	0	−4.350	−0.533
	−0.2378	0	0.928	−1.171

Unit: kcal mol^{-1}

Figure 6.36 Conformation of acetonitrile dimers. White, light-gray, and gray balls represent hydrogen, carbon, and nitrogen, respectively. Reproduced by permission of Oxford University Press, ref. 48.

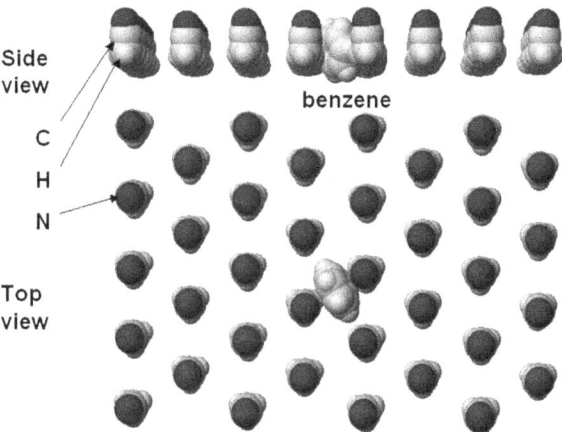

Figure 6.37 Docking of benzene on a model acetonitrile solvent phase. White, light-gray, and black balls represent hydrogen, carbon, and nitrogen, respectively.

The addition of ΔMIFS values calculated from acMIES to the MIFS values improved the correlation with the log k values. The correlation coefficient was 0.973 ($n = 18$) as demonstrated in Figure 6.38. Even though solubility prediction is impossible, the MI energy values with a model solvent phase improved the precision of the relationship between the log k values and the molecular interaction energy values. The correlation coefficients in 90 and 70% aqueous acetonitrile were 0.967 ($n = 18$) and 0.971 ($n = 17$), respectively.

The above results demonstrated the possibility of the quantitative analysis of retention times in reversed-phase liquid chromatography. In addition, a cyano-group-bonded silicone phase was constructed, and MI energy values were calculated. The sum of the MIFS and cnMIES (cyano phase) energies

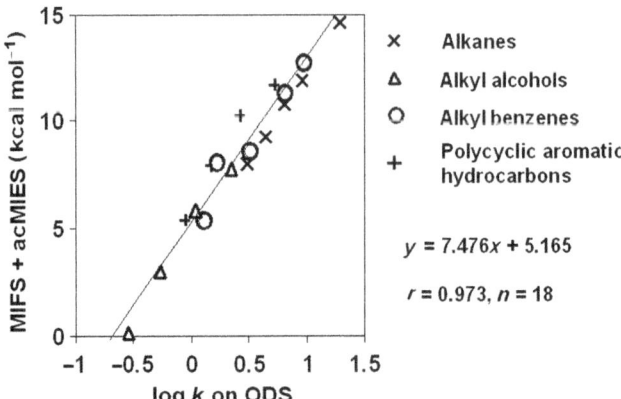

Figure 6.38   The relationship between log k and the combined calculated energy values MIFS + acMIES on an octadecyl-bonded silica gel column in 80% aqueous acetonitrile at 40 °C for some standard analytes. Reproduced by permission of Oxford University Press, ref. 48.

was correlated to log k values with a correlation coefficient of 0.952 ($n=18$). The details are not listed because the acetonitrile phase was considered to be a better model phase than the cyano-group-bonded phase.

In addition, the feasibility of the above approach was examined for log k values measured using propyl- and octyl-bonded silica gel columns, an octadecyl-bonded polyvinylalcohol copolymer gel (ODP) column, and a Capcell-pack ODS silica gel column. The main interactions were based on van der Waals forces, therefore, the van der Waals energy values were used to determine the molecular interaction between these analytes and the model phases.

The relationship between the log k values and MIVW (also MIFS) of alkanes, alky alcohols, alkyl benzenes and PAHs demonstrated the selectivity of these packing materials. The relationships for the alkanes were used as standards, and the MIVW value differences (ΔMIVW) of the other compounds were calculated. The MIVW values of the four groups were related to the acMIES values obtained from the relationships between the analytes and the model solvent phases. The ΔMIVW values were calculated using the relations calculated for each phase in different mobile phases. The obtained ΔMIVW values were combined with the MIVW value, and then correlated to the log k values.

For the octadecyl-bonded silica gel column:

$$\text{MIVW} + \Delta\text{MIVW} = 8.860 \times (\log k) + 7.072, \; r = 0.968, \; n = 18, \; \text{in } 90\% \text{ ACN (aq)},$$

$$\text{MIVW} + \Delta\text{MIVW} = 7.683 \times (\log k) + 5.284, \; r = 0.973, \; n = 18, \; \text{in } 80\% \text{ ACN (aq)},$$

$$\text{MIVW} + \Delta\text{MIVW} = 7.160 \times (\log k) + 3.539, \; r = 0.972, \; n = 18, \; \text{in } 70\% \text{ ACN(aq)}.$$

For the octyl-bonded silica gel column:

$MIVW + \Delta MIVW = 13.987 \times (\log k) + 12.277, r = 0.951, n = 18,$ in 90 % ACN (aq),

$MIVW + \Delta MIVW = 11.606 \times (\log k) + 8.219, r = 0.963, n = 18,$ in 80 % ACN (aq),

$MIVW + \Delta MIVW = 10.020 \times (\log k) + 5.350, r = 0.970, n = 18,$ in 70 % ACN (aq),

$MIVW + \Delta MIVW = 8.719 \times (\log k) + 3.239, r = 0.975, n = 17,$ in 60 % ACN (aq).

For the propyl-bonded silica gel column:

$MIVW + \Delta MIVW = 16.471 \times (\log k) + 12.369, r = 0.968, n = 18,$ in 80 % ACN (aq),

$MIVW + \Delta MIVW = 13.373 \times (\log k) + 7.753, r = 0.969, n = 18,$ in 70 % ACN (aq),

$MIVW + \Delta MIVW = 10.888 \times (\log k) + 5.038, r = 0.975, n = 18$ in 60 % ACN (aq),

$MIVW + \Delta MIVW = 8.876 \times (\log k) + 3.232, r = 0.979, n = 17,$ in 50 % ACN (aq).

For the octadecyl-bonded polyvinylalcohol copolymer column:

$MIVW + \Delta MIVW = 11.513 \times (\log k) + 11.710, r = 0.974, n = 18,$ in 90 % ACN (aq),

$MIVW + \Delta MIVW = 9.404 \times (\log k) + 6.951, r = 0.989, n = 18,$ in 80 % ACN (aq),

$MIVW + \Delta MIVW = 8.350 \times (\log k) + 4.959, r = 0.989, n = 17,$ in 70 % ACN (aq).

For the octadecyl Capcellpak column (silica gel matrix):

$MIVW + \Delta MIVW = 8.237 \times (\log k) + 7.999, r = 0.945, n = 18,$ in 90 % ACN (aq),

$MIVW + \Delta MIVW = 7.161 \times (\log k) + 6.001, r = 0.952, n = 18,$ in 80 % ACN (aq),

$MIVW + \Delta MIVW = 6.736 \times (\log k) + 4.463, r = 0.955, n = 17,$ in 70 % ACN (aq).

These results support that the above *in silico* approach for the quantitative analysis of retention times in reversed-phase liquid chromatography is feasible with high precision if ideal model phases and solubility-prediction methods are available. The liquid chromatography retention time is a combination of adsorption on a packing material surface and solvation in a mobile phase. The precision for gas chromatography is high because vaporization energies can be calculated *in silico* from analyte properties. The development of a solubility-prediction method is required to improve the precision of predicted retention times in liquid chromatography.

6.9.3   Alkyl Benzenes as Standard Compounds for Phenols

In Section 6.9.2, alkanes were used as standard compounds, because they are simple hydrophobic compounds that are suitable for, and commonly use in, gas chromatographic applications. However, alkanes are not suitable standards for liquid chromatography because of their very strong hydrophobicity and poor detectability. Therefore, the feasibility of using alkyl benzene standards was evaluated for the liquid chromatographic analysis of phenolic compounds.

The chromatographic behavior of alkanes, alkyl benzenes and alkyl- and halogenated phenols is summarized in Figure 6.39. The retention times were obtained in reversed-phase liquid chromatography using an octadecyl-bonded silica gel column. The eluent for the analysis was 70% aqueous acetonitrile containing 0.01% phosphoric acid. Chromatographic analysis was carried out at 40 °C. The model phase is shown in Figure 6.13. The measured and calculated properties are summarized in Table 6.14.

The MIHB and MIES energy values of alkanes and alkyl benzenes were close to zero, and MIFS and MIVW values were linearly related to the log k values. The relationships are outlined as follows:

For alkanes:

$$\text{MIFS} = 7.609 \times (\log k) + 3.429, \; r = 0.999, \; n = 4,$$
$$\text{MIVW} = 7.992 \times (\log k) + 3.307, \; r = 0.997, \; n = 4.$$

For alkyl benzenes:

$$\text{MIFS} = 8.838 \times (\log k) + 4.596, \; r = 0.999, \; n = 6,$$
$$\text{MIVW} = 9.041 \times (\log k) + 4.647, \; r = 0.998, \; n = 6.$$

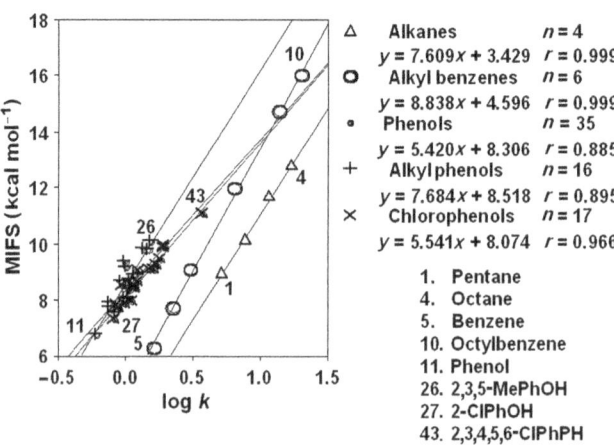

Figure 6.39   The relationship between molecular interaction energy (MIFS) and log k on a butyl-bonded phase in 80% aqueous acetonitrile.

Table 6.14 Molecular properties of standard compounds in a model acetonitrile solvent phase. Log *k* values were measured on an ODS-bonded silica gel in aqueous 70% acetonitrile containing 0.01% phosphoric acid.[49] HOMO and LUMO energy values (eV) were calculated using the PM5 program.

Compounds	log k	acFS	acHB	acES	acVW	HOMO	LUMO
Pentachlorophenol	0.5750	33.819	−1.970	44.919	−6.305	−9.5663	−1.0870
2,3,4-Trichlorophenol	0.2130	29.736	−1.882	44.701	−9.021	−9.3548	−0.5139
2,3,5-Trichlorophenol	0.2819	28.414	−1.868	43.658	−9.213	−9.6218	−0.6407
2,3,5-Trimethylphenol	0.1351	26.031	−1.479	45.999	−12.081	−8.6446	0.6278
2,3,6-Trichlorophenol	0.2271	31.194	−1.922	46.216	−8.852	−9.4494	−0.5938
2,3,6-Trimethylphenol	0.1729	25.896	−1.384	44.812	−11.260	−8.7965	0.5977
2,3-Dichlorophenol	0.0488	30.165	−1.861	45.758	−9.289	−9.3772	−0.2416
2,3-Dimethylphenol	0.0199	26.655	−1.346	44.575	−10.907	−8.7497	0.6065
2,4,5-Trichlorophenol	0.2620	28.157	−1.913	43.835	−9.607	−9.4061	−0.6117
2,4,6-Trichlorophenol	0.2869	22.744	−1.942	40.058	−10.592	−9.4147	−0.5860
2,4,6-Trimethylphenol	0.1870	24.305	−1.382	44.038	−11.785	−8.5242	0.6397
2,4-Dichlorophenol	0.1000	25.365	−1.852	43.155	−11.181	−9.2603	0.2598
2,4-Dimethylphenol	0.0318	25.507	−1.371	44.595	−11.650	−9.2603	−0.2598
2,5-Dichlorophenol	0.0748	29.576	−1.903	46.287	−10.094	−9.4201	−0.3285
2,5-Dimethylphenol	0.0318	26.409	−1.386	45.309	−11.251	−8.7269	0.6012
2,6-Dichlorophenol	0.0748	29.066	−1.956	46.162	−9.311	−9.3353	−0.2473
2,6-Dimethylphenol	0.0648	25.352	−1.376	44.252	−11.582	−8.7126	0.6133
2-Chlorophenol	−0.0830	30.386	−1.842	46.938	−9.785	−9.1713	0.1158
2-Ethylphenol	0.0519	27.443	−1.358	44.505	−10.782	−8.8304	0.5836
2-Methylphenol	−0.0773	27.204	−1.365	44.546	−10.510	−8.8189	0.5810
3,4-Dichlorophenol	0.0888	28.019	−1.459	44.165	−10.258	−9.2224	−0.1875
3,4-Dimethylphenol	−0.0348	27.510	−1.475	46.954	−11.941	−8.6557	0.6137
3,5-Dichlorophenol	0.1761	23.349	−1.475	41.098	−11.483	−9.5591	−0.3116
3,5-Dimethylphenol	−0.0070	25.755	−1.464	45.170	−11.743	−8.7977	0.6230
3-Bromophenol	−0.0079	27.147	−1.452	43.911	−10.399	−9.2247	0.1495
3-Chlorophenol	−0.0462	27.148	−1.447	43.090	−9.542	−9.2365	0.1081
3-Ethylphenol	−0.0022	26.363	−1.454	45.341	−12.387	−8.8348	0.6040
3-Methylphenol	−0.1158	27.875	−1.448	46.070	−11.096	−8.8644	0.5940
4-Bromophenol	−0.0218	27.035	−1.451	44.837	−11.437	−9.0822	0.1908
4-Chlorophenol	−0.0630	27.074	−1.451	44.897	−11.410	−9.0513	0.1535
4-Ethylphenol	0.0030	26.684	−1.477	45.625	−12.332	−8.6883	0.6061
4-Methylphenol	−0.1141	26.702	−1.465	45.559	−11.748	−8.7085	0.5991
Phenol	−0.2161	28.483	−1.464	45.658	−10.662	−8.9324	0.5697
Benzene	0.2120	32.291	0.000	47.054	−10.308	−9.4680	0.7120
Toluene	0.3570	30.200	0.000	47.086	−11.743	−9.1050	0.7030
Ethylbenzene	0.4890	31.572	0.000	47.075	−11.290	−9.0690	0.7210
Butylbenzene	0.8110	32.673	0.000	47.208	−11.246	−9.0700	0.7180
Hexylbenzene	1.1420	32.350	0.000	46.642	−11.482	−9.0730	0.7150
Heptylbenzene	1.3100	32.341	0.000	46.672	−11.755	−9.0740	0.7140

First, the difference in the MIFS values (ΔMIFS) between phenolic compounds and alkanes was related to the MI energy values resulting from the interaction between an analyte and a model acetonitrile solvent phase. These are represented by acMIFS, acMIHB, acMIES and acMIVW. The highest portion of the contributed energy was due to acMIES, while the contributions due to acMIHB and acMIVW energies were negligible. The values for acMIFS, acMIHB, acMIES and acMIVW are summarized in Table 6.14. The same approach was applied to a system in which alkanes and alkyl benzenes were used as standard compounds.

For alkane standards:

$$\Delta MIFS = 0.240 \times (acMIES) -5.425, \, r = 0.832, \, n = 35.$$

For alkyl benzene standards:

$$\Delta MIFS = 0.308 \times (acMIES) -4.369, \, r = 0.861, \, n = 35.$$

The MIHB and MIES energy values of alkanes and alkyl benzenes were close to zero, and the MIFS and MIVW values were linearly related to the log k values. The relationships are outlined as follows:

For alkanes:

$$MIFS = 7.609 \times (\log k) + 3.429, \, r = 0.999, \, n = 4,$$
$$MIVW = 7.992 \times (\log k) + 3.307, \, r = 0.997, \, n = 4.$$

For alkyl benzenes:

$$MIFS = 8.838 \times (\log k) + 4.596, \, r = 0.999, \, n = 6,$$
$$MIVW = 9.041 \times (\log k) + 4.647, \, r = 0.998, \, n = 6.$$

The MIHB and MIES energy values of alkanes and alkyl benzenes were close to zero, and the MIFS and MIVW values were linearly related to the log k values. The relationships are outlined as follows:

For alkanes:

$$MIFS = 7.609 \times (\log k) + 3.429, \, r = 0.999, \, n = 4,$$
$$MIVW = 7.992 \times (\log k) + 3.307, \, r = 0.997, \, n = 4.$$

For alkyl benzenes:

$$MIFS = 8.838 \times (\log k) + 4.596, \, r = 0.999, \, n = 6,$$
$$MIVW = 9.041 \times (\log k) + 4.647, \, r = 0.998, \, n = 6.$$

First, the $\Delta MIFS$ values between phenolic compounds and alkanes were related to the MI energy values resulting from the interaction between an analyte and the model acetonitrile solvent phase. These are indicated as acMIFS, acMIHB, acMIES and acMIVW. The highest portion of the contributed energy was due to acMIES, while the contributions due to acMIHB and acMIVW energies were negligible. The values for acMIFS, acMIHB, acMIES, and acMIVW are summarized in Table 6.14. The same approach was applied to a system in which alkyl benzenes were used as standard compounds.

For alkane standards:

$$\Delta MIFS = 0.240 \times (acMIES) - 5.425, \, r = 0.832, \, n = 35.$$

For alkyl benzene standards:

$$\Delta MIFS = 0.308 \times (acMIES) - 4.369, \, r = 0.861, \, n = 35.$$

The correlation coefficient between the MIFS and log k values for phenolic compounds ($n = 35$) was 0.885, and it was improved by the addition of acMIES values. The new correlation coefficients were 0.966 and 0.967 when using the alkane and alkyl benzene standards, respectively. This result indicated that alkyl benzenes can be used as standard compounds instead of alkanes. However, alkyl benzenes are much more hydrophobic compared to phenolic compounds. *p*-Alkyl phenols might also be suitable standards. Alkyl phenols with longer alkyl substitutes, however, were not available for study.

In addition to the above calculations, the MIVW energy values were used instead of MIFS values, since the main molecular interaction in reversed-phase liquid chromatography is hydrophobicity. The results of this calculation were identical to those obtained using MIFS energy values.

For alkane standards:

$$\Delta MIVW = 0.296 \times (acMIES) - 4.441, \, r = 0.868, \, n = 35.$$

For alkyl benzene standards:

$$\Delta MIVW = 0.239 \times (acMIES) - 5.686, \, r = 0.833, \, n = 35.$$

The correlation coefficient between the MIVW and log k values for phenolic compounds ($n = 35$) was 0.898. The correlation was improved by the addition of the acMIES values of the alkane and alkyl benzene standards to give values of 0.973 and 0.971, respectively.

The relationship between the MIFS and log k values of phenolic compounds indicated that alkyl and halogenated phenols demonstrated different behavior. The difference could be attributed to the inductive effects of the substituents. Therefore, the HOMO and LUMO energy values calculated using the PM5 program were related to the $\Delta MIFS$ energy values.

For alkane standards:

$$\Delta MIFS = -1.323 \times (HOMO) - 16.719, \, r = 0.781, \, n = 35,$$
$$\Delta MIFS = -0.858 \times (LUMO) - 4.622, \, r = 0.809, \, n = 35,$$

For alkyl benzene standards:

$$\Delta MIFS = -1.621 \times (HOMO) - 18.175, \; r = 0.773, \; n = 35,$$

$$\Delta MIFS = -1.112 \times (LUMO) - 3.338, \; r = 0.847, \; n = 35,$$

When alkanes were used as the standard compounds, the correlation co-efficients between the log k values and MIFS energy values improved from 0.885 to 0.956 and 0.962 by the addition of the HOMO and LUMO energy values. The correlation coefficients were 0.953 and 0.964 with alkyl benzenes as the standard. Similar results were obtained for MIVW.

For alkane standards:

$$\Delta MIVW = -1.612 \times (HOMO) - 18.201, \; r = 0.808, \; n = 35,$$

$$\Delta MIVW = -1.076 \times (LUMO) - 3.450, \; r = 0.861, \; n = 35.$$

For alkyl benzene standards:

$$\Delta MIVW = -1.357 \times (HOMO) - 17.302, \; r = 0.811, \; n = 35,$$

$$\Delta MIVW = -0.859 \times (LUMO) - 4.890, \; r = 0.820, \; n = 35.$$

The correlation coefficients were improved to 0.964 and 0.971 by the addition of the HOMO and LUMO energy values, respectively with alkanes as the standard. The values were increased to 0.966 and 0.969 by the addition of HOMO and LUMO energy values, respectively with alkyl benzenes as the standard.

6.10   Summary

Analyses to quantitatively relate the log k values to a combination of molecular interaction energy values, solvent effects, and/or electron localization energy values (calculated as HOMO or LUMO) in reversed-phase liquid chromatography, have been performed. The practical application of this method requires a limited number of standard compounds for column calibration.

The precision of the correlation between the log k and molecular interaction energy values was high, as long as the analyte structure was simple and flat, as demonstrated in Section 6.9.2. Specifically, such analyses are most successful when studying retention mechanisms on graphitized carbon phases (Section 6.3). This is because the most effective system for such analyses is a homogeneous and flexible model phase where the docking process may not cause errors.

The retention times measured in reversed-phase liquid chromatography were quantitatively analyzed using simple compounds like those used in gas chromatography. Alkanes were used as standards, much like Kováts retention indices. Alkyl benzenes were used as standard compounds for phenols. The main molecular interaction force between the hydrophobic surface and the analytes was hydrophobicity, which is calculated as the van der Waals energy. The desorption is due to solvation of the analytes by an organic modifier. Solvation was analyzed as the molecular interaction between a model solvent phase and the analytes. The latter interaction was mainly related to the electrostatic energy value from log k values measured in aqueous acetonitrile. The correlation coefficients between the energy values calculated *in silico* and log k were about 0.97. The log k values in reversed-phase liquid chromatography can be quantitatively calculated if an ideal model phase and a solubility-prediction method are used. These results show that this new approach to studying the quantitative analysis of chromatographic retention times in reversed-phase liquid chromatography is theoretically possible. The precision should be improved when solubility can be predicted.

References

1. T. Hanai, *HPLC: A Practical Guide*, RSC Publishing, Cambridge, 1999, pp. 109–132.
2. T. Hanai, Column types, in *Encyclopedia of Analytical Science*, Academic Press, London, pp. 2558–2567.
3. T. Hanai, Synthesis and properties of stable bonded silica gel packings and the performance, in *Advances in Liquid Chromatography*, ed. H. Hatano and T. Hanai, World Scientific, Singapore, 1996, pp. 307–327.
4. T. Hanai, New developments in liquid-chromatographic stationary phases, in *Advances in Chromatography*, ed. P. R. Brown and E. Grushka, Marcel Dekker, New York, 2000, vol. 40, pp. 315–357.
5. G. E. Berendsen and L. De Galan, Role of the chain length of chemically bonded phases and the retention mechanism in reversed–phase liquid chromatography, *J. Chromatogr.*, 1980, **196**, 21–37.
6. T. Hanai, Simulation chromatography of phenolic compounds using a computational chemical method, *J. Chromatogr., A*, 2003, **1027**, 279–287.
7. (a) T. Hanai and H. Homma, Computational chemical prediction of the retention factor of aromatic acids, *J. Liq. Chromatogr. Relat. Technol.*, 2002, **25**, 1661–1676; (b) N. Border, Z. Gabanyi and C. K. Wong, A new method for the estimation of partition coefficient, *J. Am. Chem. Soc.*, 1989, **111**, 3783–3786.
8. T. Hanai, Chromatography *in silico*, quantitative analysis of retention mechanisms of benzoic acid derivatives, *J. Chromatogr., A*, 2005, **1087**, 45–51.
9. T. Hanai, Chromatography *in silico*, quantitative analysis of retention of aromatic acid derivatives, *J. Chromatogr. Sci.*, 2006, **44**, 247–252.
10. T. Hanai, R. Miyazaki, A. Koseki and T. Kinoshita, Computational chemical analysis of the retention of acidic drugs on a pentyl-bonded

silica gel in reversed-phase liquid chromatography, *J. Chromatogr. Sci.*, 2004, **42**, 354–360.

11. T. Hanai, Design model-phases for chromatography in *silico*, presented at the 27th International Symposium in Capillary Chromatography, Riva del Garda, 2004.

12. K. J. McGraw, G. E. Hill, R. Stradi and R. S. Parker, The influence of carotenoid acquisition and utilization on the maintenance of species-typical plumage pigmentation in male American goldfinches (*Carduelis tristis*) and northern cardinals (Cardinalis cardinalis), *Physiol. Biochem. Zool*, 2001, **74**, 843–852.

13. K. J. McGraw, G. E. Hill, R. Stradi and R. S. Parker, The effect of dietary carotenoid access on sexual dichromatism and plumage pigment composition in the American goldfinch, *Comp. Biochem. Physiol., Part B: Biochem. Mol. Biol.*, 2002, **131**, 261–269.

14. T. Hanai, Separation of polar compounds using carbon columns, *J. Chromatogr., A*, 2003, **989**, 183–196.

15. T. Hanai, Analysis of mechanism of retention on graphitic carbon by a computational chemical method, *J. Chromatogr., A*, 2004, **1030**, 13–16.

16. D. D. Perrin, B. Dempsey and E. P. Serjeant, *pK_a Prediction for Organic Acids and Bases*, Chapman and Hall, London, 1981.

17. *CAChe*, Project Reader Manual from Sony–Tektronix, Tokyo, 1995. (Now Fujitsu, Tokyo).

18. T. Hanai, K. Koizumi, T. Kinoshita, R. Arora and F. Ahmed, Prediction of pKa values of phenolic and nitrogen-containing compounds by computational chemical analysis compared to those measured by liquid chromatography, *J. Chromatogr., A*, 1997, **762**, 55–61.

19. J. J. M. Stewart, personal communication, 2007.

20. T. Hanai, Quantitative structure–retention relationships of phenolic compounds without Hammett's equations, *J. Chromatogr., A*, 2003, **985**, 343–349.

21. L. Lepri, P. G. Desideri and D. Heimler, Reversed-phase and soap thin-layer chromatography of phenols, *J. Chromatogr.*, 1980, **195**, 339–348.

22. T. Hanai, Chromatography *in silico*, basic concept in reversed-phase liquid chromatography and computational chemistry, *Anal. Bioanal. Chem.*, 2005, **382**, 708–717.

23. T. Hanai, Quantitative *in silico* analysis of ion-exchange from chromatography to protein, *J. Liq. Chromatogr. Relat. Technol.*, 2007, **30**, 1251–1275.

24. F. Tazerouti, A. Y. Badjah-Hadj-Ahmed and T. Hanai, Analysis of the mechanism of retention on a modified β-cyclodextrin/silica chiral stationary phase using a computational chemical method, *J. Liq. Chromatogr. Relat. Technol.*, 2007, **30**, 3043–3057.

25. J. H. Knox, B. Kaur and G. R. Millward, Structure and performance of porous graphitic carbon in liquid chromatography, *J. Chromatogr.*, 1986, **352**, 3–25.

26. R. Tachon, V. Pichon, M. B. Le Borgne and J.-J. Minet, Use of porous graphitic carbon for the analysis of nitrate ester, nitramine and nitroaromatic explosives and by-products by liquid chromatography-at-mospheric pressure chemical ionization-mass spectrometry, *J. Chromatogr., A*, 2007, **1154**, 174–181.

27. A. V. Kiselev and D. P. Poshkus, Chromatostructural analysis (chromatography) a new method of determination of molecular structure, *Faraday Symp. Chem. Soc.*, 1980, **15**, 13–24.

28. N. S. Kulikov and M. S. Bobyleva, Molecular modeling in chromatoscopy as a new tool in the structure elucidation of novel isomers by GC/MS, *Struct. Chem.*, 2004, **15**, 51–64.

29. R. Arora, F. Ahgmed, I. Rustamov, D. Babusis and T. Hanai, Inertness and stability of newly developed wide-pore bonded silica gel, *J. Liq. Chromatog. Rel. Technol.*, 1998, **21**, 2762–2780.

30. T. Hanai, Separation of polar compounds using carbon column, *J. Chromatogr., A*, 2003, **989**, 183–196.

31. T. Hanai, H. Hatano, N. Nimura and T. Kinoshita, Computer-aided analysis of molecular recognition in chromatography, *Analyst*, 1993, **118**, 1371–1374.

32. T. Hanai, H. Hatano, N. Nimura and T. Kinoshita, Molecular recognition in chromatography aided by computational chemistry, *Supramol. Chem.*, 1994, **3**, 243–247.

33. D. J. Pietrzyk and C.-H. Chu, Separation of organic acids on Amberlite–XAD copolymers by reversed-phase high pressure liquid chromatography, *Anal. Chem.*, 1977, **49**, 860–867.

34. T. Hanai, K. Koizumi and T. Kinoshita, Prediction of retention factors of phenolic and nitrogen-containing compounds in reversed-phase liquid chromatography based on log P and pKa obtained by computational chemical calculation, *J. Liq. Chromatogr. Relat. Technol.*, 2000, **23**, 363–385.

35. Y. Arai, J. Yamaguchi and T. Hanai, Enthalpy effect in the retention of aromatic acids on an octadecyl-bonded silica gel, *J. Chromatogr.*, 1987, **400**, 21–26.

36. Y. Arai, M. Hirukawa and T. Hanai, Prediction of retention characteristics in reversed-phase liquid chromatography based on characteristics of molecules, *Nippon Kagaku Kaishi*, 1986, 969–975.

37. T. Hanai, K. C. Tran and J. Hubert, Prediction of retention times for aromatic acids in liquid chromatography, *J. Chromatogr.*, 1982, **239**, 385–395.

38. T. Hanai and J. Hubert, Chromatography of aromatic acids on ion exchangers, *J. Chromatogr.*, 1984, **316**, 261–265.

39. T. Hanai, C. Mizutani and H. Homma, Computational chemical simulation of chromatographic retention of phenolic compounds, *J. Liq. Chromatogr. Relat. Technol.*, 2003, **26**, 2031–2039.

40. T. Hanai and J. Hubert, Optimization of retention time of aromatic acids in liquid chromatography from log P and predicted pKa values, *J. High Resolut. Chromatogr. Chromatogr. Commun.*, 1984, 7, 524–528.

41. T. Hanai, New development in liquid chromatographic stationary phases, *Adv. Chromatogr.*, 2000, **40**, 315–357.
42. T. Hanai, R. Miyazaki, E. Kamijima, H. Homma and T. Kinoshita, Computational prediction of drug–albumin binding affinity by modeling liquid chromatography interactions, *Internet Electron. J. Mol. Des.*, 2003, **2**, 702–711.
43. P. N. Craig, Drug compendium, in *Comprehensive Medicinal Chemistry: the Rational Design, Mechanics Study and Therapeutic Application of Chemical Compounds*, ed. C. Hansch, P. G. Sammes and J. B. Taylor, Pergamon Press, Oxford, 1990, vol. 6, pp. 237–965.
44. T. Hanai, Y. Masuda and H. Homma, Chromatography *in silico*, retention of basic compounds on a carboxyl ion-exchanger, *J. Liq. Chromatogr. Relat. Technol.*, 2005, **28**, 3087–3097.
45. L. R. Snyder, *Principles of Adsorption Chromatography*, Marcel Dekker, New York, 1968, p. 174.
46. J. J. Kirkland, *Modern Practice of Liquid Chromatography*, Wiley Interscience, New York, 1971, p. 136.
47. T. Hanai, Quantitative *in silico* analysis of retention in normal-phase liquid chromatography II. http://www.internet–chromatography.com/html/toshihikbeitrage.html.
48. T. Hanai, Quantitative *in silico* analysis of organic modifier effect on retention in reversed-phase liquid chromatography, *J. Chromatogr. Sci.*, 2014, **52**, 75–80.
49. J. Yamaguchi and T. Hanai, Selectivity related to carbon loading and endcapping of octadecyl bonded silica gels in reversed-phase liquid chromatography of phenolic compounds, *J. Chromatogr. Sci.*, 1989, **27**, 710–715.
50. T. Hanai, Quantitative *in silico* analysis of the specificity of a graphitic carbon column, *J. Liq. Chromatogr. Rel. Technol.*, 2009, **32**, 647–655.
51. T. Hanai, Chromatography *in silico* for basic drugs, *J. Liq. Chromatogr. Rel. Technol.*, 2005, **28**, 2163–2177.

CHAPTER 7

Retention in Ion-Exchange Liquid Chromatography

7.1 Basic Concepts of Ion-Exchange Mechanisms

Molecular interaction forces are based on solubility factors. Coulombic forces are the strongest, followed by the Lewis acid–base interaction, including hydrogen bonding and charge-transfer effects, and van der Waals forces. Steric hindrance also affects the molecular interaction, and the individual interaction forces can be studied using chromatography. The main interaction force in reversed-phase liquid chromatography is the van der Waals force. The main interaction force in normal-phase liquid chromatography is the Lewis acid–base interaction, including hydrogen bonding, and that in ion-exchange liquid chromatography is the Coulombic force. Enantioseparation is achieved by the combination of these molecular interaction forces and steric hindrance as the molecular recognition of proteins. Ion-exchange is the phenomenon of ion replacement on an ion exchanger surface. Ion exchangers with hydrophobic properties can be used for reversed-phase liquid chromatography. However, the retention mechanism is not ion-exchange. Saccharides can be chromatographed on anion exchangers using an acetonitrile–water mixture as the eluent, and the separation mechanism is not ion-exchange. The ion-exchange phenomenon can be explained by the contribution of electrostatic energy. The retention mechanism for ion-exchange liquid chromatography has been quantitatively analyzed *in silico*; carboxyl and guanidine phases were selected to study the basic molecular recognition mechanism of proteins.

RSC Chromatography Monographs No. 19
Quantitative *In Silico* Chromatography: Computational Modelling of Molecular Interactions
By Toshihiko Hanai
© Toshihiko Hanai, 2014
Published by the Royal Society of Chemistry, www.rsc.org

7.1.1 Cation Exchange

A simplified model experiment *in silico* demonstrated that carboxyl groups interact with ionized aniline by Coulombic forces, and with the molecular form of aniline by Lewis acid–base interactions. The specificity can be understood from the molecular interaction energy values calculated using the molecular mechanics (MM2) program. The molecular interaction energy value (MI energy) is the energy value of a complex subtracted from the sum of the energy values of the pair of compounds, as given by the following equations. The calculated energy values are summarized in Table 7.1.

$$MIFS = FS(NH_2) + FS(COO) - FS(COO\text{–}NH_2 \text{ complex})$$
$$MIHB = HB(NH_2) + HB(COO) - HB(COO\text{–}NH_2 \text{ complex})$$
$$MIES = ES(NH_2) + ES(COO) - ES(COO\text{–}NH_2 \text{ complex})$$
$$MIHB = VW(NH_2) + VW(COO) - VW(COO\text{–}NH_2 \text{ complex})$$

$$MIFS = FS(NH_3) + FS(COO) - FS(COO\text{–}NH_3 \text{ complex})$$
$$MIHB = HB(NH_3) + HB(COO) - HB(COO\text{–}NH_3 \text{ complex})$$
$$MIES = ES(NH_3) + ES(COO) - ES(COO\text{–}NH_3 \text{ complex})$$
$$MIHB = VW(NH_3) + VW(COO) - VW(COO\text{–}NH_3 \text{complex}),$$

where FS, HB, ES, and VW are the energies of the final (optimized) structure, the hydrogen-bonding energy, the electrostatic energy, and the van der Waals energy, respectively. NH_2 and NH_3 are the molecular and ionized forms of aniline. COO is the carboxyl phase.

The contribution of the electrostatic energy was -7.937 kcal mol^{-1}, and that of hydrogen bonding was negligible because there is no hydrogen on the ionized carboxyl group in cation-exchange mode. Further studies were performed based on the difference in the atomic partial charge (apc) of the targeted atoms calculated using the MOPAC PM5 program. The results are shown in Figures 7.1 and 7.2. The apc of the target atoms is given in the

Table 7.1 Molecular interaction energies in cation-exchange mode. MIFS, MIVW, MIHB, and MIES represent the molecular interaction energy value of the final (optimized) structure, the van der Waals energy, the hydrogen-bonding energy, and the electrostatic energy (kcal mol^{-1}) of the complexes, respectively.

Analyte	MIFS	MIHB	MIES	MIVW
Molecular-form aniline (NH_2)	−6.121	−1.334	0.000	3.416
Ionic-form aniline (NH_3)	−1.048	0.000	0.000	3.496
Carboxyl group (COO)	3.370	0.000	0.000	2.954
Complex of COO–NH_2	−3.370	−1.333	−0.343	5.584
Complex of COO–NH_3	−5.691	0.000	−7.939	6.198
MI of COO–NH_2 complex	−1.127	0.001	−0.343	−0.786
MI of COO–NH_3 complex	−3.368	0.000	−7.937	−0.252

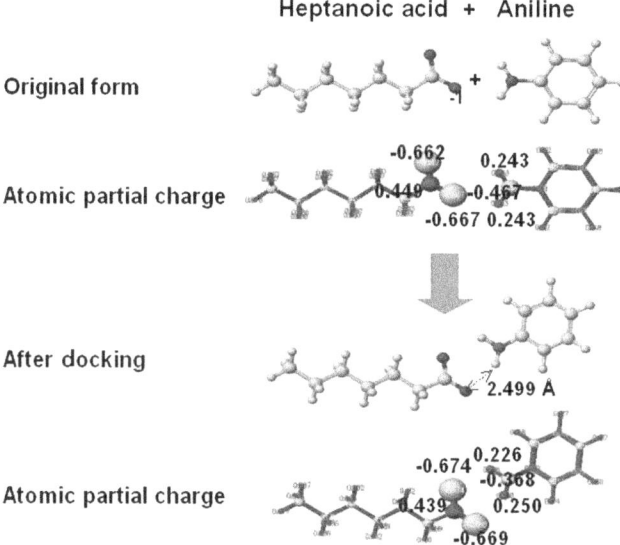

Heptanoic acid + Aniline

Figure 7.1 Model cation exchanger (heptanoic acid) and molecular (aniline) interaction.
Reproduced by permission of Taylor and Francis, ref. 15.

Heptanoic acid + Aniline⁺

Original form

Atomic partial charge

-0.662   0.229
0.449 0.227 0.111
-0.667 0.227

After docking

1.006 Å
1.326 Å

Atomic partial charge

-0.518 0.356
0.384   -0.254
0.207 0.204
-0.529

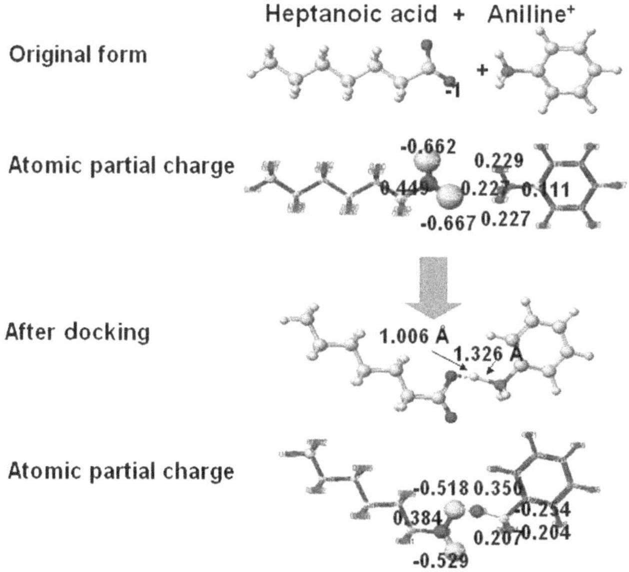

Figure 7.2 Model cation exchanger (heptanoic acid) and ion (aniline⁺) interaction.
Reproduced by permission of Taylor and Francis, ref. 15.

figures, where the gray and black indicate negative and positive charge, respectively. The larger the size, the higher the absolute value.

The atoms of the carboxyl group of heptanoic acid and the amino group of aniline were targeted for study. The apc values of the carboxyl group oxygens were −0.662 and −0.667 in the ionized form. These values increased by approximately 0.14 after complex formation with the ionized aniline. The values were −0.518 and −0.529, and they changed little when a complex was formed with the molecular form of aniline. The apc of nitrogen was −0.467 in the molecular form, and that of the ionized form was 0.111. In contrast with the oxygen, the apc of the aniline nitrogen increased by approximately 0.36 and 0.10 in the ionized and molecular forms, respectively. The apc of one hydrogen of the ionized amino group increased significantly in cation-exchange mode. The atomic distance expanded to 1.326 Å, and the hydrogen bound weakly with the oxygen of heptanoic acid. The atomic distance was 1.006 Å. These results support the contribution of Coulombic force in cation-exchange liquid chromatography.

The above findings were obtained without solvent and pH control components. The solvent and pH control components contribute to the replacement of an analyte during elution from the column, but the initial molecular interaction must occur directly between the analyte and the molecular recognition phase.

7.1.2   Anion Exchange

The anion-exchange mechanism has also been investigated using molecular interaction energy values calculated using the MM2 program. The model analyte was benzoic acid and the model phase was hexylguanidine. The molecular interaction energy values were calculated similarly to those for cation-exchange. The calculated energy values are summarized in Table 7.2.

The molecular interaction electrostatic energy value of the GUA–ionized benzoic acid complex was −5.653 kcal mol^{-1}, and that of the GUA–molecular form benzoic acid complex was −0.880 kcal mol^{-1}. This difference clearly indicates the contribution of Coulombic forces to the ion-exchange system. The molecular interaction hydrogen-bonding energy values of the GUA–molecular form of benzoic acid complex and the GUA–ionized benzoic acid complex were −8.454 and −3.925 kcal mol^{-1}, respectively. Hydrogen bonding is not the main molecular interaction force for ionized acids, but does contribute in the ion-exchange system, depending on the molecular structure of the ion exchangers. The difference can be examined based on the atomic distance and the apc of the targeted atoms. Examples are shown in Figures 7.3 and 7.4. The apc was calculated using the MOPAC PM5 program.

Table 7.2   Molecular interaction energies in anion-exchange mode.

Analyte	MIFS	MIHB	MIES	MIVW
Molecular-form benzoic acid (BAm)	−13.918	−3.459	−6.672	4.876
Ionic-form benzoic acid (BAi)	−2.549	0.000	0.000	4.743
Guanidino group (GUA)	0.036	3.824	−21.680	11.376
Complex of GUA–BAm	−35.595	−8.089	−29.232	7.810
Complex of GUA–BAi	−23.514	−0.101	−27.333	8.794
MI of GUA–BAm complex	−21.714	−8.454	−0.880	−8.442
MI of GUA–BAi complex	−21.001	−3.925	−5.653	−7.325

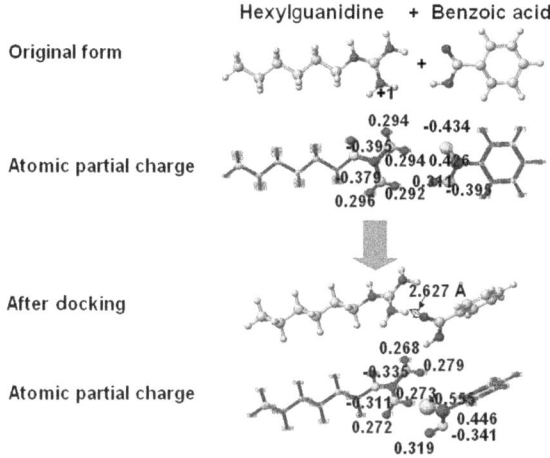

Figure 7.3   Model anion exchanger (hexylguanidine) and molecular (benzoic acid) interaction.
Reproduced by permission of Taylor and Francis, ref. 15.

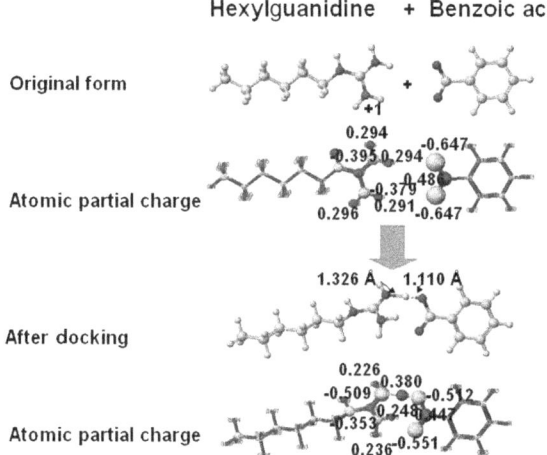

Figure 7.4   Model anion exchanger (hexylguanidine) and ion (ionized benzoic acid) interaction.
Reproduced by permission of Taylor and Francis, ref. 15.

The targeted atoms were the carboxyl group of benzoic acid and the guanidino group of hexylguanidinine. The apc values of the carboxyl group oxygen were -0.434 and -0.395 for the molecular form, and -0.647 and -0.647 for the ionized form. These values changed after complex formation. The values were -0.555 and -0.341 for the molecular form and -0.512 and -0.551 for the ionized form. The energy value change indicated an electron transfer in the ion-exchange system. The difference can also be understood from the apc of carbon and hydrogen, as well as from the apc change of the nitrogen and hydrogen of the guanidyl group. The apc of the hydrogen and nitrogen atoms of the guanidyl group was significantly reduced in ion-exchange mode, compared to that of the molecular-form interaction. The shortest atomic distance was 2.627 Å for the molecular form of benzoic acid. A guadinino group hydrogen weakly bonded with the oxygen of the benzoic acid carbonyl group, with an atomic distance of 1.110 Å. The atomic distance between the nitrogen and the hydrogen increased from 1.005 to 1.326 Å. The change in atomic distance supports a strong Coulombic force in ion-exchange liquid chromatography.

The above findings were obtained without solvent and pH control components. The solvent and pH control components, however, contribute to the replacement of an analyte for elution from the column, and the initial molecular interaction must occur directly between the analyte and the molecular recognition phase.

7.2 Retention of Ions on Graphitized Carbon

The molecular interactions between *tert*-butyl or methylphosphate ions and three model layers of graphitized carbon were calculated using the MM2 program. Polycyclic aromatic hydrocarbon (PAH) layers were constructed with 31 (PAH31) and 83 (PAH83) aromatic rings. A double layer of PAH was constructed with two layers of 22 aromatic rings (PAH22×2). The localized electronic charge of the PAH22×2 layer did not affect the adsorption of a cation as observed during *tert*-butyl–methylphosphate ion-pair formation, and had little effect on the methylphosphate ion. The van der Waals energy was the greatest contributor to adsorption, as summarized in Table 7.3.

Furthermore, the adsorption site effect of the ion–dipole type interaction was studied by adsorption of a methylphosphate ion on the larger polycyclic aromatic hydrocarbons – PAH31 and PAH83. As demonstrated before using the smaller polycyclic aromatic hydrocarbons PAH14 and PAH22, the electronic charge of the center carbons of PAH83 is less than that of PAH31. The contribution of the electrostatic energy is negligible for adsorption of such small ions at the center of PAH83, but electrostatic energy may contribute a little at the edge of PAHs. The changes in energy values by adsorption at both the center and edge of PAH31 and PAH83 are summarized in Table 7.4.

Table 7.3 Adsorption energies (kcal mol^{-1}) of $(CH_3)_4N^+$ and $CH_3HPO_4^-$ on a PAH22×2 layer. Reproduced by permission of Elsevier, ref. 16.

Energy	PAH22×2	$(CH_3)_4N^+$	CH_3HPO_4
Stretch	187.711	188.150	188.710
Stretch bend	0.000	0.146	−2.051
Improper torsion	13.258	13.258	13.258
Electrostatic	0.000	0.000	2.244
Angle	0.000	0.636	6.598
Dihedral angle	0.000	0.001	0.078
van der Waals	104.622	102.340	102.065
Hydrogen-bonding	0.000	0.000	−1.365
Final	305.5911	304.5301	309.5363

Table 7.4 Adsorption energies of $CH_3HPO_4^-$ on PAH31 and PAH83. C represents the center, and E the edge of the PAH, respectively. Reproduced by permission of Elsevier, ref. 16.

Energy	PAH31	C/PAH31	E/PAH31	PAH83	C/PAH83	E/PAH83
Stretch	—	123.067	123.080	316.552	317.233	317.136
Stretch bend	0.000	−2.041	−2.048	0.000	−2.046	−2.034
Improper torsion	0.000	0.023	0.025	0.000	0.027	0.020
Electrostatic	0.000	2.195	2.191	0.000	2.189	2.197
Angle	0.000	6.599	6.605	0.000	6.607	6.603
Dihedral angle	0.000	0.050	0.056	0.000	0.057	0.047
van der Waals	—	948.223	948.052	2426.760	2420.650	2420.623
Hydrogen-bonding	0.000	−0.849	−0.870	0.000	−0.870	−0.849
Final	—	1077.267	1077.091	2426.760	2743.847	2743.743

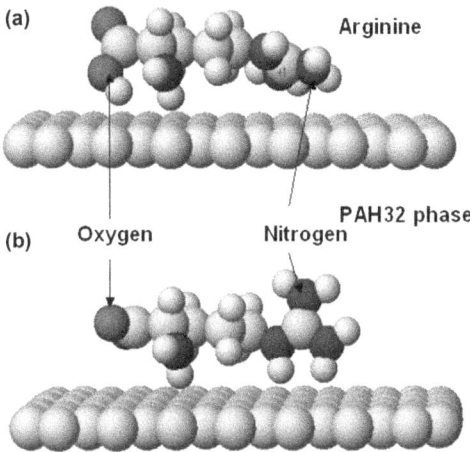

Figure 7.5 Adsorption of arginine on graphitized carbon *via* its carboxyl (a) or guanidino (b) groups. Small white, large light gray, dark gray, and black balls represent hydrogen, carbon, oxygen, and nitrogen, respectively.

The above results indicate that the adsorption site did not affect the electrostatic energy or van der Waals energy, even in cases in which the molecular sizes of the adsorbents (PAH), were quite different. Ion–dipole interactions were negligible, unlike the ion-pair formation of tetrabutyl and methylphosphate ions, with the van der Waals energy being the main contributor to adsorption.[1]

Guanidino compounds are polar and are retained on graphitized carbon layers without ion-pair reagents, and several guadinino compounds were separated in potassium citrate buffer at pH 4.5. Arginine, which was retained strongly on graphitized carbon, has two ionic groups, an anionic carboxyl group and a cationic guanidyl group. The molecular interactions between arginine and three model layers of graphitized carbon were analyzed using the MM2 program. A PAH layer was constructed with 46 aromatic rings (PAH46). Maximum retention of arginine was observed on a graphitized carbon column with potassium citric buffer, pH 4.5, and sodium phosphate buffer, pH 12.90, as the eluent.[2] Therefore, the molecular interactions were calculated using both cationic and anionic forms of arginine. Arginine adsorbs on a C32 phase. Figure 7.5(a) shows the carboxyl group of arginine interacting with the C32 phase, and Figure 7.5(b) shows the guanidyl group of arginine interacting with the C32 phase.

7.3 Cation Exchange for Basic Drugs

7.3.1 Quantitative Analysis of log *k In Silico*

Basic drugs were separated by cation-exchange liquid chromatography using a carboxyl-bonded phase. The carboxyl group mimics the cation-exchange group of a protein. The retention behavior measured in liquid chromatography was quantitatively analyzed *in silico.*[3] Using a computational chemical calculation to analyze liquid chromatographic data, the direct interaction between a model phase and an analyte was calculated in the form of energy values using the MM2 program. The quantitative analysis of retention in liquid chromatography is easy, because a homogeneous model phase can be used instead of a complicated protein model for studying docking mechanisms.

A simple model carboxyl-bonded phase was constructed to investigate basic compound–carboxyl phase interactions. The model phase consisted of 556 carbons, 48 oxygens, 957 bonds and 5448 connectors. The molecular weight was 7440. The 24 carboxyl groups were within a 390 Å^2 area. The optimized energy value was less than 0.00001 kcal mol^{-1}. The 1 : 1 adsorption form of quinine on a carboxyl phase is shown in Figure 7.6 as an example.

For the development of a quantitative structure–retention relationship (QSRR) in chromatography, the molecular interaction energy values between a model phase and an analyte can be calculated. The optimized energy value

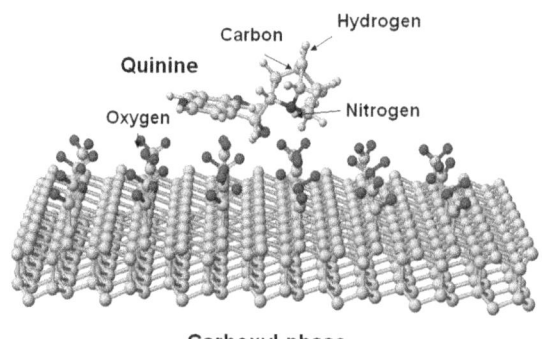

Carboxyl phase

Figure 7.6   Docking of quinine on a model carboxyl phase. Small white, light-gray, dark-gray, and black balls represent hydrogen, carbon, nitrogen, and oxygen, respectively.
Reproduced by permission of Taylor and Francis, ref. 3.

was less than 0.00001 kcal mol^{-1}. The MIFS energy is defined as the energy value of the complex subtracted from the sum of the energy values of the model phase and analyte.

$$\text{MIFS} = \text{FS}_{\text{analyte}} + \text{FS}_{\text{model phase}} - \text{FS}_{\text{complex}} \tag{1}$$

where FS is the energy value of the final (optimized) structure. The MIFS energy values of the molecular and ionized forms of the analytes are summarized in Table 7.5.

7.3.2 Inductive Effect on the pK_a of Basic Drugs

The dissociation constant (pK_a) of these basic analytes was calculated from the capacity ratios measured using eluents of pH 3.2 to 9.5, and the pH effect on the retention is given in Figure 7.7. The pK_a values obtained from the experiment were very different from those in the literature, listed in Table 7.5. Allopurinol, amoxicillin, carbamazepine, diazepam, procaine, theobromine, and theophylline did not behave as basic components during chromatography, therefore, these compounds are not included in Table 7.5, and were eliminated from further analysis. The difference was partly influenced by the methanol concentration in the eluent, but the inductive effect between the carboxyl group of the ion exchanger and the analyte might also contribute to the difference.

Table 7.5 Properties of some basic compounds. pK_a^1 values are reference values, pK_a^2 values were measured by ion-exchange liquid chromatography, pK_a^3 represents the relationship between pK_a^1 and pK_a^2, m represents the molecular form, i represents the ionized form, and hb represents the hydrogen-bonding energy (kcal mol^{-1}). Reproduced by permission of Taylor and Francis, ref. 3.

No.	Analyte	pK_a^1	pK_a^2	pK_a^3	MIFS$_m$	MIFS$_i$	hb$_m$	hb$_i$
1	Ajmaline	8.2	5.753	5.556	30.4227	31.7423	−2.317	−2.110
2	Aniline	4.69	3.818	3.408	14.2624	18.8636	−1.334	−1.334
3	Atropine	9.8	6.555	6.536	20.7190	23.0025	−3.451	−3.387
4	Dextromethorphan	8.3	6.184	5.618	27.6929	36.2590	0.000	0.000
5	Homatropine	9.9	6.413	6.597	26.0110	23.8780	−0.606	−3.306
6	Imipramine	9.5	6.231	6.352	27.3559	39.3437	0.000	0.000
7	Isoproterenol	8.6	6.005	5.801	19.1121	26.3668	−11.250	−8.581
8	Lidocaine	7.9	5.735	5.373	22.2696	28.1267	−2.406	−2.226
9	Prazosin	6.5	4.147	4.516	26.9866	24.2851	−1.257	−1.099
10	Pyridine	5.23	3.463	3.739	14.6571	17.0381	0.000	0.000
11	Quinine	8.5	5.745	5.740	29.9987	34.4376	−4.612	−4.621
12	Scopolamine	7.75	4.801	5.281	22.2733	23.4282	−4.157	−4.146
13	Terbutaline	8.8	5.912	5.924	24.2510	25.1261	−6.111	−5.812
14	Triamterene	6.2	3.851	4.332	21.7167	27.5352	−3.085	−2.844
15	Benzylamine	9.33	6.079	6.248	15.2344	21.4438	−0.973	−0.983
16	Phenylethylamine	9.82	6.650	6.548	17.5244	23.6096	−0.089	−0.089
17	N,N′-Dimethylaniline	5.15	3.935	3.690	20.9785	25.5160	0.000	0.000

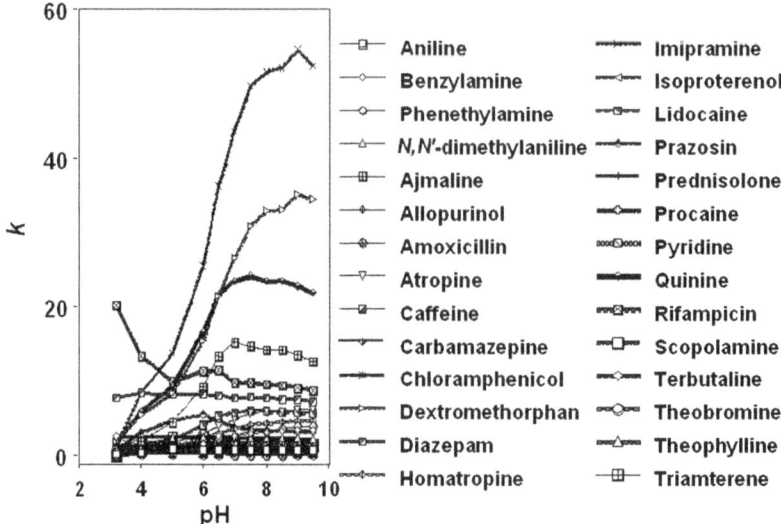

Figure 7.7   The pH effect on k for the retention of basic compounds in ion-exchange liquid chromatography using a carboxyl phase.
Reproduced by permission of Taylor and Francis, ref. 3.

The relationship between measured and reference values is given as the following equation:

$$pK_a(\text{ion-exchangeLC}) = 0.612 \times (pK_a \text{reference}) + 0.538, \; r = 0.960, n = 17 \quad (2)$$

The pK_a values of aromatic acids measured using a propylamino-bonded silica gel (NH_2) and a diethylaminoethyl-bonded silica gel (DEAE) were smaller than those measured using an octadecyl-bonded silica gel (ODS). The pK_a values measured using a sulfopropyl-bonded cation exchanger (SP) were shifted to a higher pH region.[4] The relationships are given in the following equations:

$$pK_a(NH_2) = 0.744 \times [pK_a(ODS)] + 0.446, \; r = 0.988, \; n = 29,$$
$$pK_a(DEAE) = 0.847 \times [pK_a(ODS)] + 0.268, \; r = 0.990, \; n = 29,$$
$$pK_a(SP) = 1.023 \times [pK_a(ODS)] - 0.097, \; r = 0.998, \; n = 29.$$

A total of 36 aromatic acids were used, however, 7 aromatic acids were eliminated to determine the properties of the ion exchangers. The aromatic acids eliminated were those with the possibility of intramolecular hydrogen bonding in their structures.

These data were measured under the same conditions containing 20% (by volume) acetonitrile. Therefore, the concentration effect of an organic modifier could be neglected for comparison. The effect of an organic modifier in reversed-phase liquid chromatography was approximately 0.022

pH units/% of methanol or acetonitrile, experimentally.[5,6] The concentration effect of the organic modifier depended on the buffer components, even when the pH was measured using standard buffers using a pH meter with the same concentration as the organic modifier. Furthermore, the pK_a shift in the aqueous organic solvent depended on the analyte itself.[7] This organic modifier effect was observed for carboxylic acids in reversed-phase liquid chromatography[8] and for polar aromatic compounds on cation exchanger and porous polymer gels.[9] The difficulty of standardizing the organic modifier effect is due to the difficulty in performing quantitative analysis of solvation. In this experiment, however, pH was measured before mixing with the organic modifier, therefore, the higher pH values obtained are a likely to be a property of the ion exchanger used.[4]

The slope of eqn (2) indicates the selectivity of the carboxyl phase. The inductive effect of the ion exchanger affects the shift of the pK_a values measured, and the degree of the shift might depend on the ion-exchange capacity and the ionic strength, and also the ionic strength of buffer components of the eluent. The salting-out effect at higher and lower pH regions was higher in sodium phosphate solution than that in sodium acetate solution. The pH effect on molecular interaction can be examined experimentally using liquid chromatography. The capacity ratio in a given pH eluent can be predicted from the following equation:[10]

$$k = (k_m + k_i([H^+] / K) / (1 + ([H^+] / K))\qquad(3)$$

where k_m and k_i are the capacity ratios of the molecular-form and ionized analytes, respectively, K is the dissociation constant of the analytes, and $[H^+]$ represents the hydrogen ion concentration in the eluent. k_m and k_i were replaced by molecular interaction energy values calculated using molecular mechanics. The k_m value was replaced with the MI energy value of the molecular form of the analyte ($MIFS_m$) and the k_i value was replaced by the MI energy value of the ionized form of the analyte ($MIFS_i$). The following equation is used for further discussion:

$$MIFS = (MIFS_m + MIFS_i([H^+] / K) / (1 + ([H^+] / K)\qquad(4)$$

It was necessary to first determine how to obtain the relative dissociation constants by ion-exchange liquid chromatography. The relative pK_a values measured in anion-exchange liquid chromatography were shifted to lower values compared to the reference values. The original pK_a values measured by titration were affected by the organic modifier concentration and the inductive effect of the ion-exchange groups of the bonded phase. MIFS values were calculated using eqn (4).

The hydrogen-bonding energy did not contribute to the retention, therefore further calculations were performed without the hydrogen-bonding energy values of the analytes (hb). MIFS values were calculated using $MIFS_m + hb_m$ for the molecular form and $MIFS_i + hb_i$ for the ionized form in eqn (4). The correlations between the predicted log k values and the experimental data are given in the following equations:

$$\text{MIFS}(\text{pH } 8.0) = 7.171 \times [\log k \,(\text{pH } 8.0)] + 18.090, \; r = 0.766, \; n = 17,$$

$$\text{MIFS}(\text{pH } 7.0) = 8.134 \times [\log k \,(\text{pH } 7.0)] + 18.031, \; r = 0.845, \; n = 17,$$

$$\text{MIFS}(\text{pH } 6.0) = 10.744 \times [\log k \,(\text{pH } 6.0)] + 18.879, \; r = 0.913, \; n = 17,$$

$$\text{MIFS}(\text{pH } 5.0) = 12.388 \times [\log k \,(\text{pH } 5.0)] + 21.412, \; r = 0.823, \; n = 17,$$

$$\text{MIFS}(\text{pH } 4.0) = 11.912 \times [\log k \,(\text{pH } 4.0)] + 25.040, \; r = 0.800, \; n = 17.$$

The correlation coefficient was high at pH values around the pK_a values of these compounds. However, it was low in higher pH regions, where a salting-out effect reduced the retention (Figure 7.7). The high MIFS_m and MIFS_i values indicated that the molecules were well fitted with the model phase for the computational chemical calculation.

Further calculations were performed using the original pK_a values from the literature. The pK_a values of the analytes were considered to be constant and not directly affected by ion-exchange groups and organic modifiers. The correlation coefficients are given in the following equations:

$$\text{MIFS}(\text{pH } 8.0) = 10.777 \times [\log k \,(\text{pH } 8.0)] + 18.380, \; r = 0.932, \; n = 17,$$

$$\text{MIFS}(\text{pH } 7.0) = 11.760 \times [\log k \,(\text{pH } 7.0)] + 18.643, \; r = 0.940, \; n = 17,$$

$$\text{MIFS}(\text{pH } 6.0) = 12.429 \times [\log k \,(\text{pH } 6.0)] + 19.859, \; r = 0.900, \; n = 17,$$

$$\text{MIFS}(\text{pH } 5.0) = 12.333 \times [\log k \,(\text{pH } 5.0)] + 22.287, \; r = 0.830, \; n = 17,$$

$$\text{MIFS}(\text{pH } 4.0) = 11.223 \times [\log k(\text{pH } 4.0)] + 25.621, \; r = 0.806, \; n = 17.$$

These relationships are shown in Figure 7.8. The correlation coefficients were relatively higher than those calculated using pK_a values obtained by ion-exchange liquid chromatography.

7.3.3 Three-Dimensional Model Ion Exchanger

The basic drugs used, and their molecular properties, are summarized in Table 7.6. The computational model phase is shown in Figure 7.9. One carboxyl group is surrounded by alkyl groups, and it looks like a little flower inside a flowerpot. The molecular interaction energy values of the molecular and ionized forms of the basic drugs were calculated using the MM2 program.

The molecular interaction energy values were correlated with log k values, and there was a poor correlation between the MIFS and log k values (data not shown). Electrostatic energy is the major contributor for acidic drugs, and hydrophobic interaction energy values are important for basic drugs.[11] Therefore, a new model phase was constructed to increase the contact surface area for hydrophobic interactions. The model phase was based on a model phase constructed to study the ionization effect of the docking of an acidic drug with a protein.[12] The structure is shown in Figure 7.10, where scopolamine is docked with the model phase.

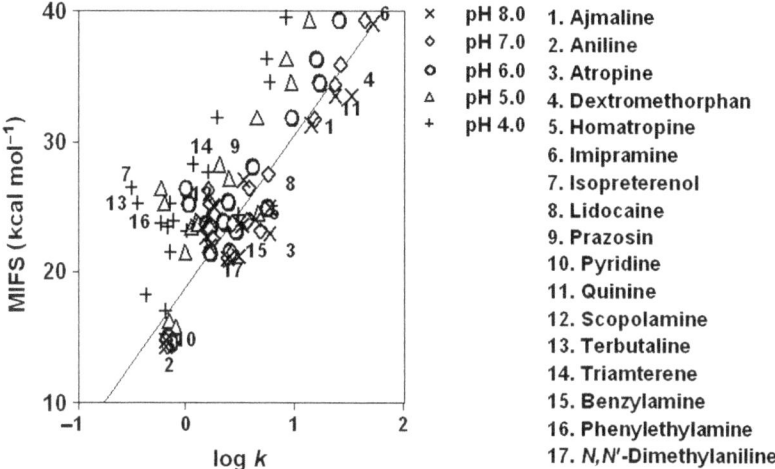

Figure 7.8 The relationship between molecular interaction energy (MIFS) and log *k* for basic compounds on a model carboxyl phase in eluents of various pH. Reproduced by permission of Taylor and Francis, ref. 3.

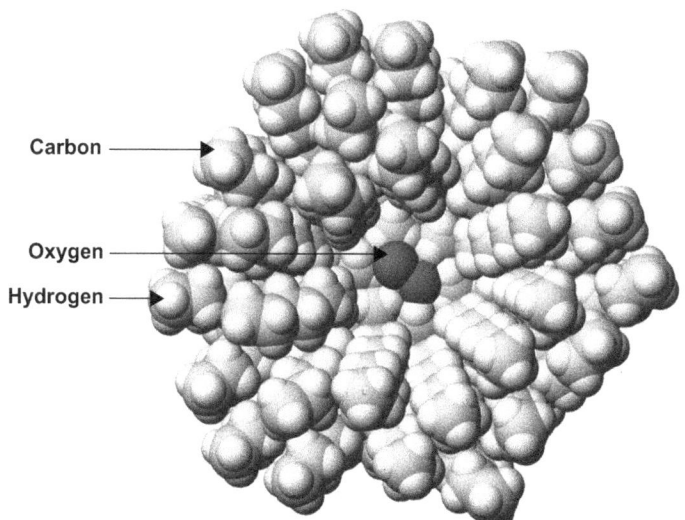

Figure 7.9 Top view of a carboxyl group surrounded by octyl groups as a model cation exchanger. Small white, light gray and black balls represent hydrogen, carbon, and oxygen, respectively.

One ionized carboxyl group was located at the center of the carbon phase and six methyl groups surrounded the carboxyl group to protect against the binding of octyl groups. There was a total of 16 octyl groups. The longer neighbor alkyl chains were bound together by hydrophobic interactions, *i.e.*, van der Waals energy, in this type of molecular modeling. The model phase

Table 7.6 Molecular properties of some basic
compounds. pK_a^1 values are reference values.

No.	Analyte	pK_a^1	pK_a	$MIFS_m$	$MIFS_i$
1	Atropine	9.8	6.56	24.170	26.389
2	Ajmaline	8.2	5.75	32.740	33.852
3	Dextromethorphan	8.3	6.18	27.683	36.258
4	Homatropine	9.9	6.41	26.617	27.183
5	Lidocaine	7.9	5.74	24.676	30.352
6	Quinine	8.5	5.75	34.611	39.058
7	Scopolamine	7.75	4.80	26.430	27.574
8	Triamterene	6.2	3.85	24.801	30.379
9	Prazosin	6.5	4.15	28.244	25.384
10	Imipramine	9.5	6.23	27.356	39.343
11	Pyridine	5.23	3.46	14.657	17.038
12	Aniline	4.69	3.82	14.596	20.197
13	Benzylamine	9.33	6.08	16.207	22.426
14	*N*-Ethylaniline	9.82	6.65	17.613	23.698
15	2,4-Dimethylaniline	5.15	3.94	20.978	25.515

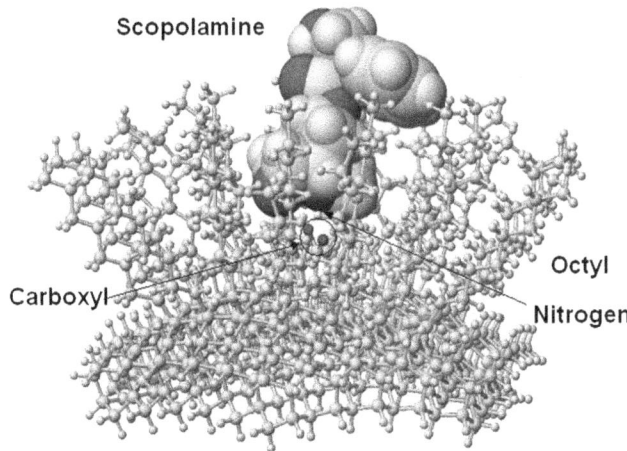

Figure 7.10   Docking of scopolamine with a carboxyl–octyl phase. Small white,
light-gray, dark-gray, and black, balls represent hydrogen, carbon,
nitrogen, and oxygen, respectively. The atom size of scopolamine is
five times that of the carboxyl–octyl phase.
Reproduced by permission of Taylor and Francis, ref. 15.

consisted of 1350 atoms including 2 oxygens, 739 hydrogens, 609 carbons,
1588 bonds, and 9121 connectors. To develop a QSRR, the MI energies were
calculated between the model phase and the analytes using eqn (1). The
optimized energy value was less than 0.00001 kcal mol^{-1}. The MI values of
the molecular and ionized forms of the analytes are summarized in
Table 7.6. The MIFS values were correlated with the capacity ratios measured
in liquid chromatography. The three-dimensional model phase demonstrated better correlation as shown in Figure 7.11.

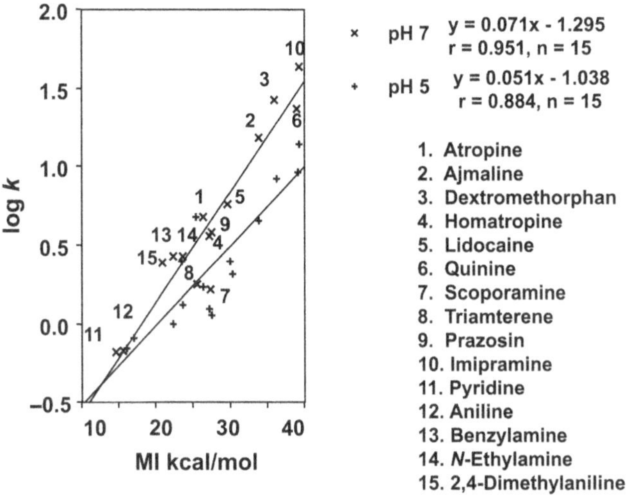

log *k*
MI kcal/mol
pH 7 y = 0.071x - 1.295
r = 0.951, n = 15
pH 5 y = 0.051x - 1.038
r = 0.884, n = 15
1. Atropine
2. Ajmaline
3. Dextromethorphan
4. Homatropine
5. Lidocaine
6. Quinine
7. Scoporamine
8. Triamterene
9. Prazosin
10. Imipramine
11. Pyridine
12. Aniline
13. Benzylamine
14. *N*-Ethylamine
15. 2,4-Dimethylaniline

Figure 7.11 The relationship between molecular interaction energy and log *k* for acidic compounds on a model carboxyl phase in eluents of various pH.

Above results indicate that molecules were dissociated based on their individual pK_a values. The original pK_a values measured by titration can be used in a theoretical approach to studying molecular interactions. Measurement of the organic modifier and ion-exchange group effects is not necessary prior to the calculation.

7.4 Anion Exchange for Acidic Compounds

7.4.1 Quantitative Analysis of Retention *In Silico*

Acidic drugs were separated by anion-exchange liquid chromatography using a guanidino phase. The guanidino group mimics an anion-exchange group on a protein. The chromatographic data were quantitatively analyzed *in silico.*[2] The acidic drugs used and their molecular properties are summarized in Table 7.7. A simple model guanidino phase was constructed to investigate acidic compound–guanidino phase interactions. The model phase contained 1117 atoms, 1470 bonds, and 8432 connectors. Twelve hexyl guanidyl groups and twelve hexyl groups were bound to a double-layer-like carbon phase that kept its rigid basic structure. The analyte was located in the center of the guanidino phase, and the complex was optimized using MM2 calculations. The optimized energy value was less than 0.00001 kcal mol^{-1}. An example of such a complex is shown in Figure 7.12, where ibuprofen is adsorbed on the guanidine phase.

A QSRR was developed using the method in Section 7.3.3. The optimized energy value was less than 0.00001 kcal mol^{-1}. The MI energy was defined by the following equation:

$$MI = FS_{analyte} + FS_{model\,phase} - FS_{complex} - hb \quad (5)$$

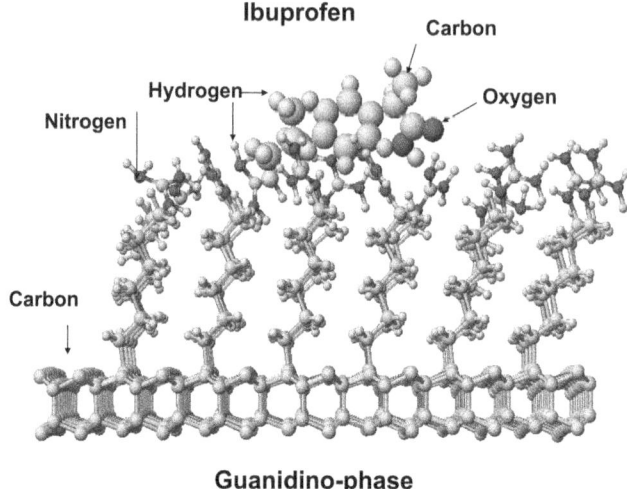

Figure 7.12 Docking of ibuprofen on a guanidino phase. White, light-gray, dark-gray, and black balls represent hydrogen, carbon, nitrogen, and oxygen, respectively. The atom size of scopolamine is 50% and the carboxyl–octyl phase is 20% of the original size.
Reproduced by permission of Taylor and Francis, ref. 15.

Table 7.7 Molecular properties of some acidic compounds. $\log nK$ represents the Human Serum Albumin (HSA)–drug binding affinity, from ref. 13. $\log nK^*$ represents the HSA–drug binding affinity measured using a two-column system, from ref. 13. pK_a^1 are reference values, from ref. 13 pK_a^2 values were measured by ion-exchange liquid chromatography. MI_m represents the molecular interaction energy calculated for the molecular-form compound, and MI_i the molecular interaction energy calculated for the ionic-form compound (kcal mol^{-1}).

No.	Analyte	$\log nK$	$\log nK^*$	pK_a^1	pK_a^2	MI_m	MI_i
1	Furosemide	5.54	5.70	4.2	5.661	23.592	31.184
2	Ibuprofen	—	6.61	5.2	6.032	24.515	32.909
3	Indomethacin	—	7.32	4.5	5.729	36.478	47.386
4	Iopanoic acid	—	7.84	—	6.051	34.234	39.888
5	Mefenamic acid	—	7.84	4.2	5.776	31.606	42.211
6	Nalidixic acid	—	4.18	6.0	6.481	23.849	28.415
7	Naproxen	5.81	5.75	4.2	5.942	22.827	32.190
8	Nicotinic acid	—	3.20	4.95	5.262	17.022	24.845
9	Phenylbutazone	5.95	6.05	4.4	5.665	28.226	42.237
10	Probenecid	—	5.29	3.4	5.537	24.142	37.505
11	Salicylic acid	4.92	4.81	3.0	5.079	21.661	30.691
12	Tolazamide	—	5.16	5.7	6.118	22.514	34.746
13	Tolbutamide	5.26	5.29	5.3	6.329	18.776	35.167
14	Warfarin	5.63	5.38	5.1	6.136	26.052	32.272

where FS is energy value of the final (optimized) structure, and hb is the hydrogen-bonding energy value of the analyte. The subtraction of hb is necessary for molecules with intramolecular hydrogen bonding. The MI energy values of the molecular and ionized forms of the analytes are summarized in Table 7.7. The MI energy values correlated with the capacity ratios measured using liquid chromatography.

7.4.2 Inductive Effect on pK_a

Although the retention times were longer for this new ion-exchange liquid chromatography system, the precision was higher for the weakly retained compounds. The negative log k values had to be neglected due to their very short retention times. With the new guanidino-bonded phase, however, retention time was reasonably long and all measured capacity ratios were analyzed.

Furthermore, the pK_a values of these analytes were calculated from the capacity ratios measured using eluents of pH 3.0 to 9.0. The pK_a values obtained from the experiment were very different to those reported in the literature (see Table 7.7). The difference was partly influenced by the methanol concentration in the eluent, but the inductive effect between the guanidino group of the ion exchanger and the analyte could also contribute to the difference. The relationship between the measured and reference values is given by the following equation:

$$p K_a (\text{ion-exchange LC}) = 0.370 \times (p K_a \text{ reference}) + 4.116,$$

$$r = 0.792, \ n = 13 \tag{6}$$

$$p K_a (\text{ion-exchange LC}) = 0.397 \times (p K_a \text{ reference}) + 4.048,$$

$$r = 0.931, \ n = 12 \tag{7}$$

The correlation coefficient improved from 0.792 to 0.931 after the elimination of nicotinic acid. The slopes of eqns (6) and (7) indicate the selectivity of the guanidino phase. The inductive effect of the ion exchanger affects the shift of the measured pK_a values, and the degree of the shift might depend on the ion-exchange capacity and the ionic strength of the phase, as well as on the ionic strength of the eluent used. The salting-out effect at higher and lower pH was higher in sodium phosphate solution than in sodium acetate solution. In the sodium acetate eluent, strong acids (*e.g.*, salicylic acid) were retained more than weaker acids. The maximum retention time was not obtained with a pH 3.22 solution due to the low dissociation constant of salicylic acid.

The difficulty in standardizing the effect of an organic modifier is due to the current obstacles in quantitatively analyzing solvation. In this experiment, however, pH was measured before mixing with an organic modifier. Therefore, the higher pH values obtained are likely to be a property of the ion exchanger used.[14]

7.4.3   Quantitative Analysis of log *k In Silico*

For the QSRR of reversed-phase liquid chromatography, the contact surface area is important, and the selection of a model phase is difficult, but a simple model phase is satisfactory for the measurement of albumin–acidic drug binding affinity in ion-exchange liquid chromatography.[12]

This led us to examine how to obtain the relative dissociation constant in ion-exchange liquid chromatography. The relative pK_a values measured by cation-exchange liquid chromatography were shifted to lower values compared to the reference values. The original pK_a values measured by titration were affected by the organic modifier concentration and the inductive effect of ion-exchange groups on the bonded phase. MIFS energy values were calculated using eqn (4).

The correlation coefficient was high at pH values near the pK_a values of these compounds. However, it was low in higher pH regions, where a salting-out effect reduced the retention.[13] The molecular interaction energy values were determined using the MM2 program, which demonstrated that retention order could be predicted, even in ion-exchange liquid chromatography with a pH-controlled eluent.

The pH effect on the molecular interactions can be examined experimentally using liquid chromatography. The capacity ratio in an eluent of a given pH can be predicted from [10] and the MI energy at a given pH can be calculated using eqn (4).

The relative pK_a values measured using this anion-exchange liquid chromatography system were shifted to higher values compared to the reference values, as given in eqns (6) and (7). The original pK_a values measured by titration were affected by the organic modifier concentration and the inductive effect of the ion-exchange groups on the bonded phase. Therefore, the MI energy at a given pH was first calculated using the measured pK_a and eqn (4). The contribution of the intramolecular hydrogen-bonding energy was high for salicylic acid, and therefore the final molecular interaction energy value was reduced by the original hydrogen-bonding energy value of the analyte, as shown in the following equations:

$$\log k(\text{pH } 3.0) = 0.065 \times (\text{MI, pH } 3.0) - 0.751, \; r = 0.830, \; n = 14,$$

$$\log k(\text{pH } 4.0) = 0.078 \times (\text{MI, pH } 4.0) - 0.945, \; r = 0.896, \; n = 14,$$

$$\log k(\text{pH } 5.0) = 0.078 \times (\text{MI, pH } 5.0) - 1.163, \; r = 0.867, \; n = 14,$$

$$\log k(\text{pH } 6.0) = 0.062 \times (\text{MI, pH } 6.0) - 1.401, \; r = 0.756, \; n = 14,$$

$$\log k(\text{pH } 7.0) = 0.071 \times (\text{MI, pH } 7.0) - 2.320, \; r = 0.824, \; n = 14,$$

$$\log k(\text{pH } 7.4) = 0.070 \times (\text{MI, pH } 7.4) - 2.450, \; r = 0.843, \; n = 14,$$

$$\log k(\text{pH } 8.0) = 0.073 \times (\text{MI, pH } 8.0) - 2.697, \; r = 0.858, \; n = 14,$$

$$\log k(\text{pH } 9.0) = 0.073 \times (\text{MI, pH } 9.0) - 2.850, \; r = 0.866, \; n = 14.$$

Further study was performed using eqns (4), (5) and (7) and the reference pK_a values listed in Table 7.7. The relationships are shown in Figure 7.13.

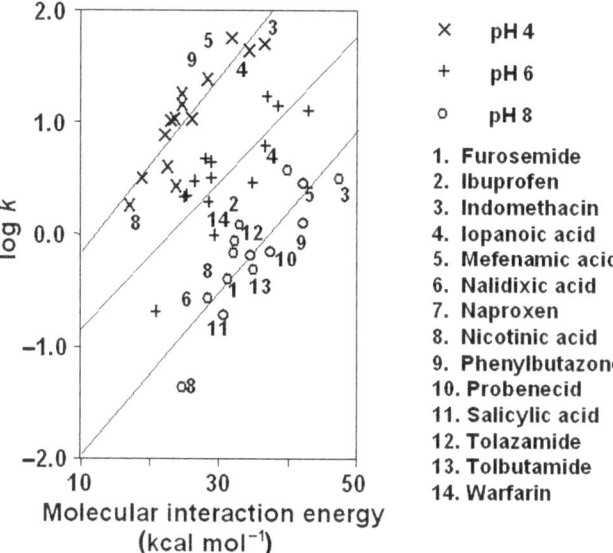

Figure 7.13 The relationship between molecular interaction energy and log *k* for acidic compounds on a guanidine phase in eluents of various pH. Reproduced by permission of Taylor and Francis, ref. 15.

$\log k(\mathrm{pH}\,3.0) = 0.065 \times (\mathrm{MI,\ pH}\,3.0) - 0.750,\ r = 0.830,\ n = 14,$

$\log k(\mathrm{pH}\,4.0) = 0.078 \times (\mathrm{MI,\ pH}\,4.0) - 0.936,\ r = 0.898,\ n = 14,$

$\log k(\mathrm{pH}\,5.0) = 0.079 \times (\mathrm{MI,\ pH}\,5.0) - 1.173,\ r = 0.902,\ n = 14,$

$\log k(\mathrm{pH}\,6.0) = 0.065 \times (\mathrm{MI,\ pH}\,6.0) - 1.503,\ r = 0.814,\ n = 14,$

$\log k(\mathrm{pH}\,7.0) = 0.071 \times (\mathrm{MI,\ pH}\,7.0) - 2.338,\ r = 0.834,\ n = 14,$

$\log k(\mathrm{pH}\,7.4) = 0.071 \times (\mathrm{MI,\ pH}\,7.4) - 2.454,\ r = 0.847,\ n = 14,$

$\log k(\mathrm{pH}\,8.0) = 0.073 \times (\mathrm{MI,\ pH}\,8.0) - 2.698,\ r = 0.859,\ n = 14,$

$\log k(\mathrm{pH}\,9.0) = 0.073 \times (\mathrm{MI,\ pH}\,9.0) - 2.850,\ r = 0.866,\ n = 14.$

These correlation coefficients were not high compared to those obtained with reversed-phase liquid chromatography. The predicted pK_a values from the reference values measured by titration can be used if a pK_a conversion equation, like eqn (7), is obtained systematically for a specific ion exchanger. This approach is still difficult, however, due to the uncontrolled inductive effect of the ion-exchange groups.

Chromatography *in silico* using a model phase is practical for studying retention mechanisms, furthermore, the elution order can be predicted even in ion-exchange liquid chromatography. It is difficult, however, to predict relative pK_a values in ion-exchange liquid chromatography. It seems that the compounds dissociate based on their pK_a, as measured by titration. A computational chemical optimization based on the molecular properties of analytes is practical in liquid chromatography, and for studying protein–drug binding affinity.

References

1. T. Hanai, Computational chemical analysis of the molecular recognition of graphitic carbon, presented at the 21st International Symposium on Capillary Chromatography and Electrophoresis, Park City, 1999.
2. Y. Inamoto, S. Inamoto, T. Hanai, M. Tokuda, O. Hatase, K. Yoshi, N. Sugiyama and T. Kinoshita, Liquid chromatography of guanidino compounds using a porous graphite carbon column and application to their analysis in serum, *J. Chromatogr., B*, 1998, **707**, 111–120.
3. T. Hanai, Y. Masuda and H. Homma, Chromatography *in silico*; retention of basic compounds on a carboxyl ion exchanger, *J. Liq. Chromatogr. Relat. Technol.*, 2005, **28**, 3087–3097.
4. T. Hanai and J. Hubert, Chromatography of aromatic acids on ion exchangers, *J. Chromatogr.*, 1984, **316**, 261–265.
5. T. Hanai and J. Hubert, Chromatographic behavior of acids on macroporous-polystyrene gels, *Chromatographia*, 1983, **17**, 633–639.
6. T. Hanai, K. C. Tran and J. Hubert, Prediction of retention times for aromatic acids in liquid chromatography, *J. Chromatogr.*, 1982, **239**, 385–394.
7. M. Rosés and E. Bosch, Influence of mobile phase acid–base equilibria on the chromatographic behavior of protolytic compounds, *J. Chromatogr., A*, 2002, **982**, 1–30.
8. J. L. van de Venne, J. L. H. M. Hendrikx and R. S. Deelder, Retention behavior of carboxylic acids in reversed-phase column liquid chromatography, *J. Chromatogr.*, 1978, **167**, 1–16.
9. T. Hanai, H. F. Walton, J. D. Navratil and D. Warren, Liquid chromatography of polar aromatic compounds on cation-exchange resins and porous polymer gels, *J. Chromatogr.*, 1978, **155**, 261–271.
10. D. J. Pietrzyk and C.-H. Chu, Separation of organic acids on Amberlite XAD copolymers by reversed-phase high pressure liquid chromatography, *Anal. Chem.*, 1977, **49**, 860–867.
11. G. Ermondi, M. Lorenti and G. Caron, Contribution of ionization and lipophilicity to drug binding to albumin: a preliminary step toward biodistribution prediction, *J. Med. Chem.*, 2004, **47**, 3949–3961.
12. T. Hanai, Molecular modeling for quantitative analysis of molecular interaction, *Lett. Drug Des. Discovery*, 2005, **2**, 232–238.
13. T. Hanai, A. Koseki, R. Yoshikawa, M. Ueno, T. Kinoshita and H. Homma, Prediction of human serum albumin–drug binding affinity without albumin, *Anal. Chim. Acta*, 2002, **454**, 101–108.
14. T. Kinoshita, T. Hanai, R. Miyazaki and J. Suzuki, *Jpn. Pat.*, JP 10160719 A2 19980619, 1998.
15. T. Hanai, Quantitative in silico analysis of ion exchange from chromatography to protein, *J. Liq. Chromatogr. Rel. Technol.*, 2007, **30**, 1251–1275.
16. T. Hanai, Separation of polar compounds using carbon columns, *J. Chromtogr. A*, 2003, **989**, 183–196.

Enantioseparation

8.1 Enantiomer Recognition in Normal-Phase Liquid Chromatography

The major interaction force for enantioseparation in normal-phase liquid chromatography is the Lewis acid–base interaction including hydrogen bonding, and steric hindrance affects stereoselectivity. Indeed, the presence of hydrogen bonding in chiral complexes is supported by NMR, IR[1] and X-ray crystallography.[2] The quantitative *in silico* analysis of chiral complexes focused on hydrogen-bonding as the docking center. The chromatographic results using so-called "Pirkle type" chiral phases were quantitatively analyzed by molecular mechanics calculations.

The model chiral phases, *N*-(*tert*-butylaminocarbonyl)-(*S*)-valylaminobutane (Phase 1) and (*R*)-1-(α-naphthyl)ethylaminocarbonyl-glycylaminobutane (Phase 2) are shown in Figure 8.1. Phase 1 was used for the enantioseparation of *N*-acetylamino acid methylesters and (*R*)- and (*S*)-4-nitrobenzoyl amino acids, but Phase 2 could not separate these enantiomers.[3] The enantiomer selectivities of *N*-(*S*)-1-(α-naphthyl)ethylaminocarbonyl-(*S*)-valylaminobutane (Phase 3), *N*-(*S*)-1-(α-naphthyl)ethylaminocarbonyl-(*R*)-valylaminobutane (Phase 4), *N*-(*R*)-1-(α-naphthyl)ethylaminocarbonyl-(*R*)-valylaminobutane (Phase 5), and *N*-(*R*)-1-(α-naphthyl)ethylaminocarbonyl-(*S*)-valylaminobutane (Phase 6),[4] which all have two chiral centers, were examined by computational chemical analysis. The structures of model Phases 3–6 are also shown in Figure 8.1.

8.1.1 Enantiomer Recognition of Phases 1 and 2

The optimized complex form of the chiral recognition molecule (CRM 1) of Phase 1 with (*R*)-acetylalanine methylester is shown in Figure 8.2. The structure optimization was performed using the MM2 program, and the

RSC Chromatography Monographs No. 19
Quantitative *In Silico* Chromatography: Computational Modelling of Molecular Interactions
By Toshihiko Hanai
Published by the Royal Society of Chemistry, www.rsc.org

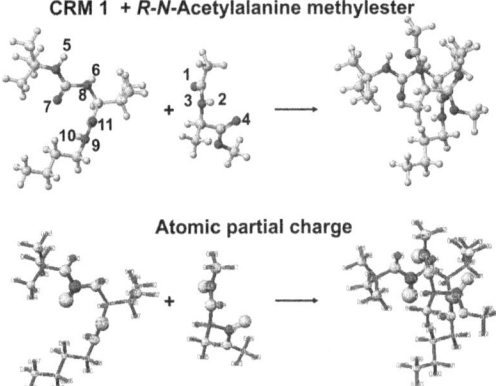

Phase 1

```
  O  OC2H5
   \ /
    Si                O^d              O^e    CH3
   / \                ||               ||      |
  O  (CH2)3--NH^a--C--CH^*1-NH^b--C--NH^c--CH- CH3
                       |                       |
                      CH                      CH3
                     /  \
                   H3C   CH3
```

Phase 2

```
  O  OC2H5
   \ /
    Si                O^d              O^e
   / \                ||               ||
  O  (CH2)3--NH^a--C--CH2-NH^b--C--NH^c--CH^*2-
                                            |
                                           CH3
```

Phase 3 - 6

```
  O  OC2H5
   \ /
    Si                O^d              O^e
   / \                ||               ||
  O  (CH2)3--NH^a--C--CH^*1-NH^b--C--NH^c--CH^*2-
                       |                       |
                      CH                      CH3
                     /  \
                   H3C   CH3
```

	*1	*2		*1	*2		*1	*2
Phase 1	(S)	-	Phase 3	(S)	(S)	Phase 5	(R)	(R)
Phase 2	-	(S)	4	(S)	(R)	6	(R)	(S)

Figure 8.1   Chemical structures of the model chiral Phases 1–6.

CRM 1 + *R-N*-Acetylalanine methylester

Atomic partial charge

Figure 8.2   Optimized stereostructure of the CRM 1 and (*R*)-acetylalanine methylester complex. White, light-gray, dark-gray, and black balls, represent
hydrogen, carbon, nitrogen, and oxygen, respectively.

atomic partial charge (apc) values were calculated using the MOPAC PM5
module of the CAChe™ program.

It is very difficult to imagine this stereostructure from Figure 8.1. The
strongest electron acceptor group is the secondary amino group of the model
phase. The apc indicates the strongest electron donor and acceptor sites
of the chiral phase and the analyte. The values of the apc were changed
positively and negatively after complex formation. The strongest electron
donor group in the complex was the carbonyl of the acetyl group of the
amino acid, and the strongest electron acceptor group was the secondary

Table 8.1 Atomic partial charges (au) of key atoms.

No.	Atom	Before complex	After complex	Δapc
Amino acid				
1	O	−0.448	−0.505	−0.059
2	H	0.272	0.284	0.015
3	N	−0.420	−0.411	−0.009
4	O	−0.414	−0.423	−0.009
Chiral phase				
5	H	0.250	0.285	0.032
6	H	0.260	0.278	0.018
7	O	−0.507	−0.521	−0.014
8	N	−0.445	−0.461	−0.016
9	N	−0.414	−0.415	−0.001
10	H	0.287	0.293	0.006
11	O	−0.472	−0.490	−0.018

amino group of the chiral phase. The apc values of key atoms, as indicated by the numbers in Figure 8.2, are summarized in Table 8.1.

The stereostructure of the CRM of Phase 2 (CRM 2) is very symmetrical, as shown in Figure 8.3. One methyl group demonstrated a small stereoselectivity, but it did not affect the enantioseparation. The apc can demonstrate the strength of the molecular interaction but not any possible enantioseparation. The possibility of enantioseparation can be studied by comparison of the energy values of the optimized structures, calculated using molecular mechanics. The molecular interaction energy value is the sum of the energy values for a pair of compounds minus the energy value of the complex, as given in the following equations:

$$MIFS = fs(amino\,acid) + fs(phase) - FS(complex),$$

$$MIHB = hb(amino\,acid) + hb(phase) - HB(complex),$$

$$MIES = es(amino\,acid) + es(phase) - ES(complex),$$

$$MIVW = vw(amino\,acid) + vw(phase) - VW(complex).$$

The calculated energy values are summarized in Tables 8.2 and 8.3. The energy value of the final (optimized) structure, the hydrogen-bonding energy, the electrostatic energy, and the van der Walls energy of the derivatized amino acids are given as fs, hb, es, and vw, respectively. FS, HB, ES, and VW are the same energy values for the complexes. MIFS, MIHB, MIES, and MIVW are molecular interaction energy values.

The energy values of the (*R*)- and (*S*)-*N*-acetylamino acid methylesters are identical, but those of their complexes are different due to steric hindrance. The MIFS energy values indicate which enantiomer eluted first. The MIFS value difference of enantiomers did not reflect the separation factor (*α*), but this type of analysis should help to study enantioseparation further, in order to design new chiral phases.[5]

The above experiments were performed for one-to-one complexes in free space. The density of bonded chiral phases is not consistent compared to

Side view   **Top view**   **Atomic partial charge**

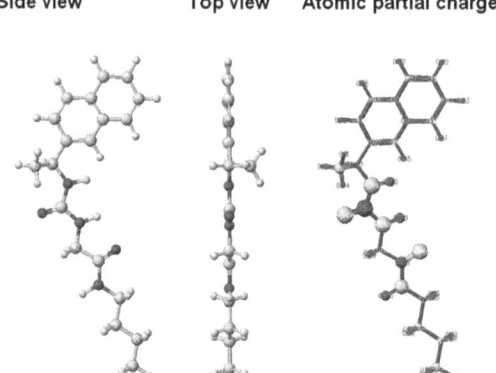

Figure 8.3   Stereostructure of the CRM of Phase 2. White, light-gray, dark-gray, and black balls, represent hydrogen, carbon, nitrogen, and oxygen, respectively.

Table 8.2   Separation factors and molecular properties of some *N*-acetylamino acid methylesters. α represents the separation factor, Eo the elution order, El the eluent (A = *n*-hexane/1,2-dichloromethane/ethanol 40:10:1 and B = *n*-hexane/1,2-dichloroethane/ethanol 100:20:1). The units for fs, he, es and vw are kcal mol^{-1}.

Amino acid	α	Eo	El	fs	hb	es	vw
(*R*)-Alanine	1.24	*R*	A	−0.857	−2.856	−4.415	4.089
(*S*)-Alanine				−0.857	−2.852	−4.414	4.091
(*R*)-Leucine	1.99	*R*	B	−2.401	−2.425	−3.650	5.349
(*S*)-Leucine				−2.401	−2.338	−3.690	5.352
(*R*)-Methionine	1.48	*R*	A	−4.792	−2.998	−6.556	4.815
(*S*)-Methionine				−4.792	−3.003	−6.560	4.810
(*R*)-Phenylalanine	1.66	*R*	B	−12.029	−2.230	−4.841	5.773
(*S*)-Phenylalanine				−12.028	−2.231	−4.840	5.774
(*R*)-Phenylglycine	1.39	*R*	B	−14.908	−3.678	−5.387	6.047
(*S*)-Phenylglycine				−14.908	−3.679	−5.384	6.045
(*R*)-Valine	1.73	*R*	B	−1.176	−2.833	−3.538	4.891
(*S*)-Valine				−1.176	−2.836	−3.538	4.891
Phase 1	—	—	—	−37.258	−5.161	−46.968	8.170

alkylsilane-bonded phases, whose carbon contents and reproducibility in chromatography are consistent. Thus, the reproducibility of retention time is not as reliable for chiral phases, and it is difficult to develop a model phase for computational chemical analysis. Untreated silanols affect the retention of analytes because hydrogen bonding is the predominant retention force for enantioseparation in normal-phase liquid chromatography. But the possibility of enantioseparation and the design of new chiral phases are feasible using computational chemical calculations for the chromatography with Pirkle-type phases.

Table 8.3 Energy values for complexes of *N*-acetylamino acid methylesters with CRM1 (FS, HB, ES, VW) and molecular interaction (MI) energies (kcal mol^{-1}).

Amino acid	FS	HB	ES	VW	MIFS	MIHB	MIES	MIVW
(R)-Alanine	−59.861	−45.381	−57.226	2.526	21.746	7.364	5.846	9.733
(S)-Alanine	−60.935	−15.019	−56.901	3.056	22.820	7.006	5.519	9.205
(R)-Leucine	−59.216	−15.155	−55.390	3.529	21.717	7.569	4.772	9.990
(S)-Leucine	−59.441	−13.850	−55.851	2.912	21.942	6.351	5.193	10.610
(R)-Methionine	−64.241	−15.143	−58.649	2.035	22.191	6.984	5.125	10.950
(S)-Methionine	−64.884	−33.620	−59.511	1.406	22.834	25.456	5.983	11.574
(R)-Phenylalanine	−72.473	−15.026	−56.165	2.768	23.186	7.635	4.356	11.175
(S)-Phenylalanine	−73.698	−21.708	−55.479	6.798	24.412	14.316	3.671	7.146
(R)-Phenylglycine	−74.357	−16.756	−57.019	3.528	22.191	7.917	4.664	10.689
(S)-Phenylglycine	−76.789	−23.041	−55.246	5.431	24.623	14.201	2.894	8.784
(R)-Valine	−60.073	−15.154	−56.467	3.065	21.639	7.160	5.961	9.996
(S)-Valine	−59.650	−13.671	−56.480	3.116	21.516	5.674	5.974	9.945

8.1.2 Enantiomer Recognition of Phases 3–6

The chiral recognition molecule *N*-(*S*)-1-(α-naphthyl)ethylaminocarbonyl- (*S*)-valylaminobutane (CRM 3) was constructed as a model *N*-(*S*)-1-(α-naphthyl)ethylaminocarbonyl-(*S*)-valine-bonded aminopropyl silica gel (Phase 3). *N*-(*S*)-1-(α-naphthyl)ethylaminocarbonyl-(*R*)-valylaminobutane (CRM 4) was constructed as a model *N*-(*S*)-1-(α-naphthyl)ethylaminocarbonyl-(*R*)-valine-bonded aminopropyl silica gel (Phase 4). *N*-(*R*)-1-(α-Naphthyl)ethylamino-carbonyl-(*R*)-valylaminobutane (CRM 5) was constructed as a model *N*-(*R*)-1-(α-naphthyl)ethylaminocarbonyl-(*R*)-valine-bonded aminopropyl silica gel (Phase 5). *N*-(*R*)-1-(α-Naphthyl)ethylamino carbonyl-(*S*)-valylamino-butane (CRM 6) was constructed as a model *N*-(*R*)-1-(α-naphthyl)ethylaminocarbonyl-(*S*)-valine bonded aminopropyl silica gel (Phase 6). These model compounds each have two chiral centers, as shown in Figure 8.1, but their chiral recognition is considered to be based on hydrogen bonding and steric hindrance.

The strongest electron acceptors and donors of each compound were determined by extended Hückel calculation using the CAChe™ program, and the results are partially summarized in Table 8.4, where the apc values of key elements are given. The net apc of hydrogen was generally about −0.05 au, however, the apc values of the hydrogens of secondary amines were about 0.2. The apc of carbon in the carbonyl groups was high, with values of 1.1 and 1.3 au compared to the typical apc values, 0.03 and −0.1, of other carbons. The apc of carbonyl oxygen was low, about −1.1 au.

The strongest electron acceptors on these molecules were the secondary amino groups, as also observed for *N*-tert-butylaminocarbonyl-(*S*)-valylaminobutane and (*R*)-1-(α-naphthyl)ethylaminocarbonylglycylamino-butane, shown as A, B and C in Figure 8.1. However, the hydrogen (B in Figure 8.1) and oxygen (D in Figure 8.1) of CRMs 3 and 4 formed intra-molecular hydrogen bonds, as shown in Figure 8.4 where the structures

Table 8.4 Net atomic partial charges (au) of key elements from extended Hückel calculations. Positions are indicated in Figure 8.1. Reproduced by permission of Elsevier, ref. 29.

Position	Atom	Net atomic partial charge (au)			
		CRM 3	CRM 4	CRM 5	CRM 6
a	H (NH)	0.2089	0.2085	0.1989	0.1989
b	H (NH)	0.2027	0.2026	0.2000	0.2000
c	H (NH)	0.1976	0.2015	0.2017	0.2013
d	O (CO)	−1.0779	−1.0799	−1.0900	−1.0697
e	O (CO)	−1.1352	−1.1336	−1.1445	−1.1444

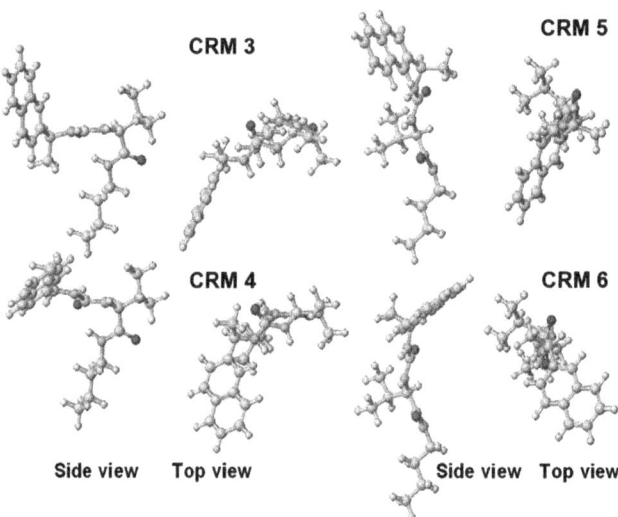

CRM 3   CRM 5
CRM 4   CRM 6

Side view   Top view   Side view   Top view

Figure 8.4 Stereostructures of CRMs 3–6. White, light-gray, dark-gray, and black balls, represent hydrogen, carbon, nitrogen, and oxygen, respectively.

optimized by MM2 calculations are shown. The electron donor groups of the analytes were the carbonyls of the acetyl and dinitrobenzoyl groups.

The complexes were, therefore, constructed with the secondary amino groups of the CRM facing a carbonyl group on the analyte for CRMs 3 and 4. Some optimized complex forms of *N*-acetylphenylalanine methylester with CRM 3 are shown in Figure 8.5, where side and top views of four complexes formed between CRM 3 and (*R*)- and (*S*)-*N*-acetylphenylalanine methylesters are shown. The minimum final energy values and the hydrogen-bonding and van der Waals energy values of the complexes are summarized in Table 8.5(a) and (b) with the energy values of the CRMs and analytes.

The energy values of complexes formed between the 'a' and 'b' hydrogens and a carboxyl group oxygen of the analyte (Figure 8.5(a) and (b)) were a few kcal higher than those of complexes formed between the 'a' and 'b' hydrogens and a carbonyl group oxygen of the analyte. Therefore, the energy values shown in Table 8.5 are those of the latter conformations. The same

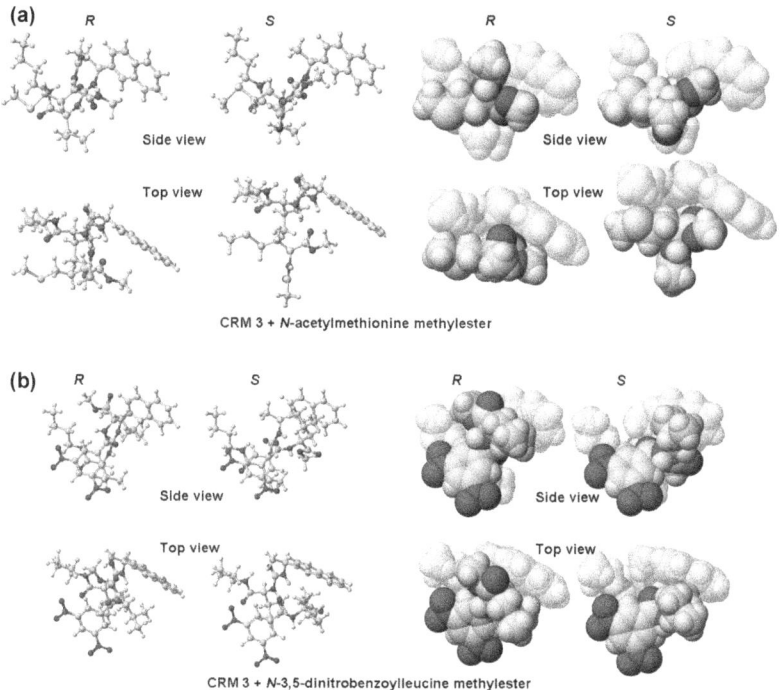

Figure 8.5 Complex forms of CRM 3 with (a) (R)- and (S)-N-acetylmethionine methylester, and (b) (R)- and (S)-N-3,5-dinitrobenzoylleucine methylester. White, light-gray, dark-gray, and black balls, represent hydrogen, carbon, nitrogen, and oxygen, respectively.

results were obtained for N-3,5-dinitrobenzoylamino acid methylesters. The final, hydrogen-bonding and van der Waals energy values of these complexes with CRM 4 are summarized in Table 8.6.

The energy values of the complexes were smaller than the sums of the CRMs and analytes. The final energy values of complexes with (R)-N-acetylamino acid methylesters were smaller than those with (S)-N-acetylamino acid methylesters except for (S)-phenylglycine. The energy values of the (S)-N-acetylphenylglycine methylester complexes (-81.83 kcal mol^{-1}) were higher than those of the (R)-form (-83.91 kcal mol^{-1}). The van der Waals energy value (18.39 kcal mol^{-1}) of the (S)-form complex was about equivalent to the sum of each energy value (18.41 kcal mol^{-1}). The MI energy of these complex conformations clearly indicated that the MIFS values of the (R)-form complexes were larger than the MIFS values of the (S)-form complexes except for (S)-N-acetylphenylglucine methylester. The MIVW energy values also supported the results of MIFS, however MIHB did not indicate the elution order. The elution order of enantiomers of N-3,5-dinitrobenzoylamino acids methylesters from CRM 3 followed the final energy values of complexes with CRM 3 and (R) and (S)-form amino acids. In the case of Phase 4, the exceptions were N-acetyl-(S)-leucine and N-3,5-dinitrobenzoyl-(R)-leucine, as

Table 8.5   (a) Molecular properties of the model phases, some *N*-acetylamino acid methylesters, and some *N*-3,5-dinitrobenzoylamino acid methylesters. Reproduced by permission of Elsevier, ref. 29.

Compound	(S)-form			Compound	(S) form		
	FS	*HB*	*VW*		*FS*	*HB*	*VW*
CRM 3	−62.77	−7.50	12.39	CRM 5	−58.65	−6.54	11.10
CRM 4	−63.88	−7.79	12.43	CRM 6	−58.58	−6.57	11.01
N-Acetylamino acid methylesters							
Alanine	−0.84	−2.93	4.08	Phenylalanine	−11.98	−2.21	5.76
Leucine	−0.10	−1.35	5.30	Phenylglycine	−14.89	−3.67	6.02
Methionine	−4.77	−2.90	4.86	Valine	−1.13	−2.50	5.00
N-3,5-Dinitrobenzoylamino acid methylesters							
Alanine	−9.39	−5.67	11.10	Phenylalanine	−17.22	−5.28	12.66
Leucine	−8.72	−5.28	11.86	Phenylglycine	−22.02	−6.91	13.11
Methionine	−12.66	−4.94	11.08	Valine	−9.79	−5.64	11.30

(b) Molecular properties of some *N*-acetylamino acid methylesters, some *N*-3,5-dinitrobenzoylamino acid methylesters, and their complexes with CRM 3.

Compound	α	Eo^a	El^a	(R)-form			(S)-form		
				FS	*HB*	*VW*	*FS*	*HB*	*VW*
Complexes of CRM 3 and *N*-acetylamino acid methylesters									
Alanine	1.17	*R*	A	−68.09	−20.99	17.13	−68.02	−20.49	16.21
Leucine	1.46	*R*	B	−66.45	−17.37	15.39	−67.68	−18.56	13.24
Methionine	1.3	*R*	A	−71.27	−18.00	14.28	−72.97	−17.13	12.97
Phenylalanine	1.33	*R*	B	−76.67	−17.70	16.93	−83.18	−23.55	15.5
Phenylglycine	1.2	*R*	B	−83.91	−22.91	15.56	−81.83	−25.50	18.39
Valine	1.43	*R*	B	−66.99	−17.62	15.15	−67.93	−17.61	14.51
Complexes of CRM 3 and *N*-3,5-dinitrobenzoylamino acid methylesters									
Alanine	2.19	*R*	A	−76.00	−22.25	18.09	−80.06	−23.77	20.07
Leucine	1.54	*R*	A	−76.49	−23.20	22.21	−78.37	−21.95	19.56
Methionine	2.04	*R*	A	−80.02	−22.27	21.74	−81.58	−21.38	18.42
Phenylalanine	1.33	*R*	A	−91.06	−27.10	20.41	−93.11	−25.31	19.29
Phenylglycine	1.29	*S*	A	−92.83	−27.48	22.06	−88.74	−22.78	23.7
Valine	1.98	*R*	A	−76.77	−22.04	17.32	−77.65	−22.51	20.66

[a]Eo: first eluted compound; El: Eluent A, *n*-hexane : 1,2-dichloroethane : ethanol (40 : 10 : 1); Eluent B, *n*-hexane : 1,2-dichloroethane : ethanol (100 : 20 : 1).

listed in Table 8.6. The difference depended on the fit between the two secondary amino groups of CRM 4 and the carbonyl group of the *N*-acetyl or 3,5-dinitrobenzoyl group and an oxygen of the methyl carboxyl group of the derivatized amino acid methylester.

CRMs 5 and 6 had no intramolecular hydrogen bonds, and therefore were considered to form more chiral complex forms than those of CRMs 3 and 4, as expected from Figure 8.4. After certain trials, two complex centers were selected. One was the same as CRMs 3 and 4, and consisted of the secondary amino groups, and the second was the carbonyl group indicated as 'd' in Figure 8.1. The optimized energy values are separately summarized in Tables 8.7 and 8.8 as C=O and N–H. The exceptions for the CRM 5

Table 8.6 Molecular properties of some *N*-acetylamino acid methylesters, some *N*-3,5-dinitrobenzoylamino acid methylesters, and their complexes with CRM 4. Reproduced by permission of Elsevier, ref. 29.

Compound	α	Eo[a]	El[a]	(R)-form FS	HB	VW	(S)-form FS	HB	VW
Complexes of CRM 4 and *N*-acetylamino acid methylesters									
Alanine	1.16	R	A	−66.98	−18.06	14.18	−67.52	−17.61	13.46
Leucine	1.74	R	B	−69.33	−20.33	13.28	−68.04	−20.53	13.98
Methionine	1.44	R	A	−71.73	−17.39	12.62	−73.14	−17.13	12.13
Phenylalanine	1.56	R	B	−79.72	−17.64	12.64	−83.55	−23.40	14.45
Phenylglycine	1.3	R	B	−84.34	−23.13	14.72	−88.73	−29.44	14.67
Valine	1.47	R	B	−67.36	−17.36	14.07	−67.90	−17.34	13.85
Complexes of CRM 4 and *N*-3,5-dinitrobenzoylamino acid methylesters									
Alanine	1.73	S	A	−78.77	−21.59	17.74	−76.11	−23.74	21.23
Leucine	1.26	S	A	−78.09	−20.31	13.88	−80.02	−21.77	17.88
Methionine	1.63	S	A	−84.58	−19.93	14.11	−83.26	−21.53	17.58
Phenylalanine	1.03	R	A	−92.24	−27.81	18.57	−92.86	−30.03	19.65
Phenylglycine	1.3	R	A	−92.08	−24.62	19.6	−93.78	−25.67	19.85
Valine	1.87	S	A	−78.58	−23.17	18.58	−78.63	−22.04	18.57

[a]Eo: first eluted compound; El Eluent A, *n*-hexane : 1,2-dichloroethane : ethanol (40 : 10 : 1); Eluent B, *n*-hexane : 1,2-dichloroethane : ethanol (100 : 20 : 1).

complexes were *N*-3,5-dinitro (*S*)-valine, (*R*)-phenylalanine and (*S*)-phenylglycine; their final energy values in these cases were not the same as their elution order. Even when their complexes having smaller van der Waals energy values were constructed, the final energy values did not support their elution order. The exceptions for CRM 6 were *N*-3,5-dintrobenzoyl-(*R*)-phenylalanine and (*S*)-leucine. The calculated energy value differences between the (*R*)- and (*S*)-form complexes of *N*-acetylamino acids were almost equivalent, but α values were larger for CRMs 5 and 6. The elution order of *N*-3,5-dinitrobenzoylamino acids was not the same for all amino acids. These results indicated that the one-to-one complex model is not adequate. The difficulty in the estimation of elution order was enhanced by the steric effects of the large naphthyl groups of the above chiral phases, and the dinitrobenzoyl groups of the analytes.

Individual analytes may form different types of complexes with the chiral phase and may not correspond to those shown in Figures 8.5–8.8. The computational chemical calculation was performed in free space, and the chromatographic separation was achieved in narrow space. In particular, the position of the naphthyl group in the chiral phase affects the steric hindrance as expected from the three-dimensional models. The hydrogen-bonding and van der Waals energy values did not directly support the chromatographic elution order.

The surface conditions of the bonded phase and the densities of the chiral recognition groups are not known. Therefore, three-dimensional structures of chiral phases were constructed using the Molecular Editor program, and

Table 8.7 (a) Molecular properties of some N-acetylamino acid methylesters and their complexes with CRM 5. FS represents the final energy, HB the hydrogen-bonding energy, and VW the van der Waals energy (kcal mol^{-1}). Reproduced by permission of Elsevier, ref. 29.

Compound	α	Eo^a	El^a	(R)-form			(S)-form		
				FS	HB	VW	FS	HB	VW
Complexes of CRM 5 (C=O) and N-acetylamino acid methylesters									
Alanine	1.21	S	A	−80.19	−15.56	5.45	−79.26	−16.39	6.2
Leucine	1.43	S	B	−80.09	−15.74	6.85	−78.33	−16.80	8.64
Methionine	1.26	S	A	−84.10	−15.87	4.93	−84.93	−16.55	4.47
Phenylalanine	1.31	S	B	−91.55	−17.12	8.68	−90.72	−16.76	7.74
Phenylglycine	1.2	S	B	−95.54	−20.66	9.61	−92.84	−18.91	9.57
Valine	1.42	S	B	−82.24	−17.95	4.84	−80.24	−16.09	6.21
Complexes of CRM 5 (N−H) and N-acetylamino acid methylesters									
Alanine	1.21	S	A	−85.84	−19.22	8.41	−85.24	−18.92	8.76
Leucine	1.43	S	B	−81.05	−17.55	4.51	−78.22	−13.60	3.89
Methionine	1.26	S	A	−85.38	−18.10	5.34	−83.65	−13.75	2.87
Phenylalanine	1.31	S	B	−91.55	−17.08	8.66	−91.49	−19.48	8.45
Phenylglycine	1.2	S	B	−95.41	−20.81	9.77	−93.59	−19.40	10.94
Valine	1.42	S	B	−79.53	−17.02	8.42	−77.02	−13.60	5.98

(b) Molecular properties of some N-3,5-dinitrobenzoylamino acid methylesters and their complexes with CRM 5.

Compound	α	Eo^a	El^a	(R)-form			(S)-form		
				FS	HB	VW	FS	HB	VW
Complexes of CRM 5 (C=O) and N-3,5-dinitrobenzoylamino acid methylesters									
Alanine	2.07	S	A	−90.99	−20.68	13.83	−88.47	−21.04	13.45
Leucine	1.56	S	A	−93.30	−19.55	11.26	−90.75	−20.88	11.6
Methionine	2.13	S	A	−94.80	−20.31	12.17	−94.70	−20.26	12.35
Phenylalanine	1.35	S	A	−100.88	−21.75	12.55	−100.69	−20.93	12.79
Phenylglycine	1.3	R	A	−108.77	−24.61	13.55	−108.64	−25.42	10.61
Valine	1.99	S	A	−91.34	−22.02	13.56	−92.45	−23.89	13.05
Complexes of CRM 5 (N−H) and N-3,5-dinitrobenzoylamino acid methylesters									
Alanine	2.07	S	A	−97.50	−22.81	10.66	−96.02	−24.08	13.93
Leucine	1.56	S	A	−94.75	−21.93	8.97	−91.09	−23.62	11.43
Methionine	2.13	S	A	−99.88	−22.01	7.44	−97.76	−23.10	9.63
Phenylalanine	1.35	S	A	−105.15	−24.72	14.79	−107.15	−25.19	10
Phenylglycine	1.3	R	A	−110.47	−24.38	8.92	−108.64	−25.42	10.61
Valine	1.99	S	A	−94.07	−23.94	11.24	−91.95	−23.82	11.07

^{a}Eo: first eluted compound; El Eluent A, n-hexane : 1,2-dichloroethane : ethanol (40 : 10 : 1);
Eluent B, n-hexane : 1,2-dichloroethane : ethanol (100 : 20 : 1).

optimized by MM2 calculations. The optimized three-dimensional structures were used to study the space between CRMs bonded to the surface of silica gels. The surface of the model silica gel was the same as that of quartz, and the basic structure consisted of 11 units of Si–O groups.[6] Figure 8.9 shows eleven CRM 3 molecules densely bonded to the surface.

Table 8.8 (a) Molecular properties of some *N*-acetylamino acid methylesters and their complexes with CRM 6. Reproduced by permission of Elsevier, ref. 29.

Compound	α	Eo^a	El^a	(R)-form FS	HB	VW	(S)-form FS	HB	VW
Complexes of CRM 6 (C=O) and *N*-acetylamino acid methylesters
Alanine	1.17	S	A	−79.86	−15.98	7.2	−79.20	−15.93	7.66
Leucine	1.71	S	B	−82.08	−19.82	7.57	−80.54	−15.12	5.75
Methionine	1.35	S	A	−85.18	−20.06	7.97	−84.54	−15.13	6.11
Phenylalanine	1.53	S	B	−91.07	−20.47	9.01	−91.53	−19.97	9.02
Phenylglycine	1.28	S	B	−95.35	−22.09	10.1	−95.31	−18.29	6.84
Valine	1.41	S	B	−80.98	−19.77	8.87	−79.45	−16.86	8.42
Complexes of CRM 6 (N–H) and *N*-acetylamino acid methylesters
Alanine	1.17	S	A	−85.54	−19.10	8.36	−85.12	−18.84	8.6
Leucine	1.71	S	B	−84.35	−19.03	8.13	−77.23	−13.73	5.15
Methionine	1.35	S	A	−86.66	−17.50	3.49	−83.03	−17.29	6.18
Phenylalanine	1.53	S	B	−93.10	−18.69	8.04	−92.68	−18.17	5.9
Phenylglycine	1.28	S	B	−98.40	−22.93	7.68	−95.47	−21.50	7.86
Valine	1.41	S	B	−84.42	−19.02	7.13	−83.59	−18.91	8.28

(b) Molecular properties of some *N*-3,5-dinitrobenzoylamino acid methylesters and their complexes with CRM 6.

Compound	α	Eo^a	El^a	(R)-form FS	HB	VW	(S)-form FS	HB	VW
Complexes of CRM 6 (C=O) and *N*-3,5-dinitrobenzoylamino acid methylesters
Alanine	1.81	R	A	−88.12	−20.20	12.51	−90.30	−21.19	11.06
Leucine	1.28	R	A	−90.14	−19.55	11.6	−91.62	−18.47	10.74
Methionine	1.61	R	A	−94.59	−21.39	9.86	−97.03	−22.11	10
Phenylalanine	1.04	S	A	−101.85	−22.65	9.86	−103.90	−25.01	12.21
Phenylglycine	1.25	S	A	−105.92	−24.57	11.54	−105.58	−22.1	10.28
Valine	1.92	R	A	−91.08	−22.14	12.14	−91.55	−21.24	10.3
Complex of CRM 6 (N–H) and *N*-3,5-dinitrobenzoylamino acid methylesters
Alanine	1.81	R	A	−92.84	−23.23	13.4	−94.51	−24.56	12.95
Leucine	1.28	R	A	−97.51	−22.58	7.18	−95.45	−23.95	11.23
Methionine	1.61	R	A	−102.31	−22.38	5.92	−106.61	−24.79	13.52
Phenylalanine	1.04	S	A	−105.48	−21.52	10.46	−109.22	−23.69	7.92
Phenylglycine	1.25	S	A	−112.36	−24.53	7.89	−109.02	−23.93	9.58
Valine	1.92	R	A	−95.09	−21.88	8.66	−95.79	−23.90	11.15

^{a}Eo: first eluted compound; El Eluent A, *n*-hexane : 1,2-dichloroethane : ethanol (40 : 10 : 1); Eluent B, *n*-hexane : 1,2-dichloroethane : ethanol (100 : 20 : 1).

The grafting rates of CRMs 3–6 are 0.37–0.53 mmol g^{-1} (ref. 3). Thus, $2.2–3.2 \times 10^{20}$ CRMs were present on the surface of one g of silica gel. The practical surface area of the silica gel was estimated to be 300 m^2 g^{-1}, therefore the area occupied by one CRM was 0.94–1.35 nm^2. If the surface area of the silica gel was 400 nm^2, the area occupied by one CRM would be 1.33–1.80 nm^2. The estimated surface areas are summarized in Table 8.9. The surface area occupied by one CRM 3 molecule was about 1 nm^2, from the results shown in Figure 8.9. The areas of CRMs 4–6 were also about 1 nm^2, and the density on the surface of the silica gel was dependent on the

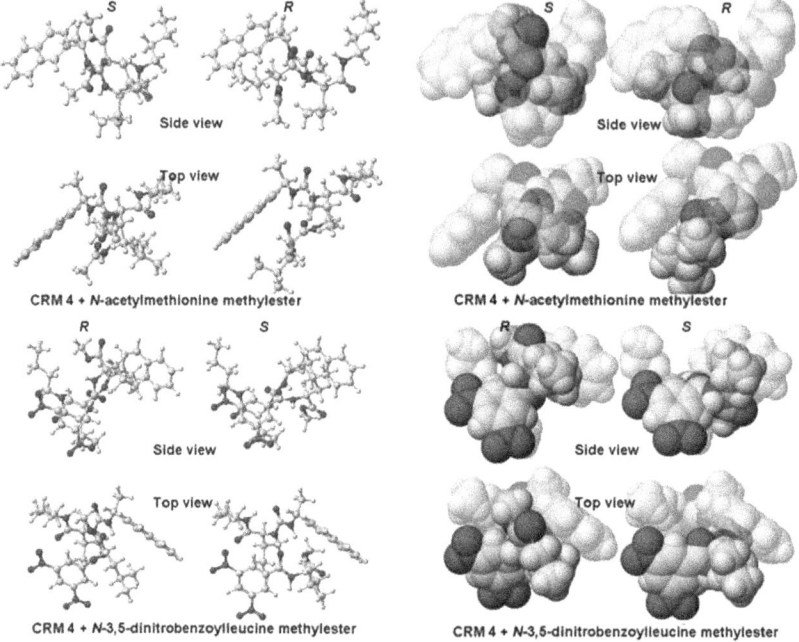

Figure 8.6   Complex forms of CRM 4 with (R)- and (S)-N-acetylmethionine methylester, and (R)- and (S)-N-3,5-dinitrobenzoylleucine methylester. White,
light-gray, dark-gray, and black balls, represent hydrogen, carbon,
nitrogen, and oxygen, respectively.

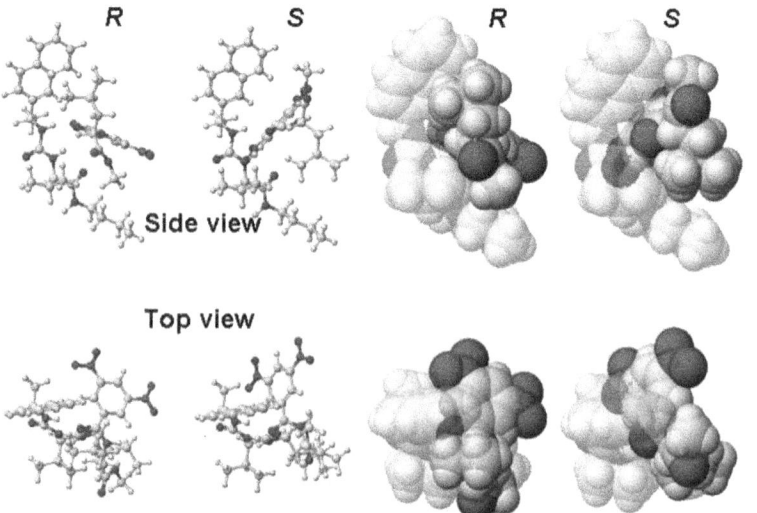

Figure 8.7   Complex forms of CRM 5 with (R)- and (S)-N-acetylmethionine methylester. White, light-gray, dark-gray, and black balls, represent hydrogen,
carbon, nitrogen, and oxygen, respectively.

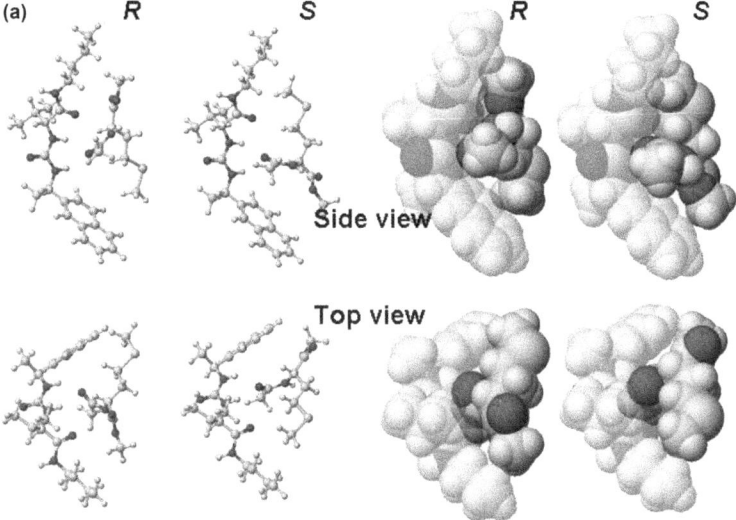

CRM 6 + *N*-acetylmethionine methylester

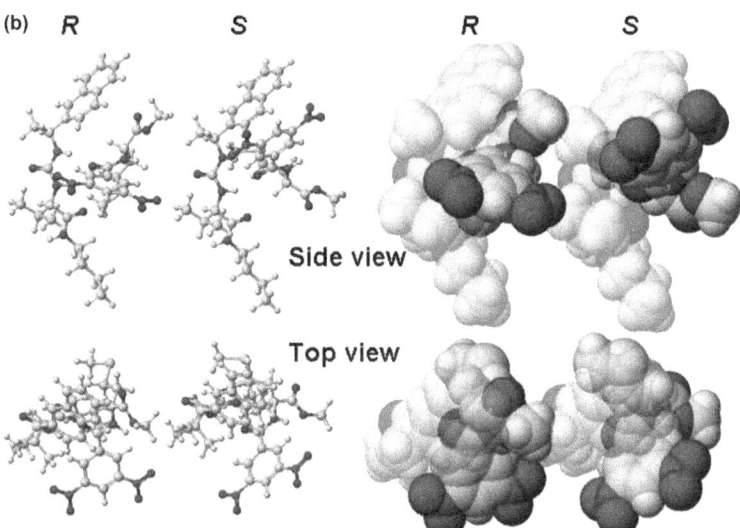

CRM 6 + *N*-3,5-dinitrobenzoylleucine methylester

Figure 8.8 Complex forms of CRM 6 with (a) (*R*)- and (*S*)-*N*-acetylmethionine methylester and (b) (*R*)- and (*S*)-*N*-3,5-dinitrobenzoylleucine methylester. White, light-gray, dark-gray, and black balls, represent hydrogen, carbon, nitrogen, and oxygen, respectively.

molecular shape, shown in Figure 8.4, where the top and side views of CRMs 3–6 are shown. A comparison of the occupied surface areas obtained experimentally, and those calculated from the molecular shape, suggested that these CRMs were densely bonded to the silica gel surface, and a larger sized analyte

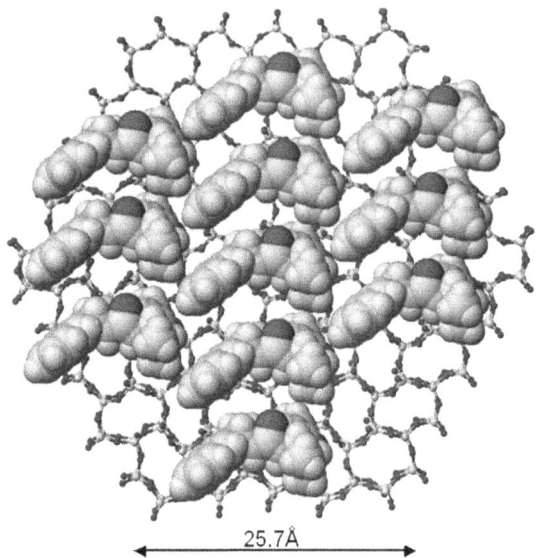

25.7Å

Figure 8.9   Top view of a model CRM 3-bonded silica gel.

Table 8.9   Occupied areas for each chiral recognition molecule. The size of one molecule in the three-dimensional structure is about 1 nm². Reproduced by permission of Elsevier, ref. 29.

CRM	Grafting rate/ mmol g⁻¹	Total CRMs×1020 per g silica	No. of CRMs per nm² [a]	Occupied area/nm² per molecule[a]	No. of CRMs per nm² [b]	Occupied area/nm² per molecule[b]
1	0.37	2.23	0.74	1.35	0.56	1.80
2	0.41	2.47	0.82	1.22	0.62	1.62
3	0.37	2.23	0.74	1.35	0.56	1.80
4	0.40	2.41	0.80	1.25	0.60	1.66

[a]Calculated from the surface area 300 m² per g of silica.
[b]Calculated from the surface area 400 m² per g silica.

may not fit into the CRM brushes to form a complex. In particular, the naphthyl groups of CRMs cover the surface of the stationary phase, meaning that the hydrogen-bonding sites, NH groups, may not be fully exposed.

This geometrical analysis indicated that one-to-one chiral complex analysis has limitations with regard to predicting the elution order. This method may be suitable when the molecular size of the CRM is quite large compared to that of analyte, and CRMs are not present at a high density on the packing material surface. However, as many CRMs as possible are bonded on commercial chiral phases. The redesign of chiral phases should help in the development of a computer-controlled enantiomer-separation system. In addition, considering the contribution of silanol groups is also important in normal-phase liquid chromatographic separation where hydrogen bonding is considered the predominant force. The prediction of elution order of enantiomers may be accurate if the analysis is performed using low-density chiral phases. The elution order was seldom different from that predicted when chromatographic results measured on high-density chiral phases were analyzed.

The balanced final energy values between the complexes with (*R*)- and (*S*)-derivatized amino acids did not correlate well with their separation factors. The final energy values, however, indicated the chromatographic elution order, and van der Waals energy values are useful in the search for the best complex conformation. The precision of prediction of chromatographic separation factors may be improved by using a powerful computer that can handle both a brush-type chiral phase as a model surface for packing, and the solvent effects.

8.2 Enantiomer Recognition in Ligand-Exchange Liquid Chromatography

Underivatized amino acids form complexes with copper(II). A variety of copper(II)-trapped phases have been developed for the enantiosepara-tion of amino acids. *N*-Salicylidene-(*R*)-2-amino-1,2-*bis*(2-butoxy-5-*tert*-butyl-phenyl)-3-phenyl-1-propanol was coated on an octadecyl-bonded silica gel column and copper sulfate solution was used as the eluent.[7] The structure was constructed using the Molecular Editor program and optimized by MM2 calculations. The optimization was performed as the energy change was less than 10^{-6} kcal mol^{-1}. The molecular weight of the chiral phase was 1449, and the final and van der Waals energies were 77.04 and -8.77 kcal mol^{-1}, respectively.

The optimized structure of the binuclear copper(II) derivative of *N*-salicyli-dene-(*R*)-2-amino-1,1-*bis*(2-butoxy-5-*tert*-butylphenyl)-3-phenyl-1-propanol is shown in Figure 8.10. Three-dimensional analysis suggested the binding site is at the front of the structure shown in Figure 8.11, due to the wide-open space compared with the other side where the site of the copper atom was very narrow. The nitrogen atom of the amino group or the oxygen ion of the carboxyl group of the analytes was bound to the copper atom that was more visible. The water molecule was not bound to the copper atom, which forms a hexadentate complex, because it did not geometrically interrupt the complex formation and it should be located at the opposite site of analyte. The complex form with (*R*)- and (*S*)-tryptophan is shown in Figure 8.12, and the calculated molecular interaction energy values are summarized in Tables 8.10 and 8.11.

This chiral phase is large and contains copper, therefore the apc could not been calculated using the computer system, but the molecular recognition center is clear. The copper and its neighboring oxygen atom form a complex with the positively charged amino acid, then the amino acid forms a complex with the copper ion in the eluent and is removed from the chiral phase. The energy values of positively charged (*R*)- and (*S*)-amino acids demonstrated small differences due to intramolecular electron localization. Complex

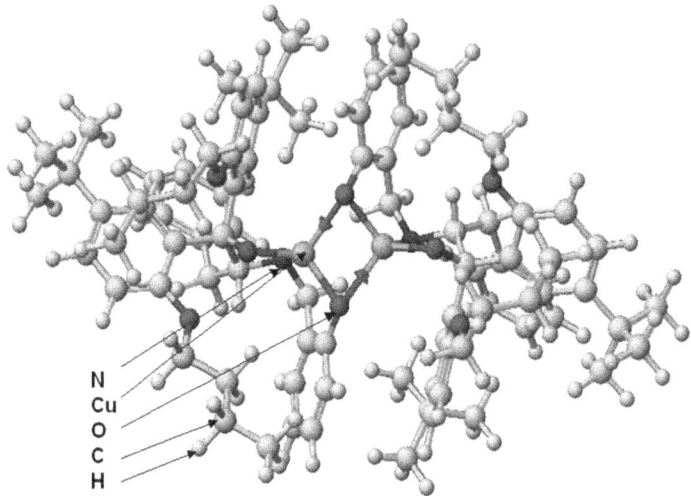

Figure 8.10 Structure of the binuclear Cu(II) derivative of *N*-salicylidene-(*R*)-2-amino-1,1-*bis*(2-butoxy-5-*tert*-butylphenyl)-3-phenyl-1-propanol. White, light-gray, dark-gray, black, and gray balls, represent hydrogen, carbon, nitrogen, oxygen, and copper, respectively.

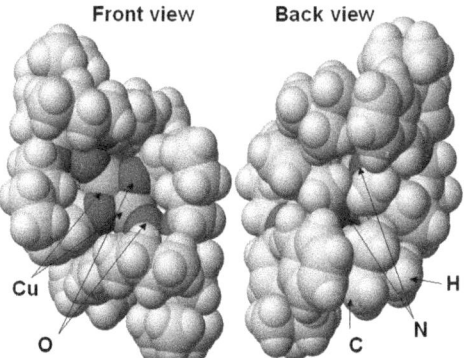

Figure 8.11 Front and back view of the binuclear Cu(II) derivative of *N*-salicylidene-(*R*)-2-amino-1,1-*bis*(2-butoxy-5-*tert*-butylphenyl)-3-phenyl-1-propanol. White, light-gray, dark-gray, black, and gray balls, represent hydrogen, carbon, nitrogen, oxygen, and copper, respectively.

formation increased the difference. The elution order can be estimated from the difference of MIFS energy values.[8]

8.3 Diastereomer Recognition in Reversed-Phase Liquid Chromatography

Analytes derivatized with a chiral derivatization reagent have been separated by reversed-phase liquid chromatography;[9–16] the reversed-phase liquid chromatography of derivatized amino acid enantiomers was analyzed using

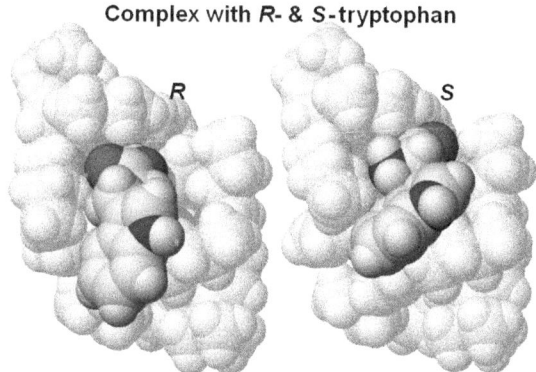

Complex with *R*- & *S*-tryptophan

Figure 8.12 Conformation of the complexes of (*R*)- and (*S*)-tryptophan with the Cu(II) derivative of *N*-salicylidene-(*R*)-2-amino-1,1-*bis*(2-butoxy-5-*tert*-butylphenyl)-3-phenyl-1-propanol. White, light-gray, dark-gray, black, and gray balls, represent hydrogen, carbon, nitrogen, oxygen, and copper, respectively.

Table 8.10 Molecular interaction energy values and separation factors of free amino acids in ligand-exchange enantioseparation. Eluent A = 1 mM copper(II) sulphate in water, B = 2 mM copper(II) sulphate in 15% aqueous acetonitrile. Reproduced by permission of Taylor and Francis, ref. 8.

Amino acid	α	Eo	El	fs	hb	es	vw
(*R*)-Aspartic acid	1.11	*S*	A	−23.3989	−7.831	−13.680	2.711
(*S*)-Aspartic acid				−23.3989	−7.833	−13.680	2.712
(*R*)-Histidine	1.18	*R*	A	−0.4312	−4.763	−5.090	0.860
(*S*)-Histidine				−0.2520	−4.664	−5.646	1.200
(*R*)-Isoleucine	1.15	*S*	A	−1.9734	−3.727	−2.584	3.045
(*S*)-Isoleucine				−2.1956	−3.678	−2.981	3.528
(*R*)-*t*-Leucine	1.34	*S*	A	−2.3856	−3.745	−3.351	3.178
(*S*)-*t*-Leucine				−3.2293	−3.901	−3.990	3.183
(*R*)-Methionine	1.30	*R*	A	−6.3321	−3.925	−4.316	2.430
(*S*)-Methionine				−5.2270	−3.735	−3.453	2.179
(*R*)-Phenylalanine	1.74	*R*	S	−18.1052	−5.234	−4.624	4.508
(*S*)-Phenylalanine				−17.7717	−5.103	−3.538	4.470
(*R*)-Phenylglycine	1.24	*R*	A	−18.6693	−5.497	−3.200	3.779
(*S*)-Phenylglycine				−18.6693	−5.497	−3.201	3.780
(*R*)-Proline	1.22	*S*	A	−4.6513	−3.867	−9.285	2.105
(*S*)-Proline				−4.6512	−3.867	−9.284	2.102
(*R*)-Serine	1.19	*R*	A	−9.5073	−4.450	−4.552	1.744
(*S*)-Serine				−9.4939	−4.316	−4.975	1.863
(*R*)-Tryptophan	2.05	*R*	B	−13.4024	−6.451	−5.097	2.841
(*S*)-Tryptophan				−12.9603	−6.125	−4.083	2.759
(*R*)-Tyrosine	2.06	*R*	B	−20.4000	−6.701	−4.735	4.453
(*S*)-Tyrosine				−20.0586	−6.571	−3.647	4.413
(*R*)-Valine	1.29	*S*	A	−3.7297	−4.031	−2.443	2.813
(*S*)-Valine				−3.7297	−4.030	−2.443	2.811
Ligand phase	—	—	—	−9.1921	0.000	−2.195	3.349

Table 8.11 Energy values and molecular interaction energies for complexes of amino acids with the binuclear copper(II) derivative of N-salicylidene-(R)-2-amino-1,1-bis(2-butoxy-5-tert-butylphenyl)-3-phenyl-1-propanol.

Amino acid	FS	HB	ES	VW	MIFS	MIHB	MIES	MTVW
(R)-Aspartic acid	−55.341	−20.937	−16.890	−4.323	22.750	13.106	1.015	10.383
(S)-Aspartic acid	−54.380	−19.873	−16.417	−4.179	21.789	12.040	0.542	10.240
(R)-Histidine	−33.085	−17.858	−7.991	−8.536	23.462	13.095	0.706	12.745
(S)-Histidine	−33.528	−17.914	−7.749	−6.882	24.084	13.250	−0.092	11.431
(R)-Isoleucine	−37.004	−17.222	−6.243	−5.440	25.839	13.495	1.464	11.834
(S)-Isoleucine	−34.208	−16.095	−5.484	−3.776	22.820	12.417	0.308	10.653
(R)-Leucine	−37.010	−16.950	−6.082	−4.817	24.256	13.013	1.463	11.197
(S)-Leucine	−35.480	−15.877	−5.283	−4.813	22.726	11.933	0.664	11.198
(R)-t-Leucine	−36.388	−16.445	−6.876	−5.619	24.810	12.700	1.330	12.146
(S)-t-Leucine	−36.342	−15.879	−7.102	−6.302	23.921	11.978	0.917	12.834
(R)-Methionine	−37.485	−16.318	−6.346	−4.482	21.961	12.393	−0.165	10.261
(S)-Methionine	−37.733	−15.888	−5.837	−5.731	23.314	12.153	0.189	11.259
(R)-Phenylalanine	−49.488	−16.155	−7.255	−3.807	22.191	10.921	0.436	11.664
(S)-Phenylalanine	−49.323	−15.448	−6.300	−4.218	22.359	10.345	0.567	12.037
(R)-Phenylglycine	−50.597	−16.780	−6.061	−5.219	22.736	11.283	0.666	12.347
(S)-Phenylglycine	−51.534	−16.273	−5.619	−5.617	23.673	10.776	0.223	12.746
(R)-Proline	−34.935	−14.552	−11.752	−5.015	21.092	10.685	0.272	10.469
(S)-Proline	−34.321	−14.605	−11.563	−4.319	20.478	10.738	0.084	9.770
(R)-Serine	−41.062	−21.823	−6.569	−3.987	22.363	17.372	−0.178	9.080
(S)-Serine	−42.663	−23.713	−7.609	−0.916	23.977	19.397	0.439	6.128
(R)-Tryptophan	−46.663	−17.589	−7.338	−6.622	24.068	11.138	0.046	12.812
(S)-Tryptophan	−46.246	−16.869	−6.795	−7.065	24.094	10.744	0.517	13.173
(R)-Tyrosine	−50.883	−15.345	−7.615	−5.239	21.291	8.644	0.685	13.041
(S)-Tyrosine	−51.621	−17.117	−6.322	−4.497	22.370	10.546	0.480	12.259
(R)-Valine	−37.265	−17.050	−6.191	−4.870	24.343	13.019	1.553	11.032
(S)-Valine	−35.861	−15.648	−5.662	−5.035	22.939	11.618	1.024	11.195

MM2 calculations, and the separation factors and capacity ratios, measured during isocratic elution, were related to the energy value changes between an analyte and a model phase. Diastereomers have very different stereo-structures. The contact surface area with bonded alkyl groups affects the elution order. Amino acids were derivatized with a chiral derivatization re-agent, such as *o*-phthalaldehyde-*N*-acetyl-ʟ-cysteine (ref. 11) and *N*-tert-butylthiocarbonyl-ʟ-cysteine, prior to the chromatographic analysis.

A model phase was constructed based on dimethylpentylsilane, consisting of 991 atoms, 1051 bonds, and 15193 connectors, containing 171 silicons, 328 oxygens, 143 carbons, and 349 hydrogens. Dimethylpentylsilane (20) and trimethylsilane (1) groups were bonded within 900 Å^2 on the polysilicondi-oxide phase. The trimethylsilane was considered to be an end-capped mol-ecule. The optimized energy value change was less than 0.00001 kcal mol^{-1}. The molecular size and alkyl chain length were determined by the calcu-lation capacity of the computer used, and the alkyl chain length effect for the hydrophobicity.[17] Optimized structures of alanine diastereomers derivatized with (+)-1-(9-fluorenyl)ethyl (FLE), the model phase, and a complex between this model phase and an enantiomer are shown in Figures 8.13–8.15,

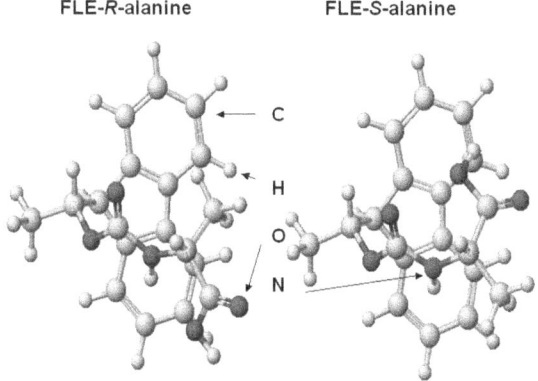

FLE-*R*-alanine   **FLE-*S*-alanine**

Figure 8.13   (+)-1-(9-Fluorenyl)ethyl chloroformate derivatized (*R*)- and (*S*)-alanine.

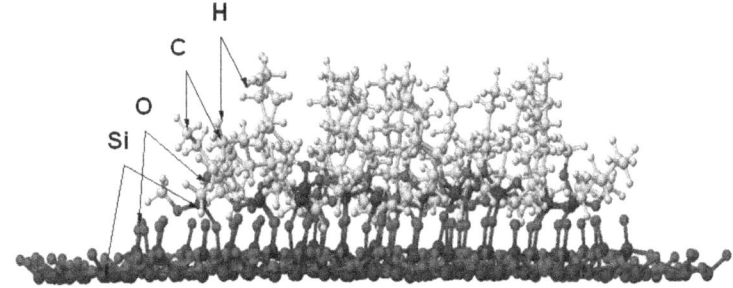

Figure 8.14   Structure of the model dimethylpentylsilane-bonded silica gel. Small white, light-gray, gray, and black balls, represent hydrogen, carbon, silicon, and oxygen, respectively.
Reproduced by permission of BioChem Press, ref. 27.

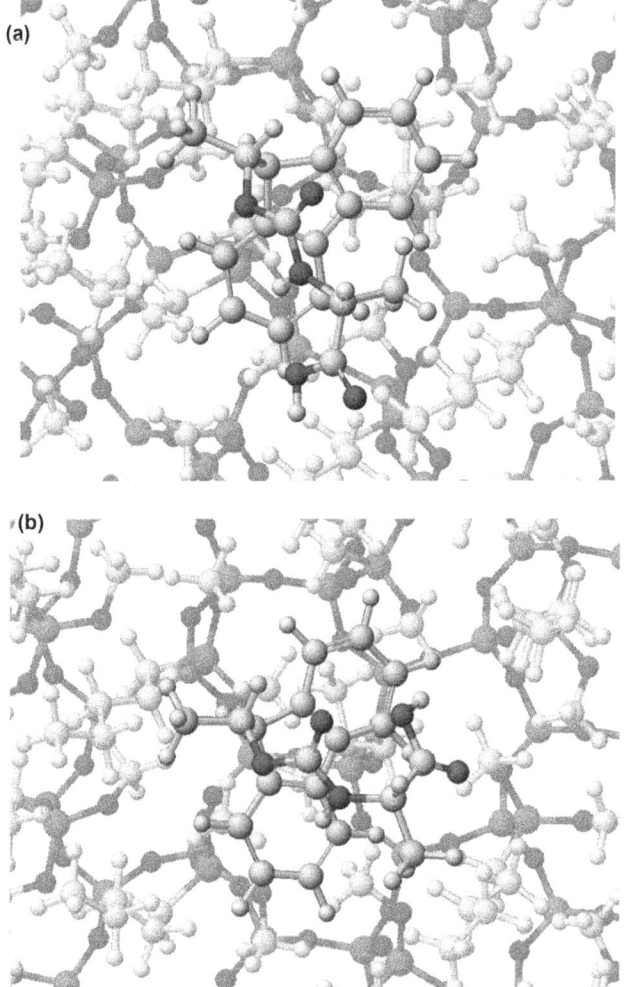

Figure 8.15   Adsorption of (a) FLE-(*R*)-alanine and (b) FLE-(*S*)-alanine on a
dimethylpentylsilane-bonded silica gel.
Reproduced by permission of BioChem Press, ref. 27.

respectively. Dimethylpentyl groups are in close proximity due to their
hydrophobic interactions. Some of them lie in free space after the molecular
interaction is optimized due to their steric hindrance.

The calculated energy values of the amino acid enantiomers derivatized
with (+)-1-(9-fluorenyl)ethyl chloroformate using the MM2 program are listed Tables 8.12 and 8.13 with chromatographic data measured in reversed-
phase from ref. 18. The energy values of individual complexes between the
model phase and an enantiomer are listed Table 8.13.

In this bonded phase, the dimethylpentyl groups surrounded one trimethyl group. Silanol groups around the trimethyl group were completely

Table 8.12 Energy values, separation factors, and capacity ratios (k) for amino acid diastereomers. Reproduced by permission of BioChem Press, ref. 27.

Amino acid	log k	α	fs	hb	es	vw
(R)-Alanine	0.751	1.10	−34.734	−15.460	−11.495	4.861
(S)-Alanine	0.792		−37.268	−17.645	−12.230	5.466
(R)-Arginine	0.549	1.14	−50.046	−14.696	−33.766	6.954
(S)-Arginine	0.605		−52.529	−8.815	−34.164	5.952
(R)-Aspargine	0.127	1.09	−65.290	−20.718	−41.294	7.529
(S)-Aspargine	0.164		−66.426	−23.673	−40.734	7.351
(R)-Aspartic acid	0.375	1.07	−60.000	−21.349	−26.546	7.460
(S)-Aspartic acid	0.405		−59.704	−24.120	−26.975	6.120
(R)-Glutamic acid	0.513	1.13	−49.242	−19.914	−19.216	4.161
(S)-Glutamic acid	0.567		−48.433	−17.487	−18.687	6.944
(R)-Glutamine	0.111	1.13	−55.929	−15.865	−30.937	6.201
(S)-Glutamine	0.164		−54.911	−17.048	−30.485	6.926
(R)-Proline	0.741	1.00	−15.360	−3.649	−10.543	7.054
(S)-Proline	0.741		−15.683	−3.735	−10.490	6.595
(R)-Serine	0.303	1.04	−35.288	−20.927	−7.579	6.052
(S)-Serine	0.320		−33.205	−19.384	−7.165	5.786
(R)-Threonine	0.468	1.13	−31.208	−11.589	−7.939	5.806
(S)-Threonine	0.520		−38.323	−24.938	−8.073	6.657
Model phase	—	—	−648.713	0.000	−403.448	−400.646

covered by alkyl groups, therefore the silanol group effect might not be a factor. The first circle of dimethylpentyl groups might not be pushed down by the presence of an analyte. The second circle of dimethylpentyl groups should support the first circle. The energy values of the amino acid enantiomers derivatized with (+)-1-(9-fluorenyl)ethyl chloroformate, calculated using the MM2 program are listed in Table 8.12, with chromatographic data measured in reversed-phase liquid chromatography from ref. 12. The energy values of individual complexes between the model pentyl phase and enantiomers are listed in Table 8.13.

For docking between an analyte and the model phase, the most hydrophobic site of the analyte faced towards the hydrophobic model phase, whose silanol activity was eliminated. The molecular interaction energy was calculated from the following equation. MI = energy value of an analyte + energy value of the model phase − energy value of a complex. The molecular interaction energy values of the final structure were related to the capacity ratios (k) of the amino acid derivatives. The correlation coefficient between MIFS and log k is given in the following equation, and the relationship is shown in Figure 8.16.

$$\text{MIFS} = 22.127 \times (\log k) + 20.692, \; r = 0.938, n = 14.$$

However, the basic amino acids (R)- and (S)-asparagine and (R)- and (S)-glutamine were not included in the above correlation due to lack of pH effects. The guanidino and amino groups of arginine and glutamine should

Table 8.13 Energy values and molecular interaction energies for complexes of amino acids with a model pentyl phase.

Amino acid	FS	HB	ES	VW	MIFS	MIHB	MIES	MIVW
(R)-Alanine	−716.373	−15.449	−414.971	−424.942	32.926	−0.011	0.028	19.435
(S)-Alanine	−718.915	−16.922	−415.852	−424.892	32.934	−0.723	0.174	18.780
(R)-Arginine	−731.092	−12.143	−438.115	−423.195	32.333	−2.553	0.901	29.503
(S)-Arginine	−736.621	−16.415	−437.406	−426.794	35.379	7.600	−0.206	32.100
(R)-Aspargine	−745.840	−25.170	−444.725	−421.302	31.837	4.452	−0.017	28.185
(S)-Aspargine	−748.306	−23.438	−444.369	−422.947	33.167	−0.235	0.187	29.652
(R)-Aspartic acid	−736.477	−18.916	−429.722	−419.922	21.138	−2.433	−0.272	26.736
(S)-Aspartic acid	−737.038	−17.503	−430.095	−425.358	28.616	−6.617	−0.328	30.832
(R)-Glutamic acid	−729.851	−11.721	−422.989	−426.524	31.896	−7.419	0.325	30.039
(S)-Glutamic acid	−731.292	−22.132	−422.767	−424.641	34.146	4.645	0.632	30.939
(R)-Glutamine	−738.278	−21.341	−436.870	−424.949	33.636	5.476	2.485	30.504
(S)-Glutamine	−738.620	−14.028	−434.048	−432.012	34.996	−3.020	0.115	38.292
(R)-Proline	−701.278	−3.615	−413.991	−429.290	37.205	0.011	0.000	35.698
(S)-Proline	−703.224	−3.660	−413.679	−431.679	38.828	−0.075	−0.259	37.437
(R)-Serine	−710.131	−18.668	−410.908	−421.484	26.190	−2.259	−0.047	26.890
(S)-Serine	−710.505	−24.279	−409.569	−418.956	28.587	4.895	−1.044	24.096
(R)-Threonine	−712.766	−14.450	−411.557	−422.818	32.845	2.861	0.170	27.978
(S)-Threonine	−719.963	−24.770	−411.591	−423.652	32.927	−0.168	0.070	29.663

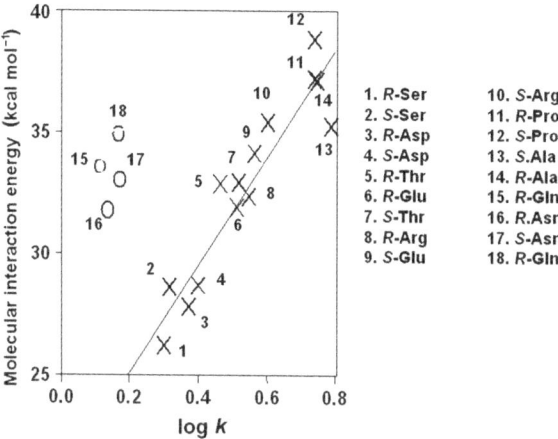

Figure 8.16   The relationship between molecular interaction energy and log k for amino acids on a dimethylpentylsilane-bonded silica gel.

be ionized and their retention time was short.[17] A one-to-one molecular interaction model might be ideal for chromatography using a Pirkle-type bonded phase, however this model can also be applied to the reversed-phase liquid chromatography of enantiomers, such as phenolic compounds.

Under these liquid chromatographic conditions, the elution order of amino acids was Asn < Gln < Ser < Asp < Thr < Glu < Arg < Ala < Pro. The interaction energy values of Asn and Gln, however, were high. The elution order of amino acids derivatized with orthophthaldialdehyde was Asp < Glu < Cys < Asn < Ser < Gln < Thr < Arg < Ala.[19] The difference in elution order depends on the organic modifiers used in the eluent. The former system[12] used tetrahydrofuran, and the latter system used acetonitrile and methanol. Tetrahydrofuran is a selective organic modifier in reversed-phase liquid chromatography. The elution order of analytes did not have a linear relationship with the octanol–water partition coefficient (log P) in reversed-phase liquid chromatography using tetrahydrofuran as an organic modifier.[20] Acetonitrile is the most suitable organic modifier for obtaining a high correlation coefficient between log k and log P. Therefore, if the purpose is to predict the elution order, the capacity ratios should be measured in reversed-phase liquid chromatography using acetonitrile as the organic modifier. In the chromatographic results, (R)-amino acids eluted first followed by the (S)-amino acids, however the interaction energy value of (S)-alanine was smaller than that of (R)-alanine (Table 8.12). More precise calculations obtained using a better model phase and solvent effect might resolve this discrepancy. Even though the solvent effect is not predicted by the above calculations, computational chemical simulation of reversed-phase liquid chromatography is useful for determining enantioseparation.

The capacity ratios and the separation factors of amino acid enantiomers derivatized with (+)-1-(9-fluororenyl)ethyl chloroformate measured in isocratic elution on reversed-phase liquid chromatography were related to energy value changes between an analyte and a model phase. The energy values of their final structures correlated well to the log k values of the amino acid derivatives.

8.4 Enantiomer Recognition of Derivatized β-Cyclodextrin

The inclusion complexes formed between β-cyclodextrin (β-CD) and various optically active solutes (drugs and herbicides, Table 8.14) were modeled and refined using molecular modeling methods. The interaction energies of the complexes formed were calculated for both enantiomers and were correlated with the experimental retention data measured by normal-phase HPLC.[21] A model cyclodextrin containing 441 atoms, 469 bonds, and 3513 connectors

Table 8.14 The optically active solutes (drugs and herbicides) used to form complexes with a cyclodextrin model phase. Reproduced by permission of Taylor and Francis, ref. 28.

No.	Analyte
1	1,1′-Bi-2-naphthol
2	2,2-Dimethoxy-1,1′-binaphthalene
3	2-Naphthylethanol
4	*trans*-2-Phenyl-3-(4-chlorophenyl)-oxirane
5	*trans*-2,3-Diphenyloxirane
6	*N*-(3,5-Dinitrobenzoyl)-2-aminomethylpropanoate
7	*N*-(3,5-Dinitrobenzoyl)-2-amino-3-methylmethylbutanoate
8	*N*-(3,5-Dinitrobenzoyl)-2-amino-3phenylmethylpropanoate
9	*N*-(3,5-Dimethylbenzoyl)-2-aminomethylpropanoate
10	9-Anthryl-2,2,2-trifluoroethanol
11	*N*-(3,5-Dinitrobenzoyl)-1-phenylethanamine
12	*O*-(3,5-Dinitrobenzoyl)-1-(2-naphthyl)-ethanol
13	*N*-(3,5-Dinitrobenzoyl)-1-(1-naphthyl)-ethanamine
14	Troger base
15	Butylfluazifop
16	Ethoxyethylhaloxyfop
17	Ethoxyethylchlorazifop
18	Methyldichlofop
19	Propranolol
20	Indapamide
21.	Phenoxypropionic acid
22	2-Chlorophenoxypropionic acid
23	3-Chlorophenoxypropionic acid
24	4-Chlorophenoxypropionic acid
25	Dichloprop
26	Silvex
27	Mecoprop
28	Ibuprofen
29	Mandelic acid

was first constructed as a model phase, and the molecular interaction energy between this model phase and a standard compound was calculated using the MM2 module from version 5 of the CAChe™ program.

The calculations were performed with the small opening 'locked' using phenylcarbamate groups, while the large opening of the CD ring was unlocked, allowing for inclusion of the guest solute. Figure 8.17 shows an example of the optimized analyte structure ('6R') and a partly locked derivatized CD complex. A completely unlocked chiral stationary phase, in which the structure is deformed, was not suitable for this approach, and the initial condition before docking affected the final structure. The complex conformations of (R)-N-(3,5-dinitrobenzoyl)-2-amino-methylpropanoate and (S)-N-(3,5-dinitrobenzoyl)-2-amino-methylpropanoate are shown in Figure 8.18(a) and (b). These structures indicated that the twisted structure may contribute to the different fitting, where the analytes screwed into the cyclodextrin. The optimized (spiral) structures of (R)-N-(3,5-dinitrobenzoyl)-2-amino-methylpropanoate and (S)-N-(3,5-dinitrobenzoyl)-2-amino-methylpropanoate, with electron density are shown in Figure 8.19.

The calculated energy values are summarized in Table 16 of the Appendix (p. 313). The calculated values were the lowest energy values of each compound and complex. MIHB, MIES, and MIVW were used to study the contribution to the molecular interaction. As a preliminary study, we compared the different interaction energy values for each enantiomer couple with its selectivity (α).

A comparison of the calculated interaction energy values with the measured capacity ratios and selectivities indicated that the best correlation was observed between MIFS and α (selectivity), as shown in Table 17 of the Appendix (p. 314). Thus, when MIFS is significant, enantioseparation occurs regardless of the values of the other energy parameters. On the other hand,

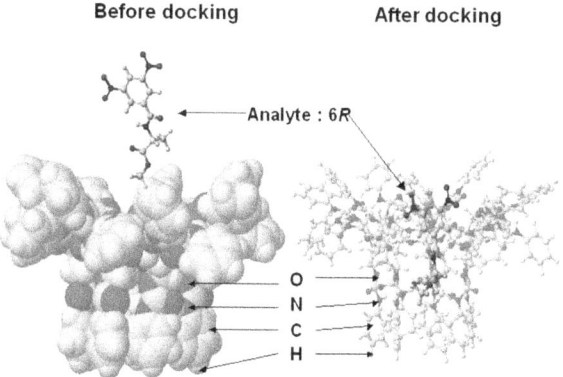

Figure 8.17   Optimized structure of an analyte (6R) and partly locked derivatized cyclodextrin complex. Small white, gray, dark-gray, and black balls represent hydrogen, carbon, nitrogen, and oxygen, respectively. Reproduced by permission of BioChem Press, ref. 27.

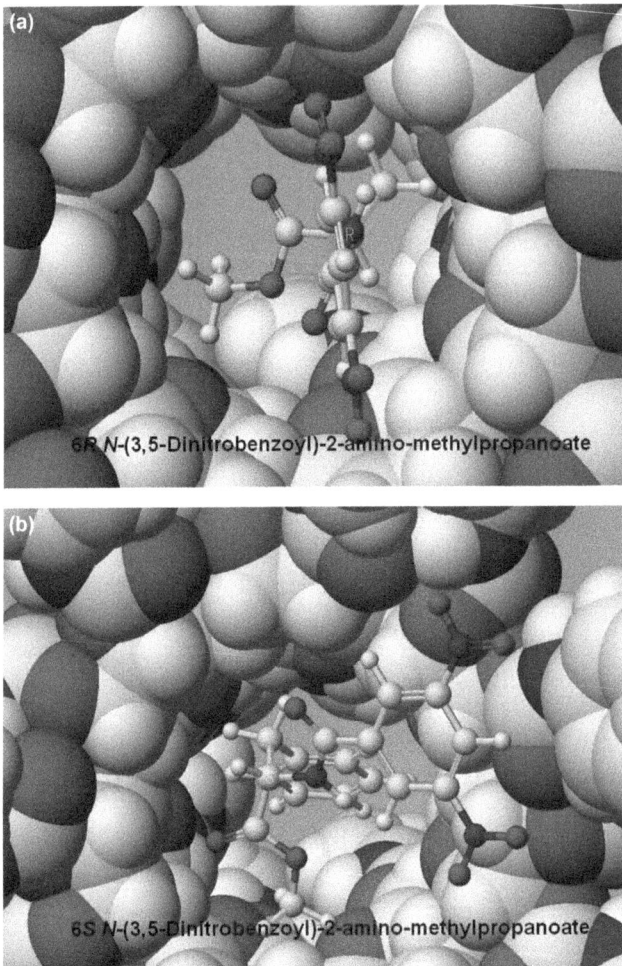

Figure 8.18   Top view of complexes of (a) (*R*)-*N*-(3,5-dinitrobenzoyl)-2-amino-methylpropanoate and (b) (*S*)-*N*-(3,5-dinitrobenzoyl)-2-amino-methyl-propanoate Small white, light-gray, dark-gray, and black balls represent hydrogen, carbon, nitrogen, and oxygen, respectively. Reproduced by permission of Taylor and Francis, ref. 28.

when both enantiomer complexes give close MIFS values, no resolution is observed in HPLC, even if MIHB, MIES, and MIVW are quite different (*e.g.*, Solutes 14 and 16). This correlation explains why we considered the MIFS parameter to be the most significant molecular interaction energy value. We noticed that the greater the difference between the MIFS values, the higher the measured calculated α (selectivity) for most racemates. To evaluate the threshold limit of enantioseparation in terms of MIFS energy, we correlated the difference (ΔMIFS) for each couple with its α (selectivity) value, as shown in Table 17 of the Appendix (p. 314).

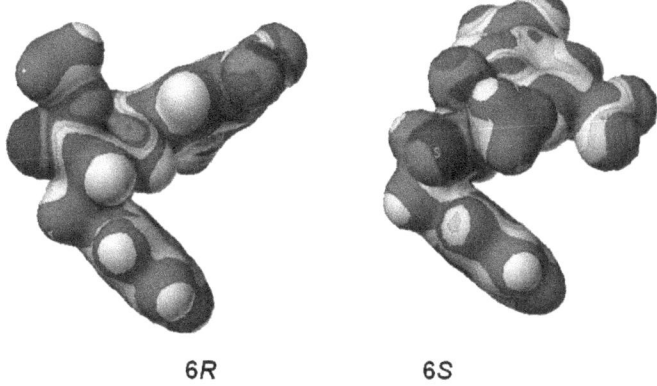

6R   6S

N-(3,5-dinitrobenzoyl)-2-amino-methylmethylpropanoate

Figure 8.19 Optimized structures of (R)-N-(3,5-dinitrobenzoyl)-2-amino-methyl-propanoate and (S)-N-(3,5-dinitrobenzoyl)-2-amino-methylpropanoate showing electron density.

For neutral compounds (Solutes 1–19), for example, among the 19 compounds tested on the studied chiral stationary phase, 17 had a good correlation between ΔMIFS and α (selectivity). When ΔMIFS was greater than 0.29, enantioseparation occurred ($\alpha > 1$). When ΔMIFS was less than 0.29, no separation occurred ($\alpha = 1$). For solutes 10 and 16, the same ΔMIFS value was obtained, and while the 12th couple was separated, the 19th couple was not. Thus, the ΔMIFS limit value seems to be approximately 0.29, which was confirmed with Solute 6 (ΔMIFS = 0.436, $\alpha = 1.09$).

A similar conclusion was drawn for acidic compounds (Solutes 20–28): among the nine analytes studied, eight compounds had a good correlation between ΔMIFS and α. The only exception was Solute 26, for which ΔMIFS = 2.3447, and no separation occurred. The threshold value ΔMIFS should be in the range 0.5 to 1.2. For this type of solute, the effect of the solvent on the calculations is very important. The obtained energy values were too high and unrealistic; even after more than 10 conformation interactions, the calculations were not believable. More comprehensive results were obtained when the effect of trifluoroacetic acid complexed with the analyte was taken into account. Figure 8.20 shows examples of optimized structures of mandelic acid ('29R' and '29S') and trifluoroacetic acid (TFA) complexes with their electron density, demonstrating the different types of screws. Figure 8.21 shows the initial condition before docking and after docking of a complex of (R)-mandelic acid–trifluoroacetic acid with the derivatized β-cyclodextrin. The optimized (R)- and (S)-mandelic acid–trifluoroacetic acid complexes inside the derivatized β-cyclodextrin are shown in Figure 8.22.

Molecular mechanic calculations can be used to predict elution order. MM2 calculations, however, do not provide information about the nature of the

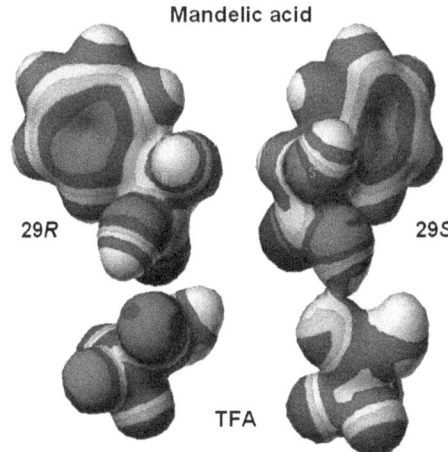

Figure 8.20 Optimized structures of (*R*)- and (*S*)-mandelic acid–trifluoroacetic acid (TFA) complexes showing electron density.
Reproduced by permission of Taylor and Francis, ref. 28.

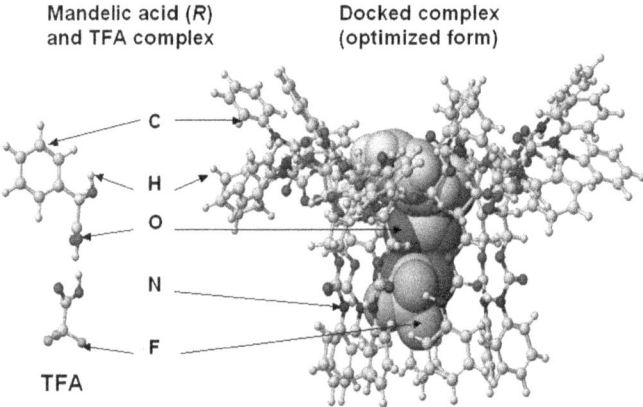

Figure 8.21 Optimized structures of initial and docked (*R*)-mandelic acid–trifluoroacetic acid, and its complex with derivatized β-cyclodextrin. White, light-gray, dark-gray, gray, and black balls represent hydrogen, carbon, nitrogen, fluorine, and oxygen, respectively.

selectivity nor the absolute magnitude of the selectivity. The hydrophobic effect is a major driving force in the formation of inclusion complexes. Molecular modeling results suggest that the CD cavity is in fact highly polar and well suited for coordinating with the ammonium and carboxylic groups of amino acids.[22] This β-CD chiral stationary phase does not appear to have a specific interaction site inside its cavity, and the molecular interaction likely occurs inside the cavity. The most important function should be steric hindrance. For enantioseparation on CD derivatives, molecular mechanics calculations

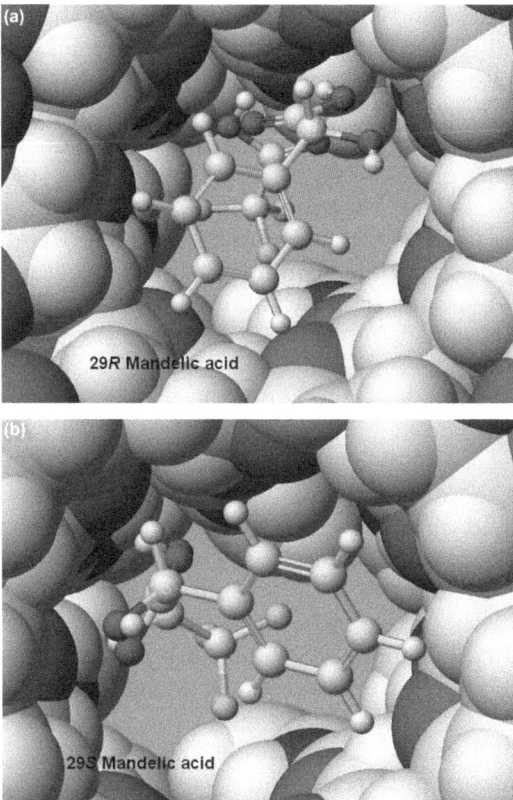

Figure 8.22 Optimized structures of (a) (*R*)- and (b) (*S*)-mandelic acid– trifluor-oacetic acid inside the derivatized β-cyclodextrin.

indicate that van der Waals interactions are the most reliable for both association and chiral discrimination.[23–25]

The properties of CD multimodel inclusion complexes were correctly reproduced by MM2 calculations. This method allows us to correctly determine the differences in the complexation of enantiomeric forms. Special care should be taken when selecting the molecules to model, and while analyzing the final geometries and interactions.[26]

References

1. Y. Dobashi, A. Dobashi, H. Ochiai and S. Hara, New, rational molecular design for chiral recognition involving application of dual hydrogen bond association, *J. Am. Chem. Soc.*, 1990, **112**, 6121–6123.
2. Y. Dobashi, S. Hara and Y. Iitaka, Dual hydrogen bond association of (*R,R*)-*N,N'*-diisopropyltartamide with (*S,S*)-9,10-dimethyl-9,10-dihydro-phenanthrene-9,10-diol, *J. Org. Chem.*, 1988, **53**, 3894–3896.

3. N. Oi, H. Kitahara and R. Kira, Elution orders in the separation of enantiomers by high-performance liquid chromatography with some chiral stationary phases, *J. Chromatogr.*, 1990, **535**, 213–227.

4. N. Oi, Recent progress of chiral stationary phases for HPLC, in *Advances in Liquid Chromatography*, ed. T. Hanai and H. Hatano, World Scientific, Singapore, 1996, pp. 213–230.

5. T. Hanai, H. Hatano, N. Nimura and T. Kinoshita, Computational chemical analysis of the separation of derivatized (*R*)- and (*S*)-amino acid enantiomers of *N*-(*tert*-butylaminocarbonyl)-(*S*)-valylaminopropylsilica gel and (*R*)-1-(α-naphthyl)ethylaminocarbonyl-glycylaminopropylsilica gel by liquid chromatography, *J. Liq. Chromatogr. Relat. Technol.*, 1996, **19**, 1189–1204.

6. T. Hanai, H. Hatano, N. Nimura and T. Kinoshita, Analysis of chemically bonded silica gel by computational chemistry, *J. Liq. Chromatogr.*, 1993, **16**, 109–114.

7. N. Oi, H. Kitahara and F. Aoki, Enantiomer separation by high-performance liquid chromatography with copper(II) complexes of Shiff bases as chiral stationary phases, *J. Chromatogr.*, 1993, **631**, 177–182.

8. T. Hanai, H. Hatano, N. Nimura and T. Kinoshita, Computational chemical analysis of the chiral recognition of binuclear copper(II) of *N*-salicylidene (*R*)-2-amino-1,2-*bis*(2-butoxy-5-*tert*.butylphenyl)-3-phenyl-1-propanol in liquid chromatography, *J. Liq. Chromatogr.*, 1994, **17**, 4327–4334.

9. N. Nimura, A. Toyama, Y. Kasahara and T. Kinoshita, Reversed-phase liquid chromatographic resolution of underivatized D,L-amino acids using chiral eluents, *J. Chromatogr.*, 1982, **239**, 671–675.

10. D. W. Aswad, Determination of D- and L-aspartate in amino acid mixtures by high-performance liquid chromatography after derivatization with a chiral adduct of *o*-phthaldialdehyde, *Anal. Biochem.*, 1984, **137**, 405–409.

11. N. Nimura and T. Kinoshita, *o*-Phthalaldehyde-*N*-acetyl-L-cysteine as a chiral derivatization reagent for liquid chromatographic optical resolution of amino acid enantiomers and its application to conventional amino acid analysis, *J. Chromatogr.*, 1986, **352**, 169–177.

12. S. Einarsson, B. Josefsson, P. Moller and D. Sanchez, Separation of amino acid enantiomers and chiral amines using precolumn derivatization with (+)-1-(9-fluorenyl)ethyl chloroformate and reversed-phase liquid chromatography, *Anal. Chem.*, 1987, **59**, 1191–1195.

13. R. Schuster, Determination of amino acids in biological, pharmaceutical, plant and food samples by automated precolumn derivatization and high-performance liquid chromatography, *J. Chromatogr.*, 1988, **432**, 271–284.

14. C. Carducci, M. Birarelli, V. Leuzzi, G. Santaga, P. Serafini and I. Antonozzi, Automated method for the measurement of amino acids in urine by high-performance liquid chromatography, *J. Chromatogr., A*, 1996, **729**, 173–180.

15. H. Bruckner and M. Wachsmann, Design of chiral monochloro-S-trizine reagents for the liquid chromatographic separation of amino acid enantiomers, *J. Chromatogr. A.*, 2003, **998**, 73–82.
16. N. Todoroki, K. Shibata, T. Yamada, Y. Kera and R.-H. Yamada, Determination of *N*-methyl-D-spartate in tissues of bivalves by high-performance liquid chromatography, *J. Chromatogr., B*, 1999, **728**, 41–47.
17. T. Hanai, New developments in liquid chromatographic stationary phases, *Adv. Chromatogr.*, 2000, **40**, 315–357.
18. S. Einarsson, B. Josefsson, P. Moller and D. Sanchez, Separation of amino acid enantiomers and chiral amines using precolumn derivatization with (+)-1-(9-fluorenyl) ethyl chloformate and reversed-phase liquid chromatography, *Anal. Chem.*, 1989, **59**, 1191–1195.
19. Y. Ishida, Amino acids, in *Liquid Chromatography in Biomedical Analysis*, ed. T. Hanai, Elsevier, Amsterdam, 1991, pp. 47–80.
20. T. Hanai, *HPLC: A Practical Guide*, RSC Publishing, Cambridge, 1999, pp. 57–65.
21. F. Tazerouti, A. Y. Badjah-Hadj-Ahmed, B. Y. Meklati, F. Pilar and C. Minguillon, Enantiomeric separation of drugs and herbicides on a β-cyclodextrin-bonded stationary phase, *Chirality*, 2002, **14**, 59–66.
22. J. Ramirez, S. Ahn, G. Grigorean and C. B. Lebrilla, Evidence for the formation of gas-phase inclusion complexes with cyclodextrins and amino acids, *J. Am. Chem. Soc.*, 2000, **122**, 6884–6890.
23. M-Y. Nie, L.-M. Zhou, Q.-H. Wang and D.-Q. Zhu, Enantiomer separation of mandelates and their analogs on cyclodextrin derivative chiral stationary phases by capillary GC, *Anal. Sci.*, 2001, **17**, 1183–1187.
24. C. Bicchi, C. Brunelli, G. Cravotto, P. Rubiolo, M. Galli and F. Mendicuti, Cyclodextrin derivatives in enantiomer GC separation of volatiles. Part XXI: Complexation of some terpenoids with 2-*O*-acetyl-3-*O*-methyl- and 2-*O*-methyl-3-*O*-acetyl-6-*O*-*t*-hexyldimethylsilyl-γ-cyclodextrins: Molecular mechanics and molecular dynamics, *J. Sep. Sci.*, 2003, **26**, 1479–1490.
25. C. Wensheng, Y. Yanmin and S. Xueguang, Chiral recognition of aromatic compounds by β-cyclodextrin based on bimodal complexation, *J. Mol. Model.*, 2005, **11**, 186–193.
26. F. Perez, C. Jaime and X. Sanchez-Ruiz, MM2 calculations on cyclodextrins: Multimodal inclusion complexes, *J. Org. Chem.*, 1995, **60**, 3840–3845.
27. T. Hanai, Computational chemical analysis of enantiomer separations of derivatized amino acids in reversed-phase liquid chromatography, *Internet Electronic J. Molecular Design*, 2004, **3**, 379–386.
28. F. Tazerouti, A. Y. Badjah-Hadj-Ahmed and T. Hanai, *J. Liq. Chromatog. Rel. Technol.*, 2007, **30**, 3043–3057.
29. Computational chemical analysis of chiral recognition in liquid chromatography, selectivity of *N*-(*R*)-1-(α-naphthyl)ethylamino carbonyl-(*R* or *S*)-valine and *N*-(*S*)-1-(α-naphthyl)ethylamino carbonyl-(*R* or *S*)-valine bonded aminopropyl silica gels, *Anal. Chim. Acta*, 1996, **332**, 213–224.

Human Serum Albumin–Drug Binding Affinity Based on Liquid Chromatography

9.1 Introduction

The discovery of new drugs has been accelerated by combinatorial chemistry. The fast screening of drug candidates is very important, and the octanol–water partition coefficient ($\log P$) and the dissociation constant (pK_a) are easily measured. Human serum albumin (HSA) is a 66 500 Da protein, is the most common and abundant plasma protein, and is considered to be a multifunctional plasma transport protein. It constitutes approximately 60% of the total serum protein and is a small globular protein with a high electrophoretic mobility; it acts to maintain homeostasis in the body, providing a protein reservoir. HSA displays the property of conformational adaptability, which allows the binding of ligands including bilirubin, fatty acids, tryptophan, and many drugs,[1] and binding to HSA can prolong the *in vivo* half-life of a drug. The physiological function of HSA does not depend on specific interactions but on the broad non-specific physico-chemical character of the protein. The binding of lead compounds as drug candidates with HSA represents a major challenge in drug discovery while a degree of albumin binding may be desirable in helping to solubilize compounds. An excessively high affinity of a drug for a protein (>95% bound) requires correspondingly high doses to achieve an effective concentration *in vivo*. Thus, the binding of drugs to HSA is one of the most important factors determining their pharmacokinetics.

HSA is the most abundant plasma protein, and accounts for the entire drug binding in plasma.[2–4] The binding of drugs to HSA influences

RSC Chromatography Monographs No. 19
Quantitative *In Silico* Chromatography: Computational Modelling of Molecular Interactions
By Toshihiko Hanai
© Toshihiko Hanai, 2014
Published by the Royal Society of Chemistry, www.rsc.org

drug pharmacokinetics and causes pharmacological effects.[5] The pharmacological effect is directly related to the free, rather than the total, concentration of a drug in plasma, Since drug binding to HSA is readily reversible, the drug–HSA complex serves as a circulating drug reservoir that releases more of a drug as the free drug is biotransformed or excreted. HSA binding thus decreases the maximum intensity, but increases the duration of action of many drugs. The drug–HSA interaction is influenced by the concentration of the drug, and that of HSA. Changes in the free fraction of a drug in the serum can have major clinical consequences.

Under physiological conditions, glucose reacts non-enzymatically with a wide variety of proteins to form stable adducts.[6–8] It has been hypothesized[9] that non-enzymatic glycosylation (NEG) of albumin may induce changes in its chemical, physical, and ultimately, biological properties, leading eventually to the pathological sequence of diabetes. This hypothesis is supported by a growing body of evidence from clinical experience[10,11] and animal model studies,[12] which indicate that chronic, subclinical hyperglycemia, rather than insulin deficiency itself, may be the major factor contributing to the progressive, secondary complications of diabetes.

Although a direct correlation between an increase on glycosylated proteins and the pathophysiology of diabetes has not been found, alterations to the biochemical properties of HSA subsequent to glycosylation have been described. Early glycosylation products will change the isoelectric point of the protein. As a result of these changes to its charge, the molecule can undergo conformational changes.[13] Glycosylated HSA (GHSA) is catabolized faster and accumulated more rapidly by the liver than native HSA[14] and is also deposited on the walls of the microvasculature.[15] NEG of HSA is elevated two- to three-fold in diabetes[16] and substantial evidence suggests that the end-products of advanced NEG may be responsible for many of the long-term complications associated with the disease.[6,17] Thus, glycosylation seems to play a role in both, aging and the pathological complications of patients with diabetes mellitus.

NEG of HSA has a severe effect on the structure of proteins, changing their functional and biological properties. For example, glycosylation is reported to reduce the binding of *cis*-parinaric acid and bilirubin[18] and some drugs[18–22] whilst hemin[18] and palmitate[23] remained unaltered. NEG of HSA may also alter its drug-binding capacity and influence the distribution and elimination of compounds, as well as the duration and intensity of their pharmacologic effect.[24–27] The glycosylation of binding sites or regions may alter the capability of HSA.

When glycosylation occurs, the binding affinity of HSA decreases, perhaps due to a conformational change or steric hindrance. Advanced Maillard reaction products increase the binding affinity compared to native HSA, due to their hydrophobicity. That is, the amount and degree of GHSA must influence the binding of drugs to HSA. Quantitative analyses of advanced GHSA, and the identification of the structure of advanced GHSA will elucidate the effect of glycosylation on drug binding.

How glycosylation inhibits or accelerates drug binding is an important subject for therapeutic purposes. New knowledge about NEG of HSA may lead to significant advances in both diagnosis and therapy. The effect of the NEG of HSA on the binding of other strongly bound ligands, among these a great number of drugs, has apparently not yet been examined. However, no one technique can handle all requirements to measure protein binding, and a suitable technique must be selected for a given protein–drug pair, mainly due to the solubility of drugs. Numerous methods such as equilibrium dialysis, spectroscopic methods, ultra-filtration, ultra-centrifugation, liquid chromatography and electrophoresis have been used to study the binding of drugs to proteins.

Equilibrium dialysis and ultra-filtration have been used as standard methods to measure protein–drug binding, because of their simplicity and general applicability to many different systems *in vitro* and *in vivo*. Equilibrium dialysis is based on the establishment of an equilibrium state between a protein compartment and a buffer compartment that are separated by a membrane, which is permeable only to a low-molecular-weight ligand. This method is the theoretically the most accurate way to determine free and bound ligands because the equilibrium is not shifted when aliquots are taken from both sides of the dialysis membrane. However, the disadvantage of equilibrium dialysis is the long equilibrium time, volume shifts, Donnan effects, hindering of the passage of free ligand, non-specific adsorption to dialysis apparatus and difficulty in the control of the pH of the dialysate.[28]

Ultra-filtration with semipermeable membranes has also been used for routine free drug monitoring in clinical laboratories; it is a simple and fast method and involves no dilution effects or volume shifts. The stability of the binding equilibrium during the separation process and the necessity of a large number of data points are problematic aspects of this method. It is therefore advisable to validate, especially in case of low-affinity interactions, the basic assumption that the binding ratio of protein-bound drugs to free drugs remains constant.

The development of chromatographic technology has allowed shorter analysis times, the consumption of fewer chemicals, and higher precision and reproducibility than conventional techniques, such as equilibrium dialysis and ultra-filtration. It also provides the possibility to detect very small differences in the binding affinity of ligands. Several variants of high-performance size-exclusion chromatographic techniques for binding interactions have been developed, *i.e.*, the Hummel–Dreyer method,[29,30] frontal analysis,[31,32] the vacancy peak method,[33] retention analysis[34] and the immobilized protein column method.[35] The measurement of retention time using an immobilized HSA column seemed simple, but the capacity ratios did not correlate well with HSA–drug binding affinity measured by free solution methods. The method requires a specific standard for the measurement of the pharmacokinetics of new chemicals.

The measurement of the percentage concentration of bound drugs is simple, but how this can replace the orthodox binding affinity log nK value is not clear. Computational chemical prediction methods require further study before any practical applications. In addition, the measurement of the percentage concentration of bound drug is simple, but the type of value that can reflect the real pharmacokinetic value is more important than a simple measurement.

Capillary electrophoresis and ultra-centrifugation methods are also not sufficient due to poor reproducibility and the requirement for a large quantity of proteins, respectively. These are fundamental problems in protein–drug binding measurements. Reference log nK values vary significantly, probably due to the different qualities of HSA and the different analytical systems used.

Drug–albumin binding sites have been studied, but albumin also functions as a scavenger. This indicates that the albumin structure has the flexibility to carry a variety of compounds and the affinity effect may not be specific. The main binding forces are hydrophobic interactions and ion–ion interactions, and specific steric effects may not be important. Ion-exchange liquid chromatography with newly developed ion exchangers that mimic ion exchangers of HSA seems to be practical. Acidic drug–HSA and basic drug–HSA binding affinities have been successfully determined by a combination of reversed-phase and ion-exchange liquid chromatography.[36,37] The guanidyl groups of arginine should work as anion-exchange groups, and the carboxyl groups of aspartic and glutamic acids should work as cation-exchange groups. The chromatographic behavior of acidic and basic drugs was studied using these columns, and their capacity ratios were correlated with their log nK values measured by the modified Hummer–Dreyer method. A pentyl-bonded silica gel was more stable than a butyl-bonded silica gel,[38] and was a simpler method for estimating the binding affinity compared to the previous study with a butyl-bonded silica gel column.[37] Such simple liquid chromatography may be useful for measuring the albumin–drug binding affinity without albumin. However, a faster analytical method is required. The retention times of acidic drugs were measured using a guanidino phase with pH-controlled eluent to determine their molecular form in pH 7.4 eluent. Their log nK values were investigated with a computational chemical analysis using a molecular mechanics calculation program (MM2).

9.2 Measurement of HSA–Drug Binding Affinity using HSA

The measurement of HSA–drug binding affinity (log nK) is time consuming. Ways of shortening the analytical period have been extensively studied. Equilibrium dialysis requires a large amount of HSA and a long time. A frontal analysis requires a large amount of HSA too. The Hummel–Dreyer method requires little HSA. The scaling down of these methods reduced the amount of HSA, but not the time needed. Using an ultra-centrifuge shortens the analytical time, but the procedure is expensive and requires a certain amount of pure HSA. A reverse nuclear Overhauser effect (NOE) pumping technique for NMR was proposed to clarify binding and non-binding signals.[39] NMR diffusion experiments are a simple way to quantitatively describe ligand–protein interactions without prior knowledge of the number of binding sites, or the binding stoichiometry. Interligand NOE detected in the diffusion analysis of protein solution containing both ligands provided insight into the conformations adopted by these ligands while bound in common HSA-binding pockets.[40] The binding of a ^{13}C-carboxyl-labeled palminate to HSA in the presence and absence of competitor ligands, drugs, and endogenous ligands such as hemin, phenylbutazone, propofol, and diazepam, was studied using NMR spectroscopy to complete the correlation of NMR chemical shifts with specific structural binding sites.[41] A dinuclear gadolinium(III) complex of an amphiphilic chelating ligand was synthesized and evaluated as a potential magnetic resonance imaging contrast agent. The fitted value of the binding constant to HSA (K_a) was 10^4 M^{-1}. It did not exhibit necrosis avidity despite the binding to HSA. Binding to albumin is not a key necrosis-targeting property.[42] Interaction between hesperetin and HSA was revealed by spectroscopic methods. The binding affinity (K_a) was 8.11×10^4 M^{-1}. A partial unfolding of HSA in the presence of the drug was observed.[43] An aqueous solution of heptamethine cyanine was used to study the stoichiometry of the dye–albumin complex.[44]

A HSA-immobilized adsorbent was developed, and a column chromatographic method was used to measure the binding constant. An HSA-immobilized column modified using disulfiram demonstrated monitoring of disulfiram with potentially co-administered drugs, and enantioselectivity.[45] An HSA-immobilized column was used to study the thermodynamics of sulindac's binding to HSA at different temperatures,[46] and to study the binding of phenytoin. The association constants for phenytoin at the indole–benzodiazepine and digitoxin sites were 1.04 $(\pm 0.05) \times 10^4$ M^{-1} and 6.5 $(\pm 0.6) \times 10^3$ M^{-1}, respectively. Both allosteric interactions and direct binding for phenytoin appear to take place at the warfarin–azapropazone and tamoxifen sites.[47] Affinity capillary electrophoresis was also used to measure binding constants.[48]

Furthermore, the retention times of a variety of drugs were measured using an immobilized-HSA column and a proposed log K_{HSA} (ref. 49) or simply the percentage concentration of bound drugs.[50] Mimic ion exchangers were developed to measure the HSA–drug binding affinity. The log k values were compared with the HSA–drug binding affinity to evaluate practical use.[37] Computational chemical prediction was proposed *via* Quantitative Structure Activity Relationship (QSAR) analysis using 53 descriptors,[49] E-state topological structure representations,[52] or direct calculation of molecular interaction energy values using model phases.[53] In this chapter, these simple liquid chromatographic methods were analyzed by comparison with reference HSA binding affinity values measured using a free solution of HSA. The feasibility of computational chemical analytical methods was evaluated.

The modified Hummel–Dreyer method requires a small amount of protein.[54] The protein–drug binding affinity was determined using the modified Hummel–Dreyer and ultra-filtration methods[54] to obtain the reference log nK values of several drugs.[55] An example of the chromatograms obtained using the modified Hummel–Dreyer method is shown in Figure 9.1, where four chromatograms are overlapped. Examples of the Scatchard plots obtained by the Hummel–Dreyer method are shown Figure 9.2.

The experimental data obtained by the modified Hummel–Dreyer method were analyzed using Scatchard plots. Binding affinity values were calculated from a Langmuir-type equation. When a compound binds to only one binding site on a protein, the model can be represented by the following equation:

$$r = \frac{C_b}{P} = \frac{nKPC_f}{1 + KC_f} \tag{1}$$

where r is the mean number of moles of drug bound per mole of protein, C_b is the concentration of drug bound to protein, C_f is the free drug concentration, P is the total protein concentration, n is the number of binding sites, and K is the equilibrium association constant for the class of site. To determine the binding affinity from a plot of r/C_f *versus* r, non-linear regression analysis was carried out by the least-squares method. The log nK values are listed in Table 18 of the Appendix (p. 315). The measured log nK values of furosemide, naproxen, procaine, phenylbutazone, salicylic acid, sulfamethoxazole, tolbutamide and warfarin are summarized with the reference values in Table 18 of the Appendix (p. 315).

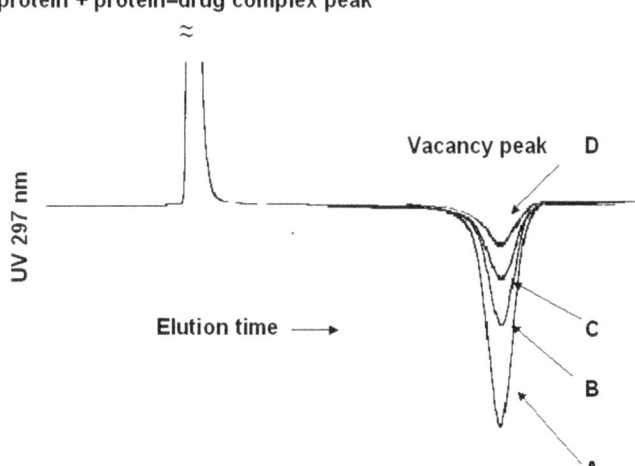

Drug concentration: (A) 50 μM, (B) 25 μM, (C) 12.5 μM, (D) 6.25 μM

Figure 9.1   Chromatograms obtained using the modified Hummel–Dreyer method. Reproduced by permission of John Wiley & Sons, ref. 55.

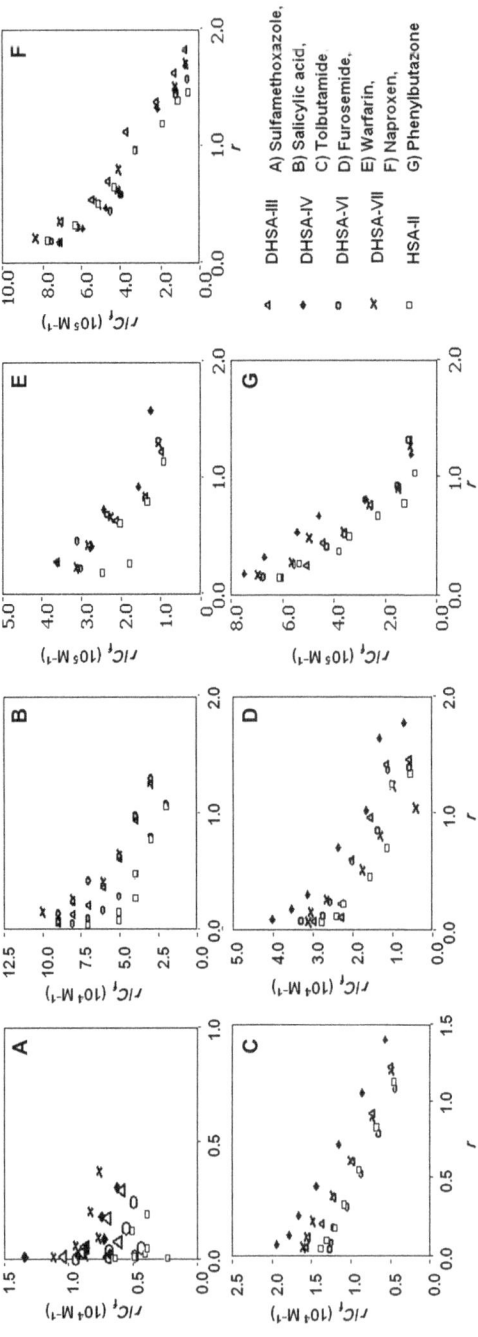

Figure 9.2 Scatchard plots obtained using the modified Hummel–Dreyer method with HSA containing various percentages of glycosylated human serum albumin (GHSA). The GHSA contents of DHSA-II, DHSA-IV, DHSA-VI, DHSA-VII, and HSA II are 33.6, 3.23, 36.9, 63.0, and 59.9%, respectively.

9.3 Measurement of HSA–Drug Binding Affinity using Liquid Chromatography

The retention time of drugs was measured using an immobilized-HSA column, and the logarithm of the capacity ratio was defined as log k_{HSA} where NaNO$_3$ was used as the t_0 (void volume) marker.[49] Another method is the measurement of capacity ratio using a guanidyl-bonded silica gel for acidic drugs and a carboxyl-bonded silica gel for basic drugs. These bonded phases were considered to be mimic ion exchangers of proteins. The logarithm of capacity ratio was correlated with the HSA–drug binding affinity measured using purified HSA.[55] The log k values, k_A and k_B, are summarized with the binding affinity measured using HSA, nK$_3$, together with nK$_1$ (log K_{HSA}), nK$_2$ (predicted log K_{HSAhsa}), nK$_4$ and nK$_4$ from the literature, PB$_1$ and PB$_2$ (the % concentration of bound drugs from the literature), and pK_a and log P values can be found in Table 18 of the Appendix (p. 315).[37,48–53,55–58]

The measurement of retention time using an immobilized-HSA column is simple if the HSA was purified before the immobilization and the purity is guaranteed, and the properties of HSA were not damaged during the immobilizing process. The log k values (nK$_1$) measured on an immobilized HSA column (ThermoHypersil, 150×4.6 mm i.d.) with 25 mM Na$_2$HPO$_4$–KH$_2$PO$_4$ buffer (pH 7.0)/acetonitrile (85:15, v/v)[49] were correlated with nK$_3$ values measured using purified HSA[55] and four sets of reference values. The correlation is shown in the following equations:

$$nK_3 = 0.593 \times (nK_1) + 4.901, \quad r = 0.198, \quad n = 9,$$

$$nK_4 = 1.980 \times (nK_1) + 5.540, \quad r = 0.731, \quad n = 9,$$

$$nK_5 = 0.753 \times (nK_1) + 3.515, \quad r = 0.884, \quad n = 9,$$

$$nK_6 = 1.309 \times (nK_1) + 5.770, \quad r = 0.756, \quad n = 11,$$

$$nK_7 = 0.103 \times (nK_1) + 4.882, \quad r = 0.058, \quad n = 34.$$

where nK$_3$ was measured by a modified Hummel–Dryer method using purified HSA (purity > 97%), and nK$_4$ and nK$_5$ were derived from the relationship between molecular interaction energy values and capacity ratios measured on a guanidyl or carboxyl-group bonded silica gel: nK$_4$ for acidic and nK$_5$ for basic drugs, respectively. The nK$_6$ values were taken from ref. 57 and nK$_7$ values were collected from the literature. nK$_6$ and nK$_7$ were measured using a free solution of HSA.

The relationship between chromatographic retention and nK$_3$ was far from perfect. The poor correlation ($r = 0.198$) indicated that that the log k values measured on an immobilized-HSA column could not be correlated with the binding affinity values measured with a modified Hummel–Dreyer method using purified HSA. Lidocaine, quinine, and scoporamin were outliers. Possible reasons are the immobilization deformed the binding sites or diminished the flexible structure of albumin, and the purity of HSA was poor. A marketed HSA contained more than 30% of glycosylated HSA, furthermore the purification was tedious. The glycosylated albumin affected the drug binding affinity.[55] Therefore, the purity of HSA should be mentioned when HSA is used directly or in the immobilized form.

The correlation was also very poor ($r = 0.058$) with nK_7 measured by free solution methods, even when only the most suitable values were selected. An acceptable correlation was obtained with nK_4 for acidic compounds measured using a guanidyl-group silica gel, and nK_5 for basic compounds measured using a carboxyl-bonded silica gel.[53] Retention also correlated with nK_6 values from ref. 57.

9.4 Measurement of HSA–Drug Binding Affinity without HSA: Part 1

Acidic drugs are highly bound to albumin. Despite the fact that albumin itself carries a net negative charge at physiological pH (pH 7.4), it is frequently proposed that albumin binds negatively charged ligands in preference to ligands carrying a positive charge. Moreover, even when reference values of binding affinity were measured by the same method, they varied due to the existence of endogenous substances such as fatty acids and bilirubin.[24–27] One reason for the variation in reference values is the existence of glycosylated albumin in HSA. The concentrations and the Maillard reaction stages of glycosylated albumin were different in all batches of commercial HSA tested.[55] Furthermore, the purity of HSA is not always mentioned with the binding affinity data of drugs. Therefore, reference values collected from literature were not suitable for studying binding mechanisms.

After the purification of a commercial HSA (Fraction V) by affinity liquid chromatography using borate resin, a purified albumin containing only 3.2% GHSA was used to measure drug binding affinities. The correlation of the binding affinity of several drugs such as sulfamethoxazole, salicylic acid, tolbutamide, furosemide, warfarin, naproxen and phenylbutazone with their respective log *P* values was examined; the correlation coefficient was 0.972 ($n = 7$).[55] Thus, the log *P* values were related to the log *k* values measured as the molecular form in reversed-phase liquid chromatography. The surface of a packing material is not like a protein, even if the three-dimensional structure of the protein is neglected in the analysis. The question arises as to why such constants measured as the ionic form were related to the log *P* values. It therefore seems that the ion–ion interaction was negligible.

The anion-exchange groups of albumin are the amino groups of lysine and the guanidino group of arginine. Their dissociation constants are 10.53 and 12.48, respectively. The amino group contributed to the glycosylation of albumin, but the guanidyl group did not.[59] The ionic strength of a propylamine-bonded phase was weaker than that of a diethylaminoethyl-bonded phase for the retention of acids.[60] Therefore, the basic study of acidic drug–protein binding cannot be performed without studying the work of the guanidino group, because the guanidyl group of arginine is an important anion-exchange group on proteins for acidic drugs. A guanidyl-group-bonded silica gel and a guanidyl-group-bonded polyvinylalcohol gel[61] were therefore used in the present study to clarify the ion–ion interaction between guanidino groups and acidic drugs in liquid chromatography. The steric effect of proteins cannot be examined, but the ion–ion interaction between acidic drugs and guanidyl groups can be examined on the bonded-silica gel. A strong anion exchanger (triethylphenylammonium-bonded silica gel) was developed for comparison of their ion-exchange capabilities.

The capacity ratios of molecular-form compounds measured in reversed-phase liquid chromatography can be related to their log P values, given by the following equation (2):

$$\log k = a\log P + b \tag{2}$$

where a and b are constant under certain conditions. The logarithm of the capacity ratio (log k) of the molecular form of analytes can be predicted from eqn (2) if their log P values are known.[62] The k_{max} in eqn (3) is directly related to log P in eqn (2). The log k values in reversed-phase liquid chromatography measured at pH 7.5 may be related to their hydrophobicity (log P values), but acidic drugs are completely ionized at this pH. The capacity ratios of some acidic drugs were measured on a butyl-bonded silica gel column (50×2.1 mm i.d.) in 25 mM phosphate buffer (pH 2.0–8.0) containing 50% methanol using a flow rate of 0.4 mL min^{-1} at 37 °C. The order of the strength of hydrophobic interaction of these drugs with the butyl phase at pH 7.5 was diazepam > indomethacin > phenylbutazone > tolazamide > tolbutamide > naproxen > warfarin > furosemide > salicylic acid > benzoic acid > sulfamethoxazole, as shown in Table 9.2. The correlation coefficient between the binding affinity of these drugs with HSA (log nK value)[55] and log k values at pH 7.5 in reversed-phase liquid chromatography using a butyl column was 0.937 ($n = 7$). In this calculation, the log nK values of sulfamethoxazole, salicylic acid, tolbutamide, warfarin, furosemide, naproxen, phenylbutazone were used, because the log nK values were measured under the same conditions using the same albumin containing only 3.2% glycosylated albumin. The other reference data in Table 18 of the Appendix (p. 315) were measured using albumin whose purity was not given. The relationship between log nK and log k is given by the following equation:

$$\log nK = 0.835 \times [\log k(R)] + 5.526, \quad r = 0.937, \quad n = 7.$$

where log k(R) is log k measured by reversed-phase liquid chromatography using a butyl column (C4 in Table 9.1). The value was smaller than the correlation coefficient (0.972) between the log nK and log P calculated using MaclogP™ based on Hansch n constants, ClogP, values.[55] Thus, the difference between 0.937 and 0.972 could not be explained by only the hydrophobicity of the drugs, because their ClogP values are predicted for the molecular form, not the ionic form. The gap between 0.937 and 0.972 can be

Table 9.1 Capacity ratios measured on different columns. C4 represents butyl-bonded silica, SG represents guanidino-bonded silica, SA represents triethylphenylammonium-bonded silica (IX$_1$), PG represents a guanidyl-group-bonded polyvinylalcohol copolymer (IX$_2$), and Ph represents phenylpropyl-bonded silica. Reproduced by permission of Elsevier, ref. 36.

| | *log* k | | | | |
| | C4 | SG | SA | PG | Ph |
Drug/Column pH	*7.5*	*7.5*	*7.0*	*7.5*	*7.5*
Benzoic acid	−0.772	—	—	—	—
Diazepam	0.974	−0.087	−0.253	−0.690	1.376
Furosemide	−0.330	−0.565	1.271	−0.046	−0.155
Indomethacin	0.789	−0.433	1.245	−0.249	0.936
Naproxen	0.125	−0.313	0.901	−0.346	0.086
Phenylbutazone	0.527	−0.527	1.078	−0.887	0.386
Procaine	0.359	0.503	−1.267	−0.860	0.577
Salicylic acid	−0.638	−1.287	1.168	−0.222	−1.046
Sulfamethoxazole	−1.721	−1.211	0.487	−0.565	−1.699
Tolazamide	0.248	−0.153	0.697	−0.855	0.204
Tolbutamide	0.191	−0.320	0.794	−0.774	0.072
Warfarin	−0.011	−0.474	1.265	−0.451	−0.051

adjusted by the addition of the values of the ion–ion interaction to the relationship between log nK and log k measured at pH 7.5 in reversed-phase liquid chromatography using a butyl column, log k(HP) (log k measured using hydrophobic phase). Otherwise, such a high coefficient of determination cannot be obtained without the combination of hydrophobic and ion–ion interactions.

The ion–ion interactions of acidic drugs were measured by anion-exchange liquid chromatography. The guanidyl-group-bonded silica gel column (150×2.0 mm i.d.) was used in a 50 mM sodium phosphate buffer with a flow rate of 0.2 mL min^{-1} at 37 °C, and the capacity ratios measured at pH 7.5 are given in Table 9.1. The capacity ratio related to the pH of the eluent was not like the theoretical relationship given by in eqn (3) for ion-exchange liquid chromatography, because the K values were not the same as those measured by reversed-phase liquid chromatography.[60]

$$k = \frac{k_{max} + k_{min}(K/[H^+])}{1 + (K/[H^+])} \qquad (3)$$

where k is the capacity ratio at a given pH, and k_{max} and k_{min} are maximum and minimum k values for an analyte. K is derived from the dissociation constant pK_a, and H$^+$ is the concentration of hydrogen ions in the eluent.[63] According to eqn (3), the k values are constant in eluents of both low and high pH. The chromatographic capability of the guanidino column for acidic drugs was compared to that of a strong anion exchanger (triethylphenylammonium-bonded silica gel; 50×4.6 mm i.d.). The k values measured at pH 7.0 are given in Table 9.1. The eluent was 25 mM phosphate buffer (pH 2.0–7.0) containing 70% methanol with a flow rate of 1.0 mL min^{-1} at 37 °C. However, the k values increased with increasing eluent pH, whereas k values decreased with increasing eluent pH in the conditions where the k values measured in other pHs are not given in this text due to the consideration of physiological pH. This means that the sodium phosphate buffer excluded acidic drugs by ion-exclusion effects at higher pH. This phenomenon is commonly observed in ion-exchange liquid chromatography using a weak ion-exchanger and a buffer solution with strong ionic strength.

The k values of acidic drugs were very small on the guanidyl-group-bonded column in the 50 mM sodium phosphate buffer at pH 7.4, which was used for the measurement of drug–protein binding. It seemed that the ion–ion interaction between the guanidine phase and the acidic drugs can be neglected.[59] However, the results obtained using the triethylphenylammonium column demonstrated selectivity related to ion–ion interaction at pH 7.0. The order of the strength of the ion–ion interaction of acidic drugs on the strong anion exchanger was furosemide > warfarin > indomethacin > salicylic acid > phenylbutazone > naproxen > sulfamethoxazole > tolbutamide > tolazamide > benzoic acid > diazepam. The reason the difference in ion–ion interactions for acidic drugs could not be obtained on the guanidine phase, may be due to its relatively polar matrix. For this type of matrix, 100% aqueous buffer solutions should be used to measure the pH effect on the retention behavior of acidic drugs.

The addition of log k values at pH 7.0 in ion-exchange liquid chromatography using a triethylphenylammonium phase (IX_1) to log k values of the above results improved the correlation coefficient to 0.959 ($n = 7$). The relationship is given by the following equation:

$$\log nK = 0.702 \times [1.00 \times \log k(R) + 0.90 \times \log k(IX_1)] + 4.862,$$
$$r = 0.959, \quad n = 7$$

where log $k(IX_1)$ was measured by ion-exchange liquid chromatography. A combination of the values of the hydrophobic and ion–ion interactions measured in liquid chromatography demonstrated that this liquid chromatographic method can predict log nK values. However, the ion-exchange group was far from a model of a protein. Generally, a hydrophobic matrix for an ion exchanger improves the ionic strength of ion-exchange group. Therefore, another guanidino-phase column made with a vinylalcohol copolymer gel (IX_2) was used to measure the log k values of these drugs at pH 7.5, and the values are listed in Table 9.1. The guanidino column made with polyvinylalcohol copolymer (150×4.6 mm i.d.) was used in 70% aqueous methanol containing 15 mM sodium phosphate buffer (pH 7.5) with a flow rate of 0.5 mL min^{-1} at 37 °C. The log k values were used for the further estimation of log nK values, instead of using the log k values measured on the triethylphenylammonium phase (IX_1). The correlation coefficient improved to 0.967 ($n = 7$). The relationship is given the following equation:

$$\log nK = 0.903 \times [1.00 \times \log k(R) + 0.70 \times \log k(IX_2)] + 5.841,$$
$$r = 0.967, \quad n = 7$$

where log $k(\mathrm{IX}_2)$ is log k measured on the guanidino phase with a pentacthylenehexamine-bonded vinylalcohol copolymer gel.[61] The phenylpropyl-bonded silica gel column (50×2.1 mm i.d.) was applied to investigate the π–π interactions of acidic drugs. The eluent was 25 mM phosphate buffer (pH 2.0–8.0) containing 50% methanol with a flow rate of 0.4 mL min^{-1} at 37 °C, and the capacity ratios at pH 7.5 are listed in Table 9.1. The capacity ratios of acidic drugs on the phenylpropyl column were larger than these on the butyl column, because the hydrophobicity of the phenylpropyl phase is stronger than on the butyl phase. In the results of the phenylpropyl and the butyl columns, the correlation coefficients between log k in pH 7.5 and ClogP were 0.760 and 0.721 ($n = 11$), respectively. This demonstrates that π–π interactions contribute to a phenylpropyl column, in addition to hydrophobic interactions. Furthermore, the k values of acidic drugs on the phenylpropyl phase were compared with those on the butyl phase, sulfamethoxazole, furosemide and phenylbutazone were retained more strongly on the phenylpropyl phase by a strong π–π interaction at pH 2.0, and furosemide, warfarin and diazepam were retained more strongly on the phenylpropyl phase at pH 7.5. Therefore, the π–π interactions should be considered for the analysis of drug–protein binding. The log k values measured on the phenylpropyl phase, log $k(\pi)$, were used further estimation of log nK values. The relationship is given in the following equation:

$$\log \mathrm{n}K = 1.107 \times [1.00 \times \log k(R) + 0.70 \times \log k(\mathrm{IX}_2) - 0.20 \times \log k(\pi)] + 5.886,$$
$$r = 0.953, \quad n = 7.$$

The results demonstrated that the addition of π–π interactions improved the correlation coefficient to 0.953.

The above results demonstrate that a combination of reversed-phase and ion-exchange liquid chromatography is a new simple method to estimate the binding affinity of drugs and proteins with high reproducibility. Such an experiment usually requires methanol as an organic modifier for the butyl- and phenylpropyl-bonded silica columns, and the guanidino column made with polyvinylalcohol copolymer. The hydrophobicity of the guanidine phase made with silica gel should be lower than that of albumin, but the hydrophobicity of the triethylphenylammonium phase and the guanidino column made with polyvinylalcohol copolymer was too high to be a model of albumin. The further development of a guanidino phase suitable for measuring ion–ion interactions on albumin is required. Moreover, the further development of packing materials suitable for the measurement of hydrophobicity and π–π interactions is necessary.

9.5 Measurement of HSA–Drug Binding Affinity without HSA: Part 2

Simplification of the above method was achieved using a pentyl-bonded silica gel column for measuring hydrophobic interactions, and guanidyl- and carboxyl-bonded silica gel columns[56] for measuring ion–ion interactions in the same eluent at pH 7.40. These k values are summarized in Table 9.2. The log k values of acidic drugs did not show good correlation with either their log P values or the log nK values obtained from the literature. Albumin–drug binding affinity could be determined from the k values measured by reversed-phase and ion-exchange liquid chromatography for some standard drugs. The standard acidic drugs were furosemide, naproxen, phenylbutazone, salicylic acid, sulfamethoxazole, tolbutamide and warfarin, and the basic drugs were scopolamine, lidocaine, quinine, dextromethorphan and imipramine.

The k values of acidic and basic drugs measured using the above systems were correlated to their binding affinity log nK values. The log nK values of acidic drugs have been measured previously,[55] and those of basic drugs were measured by the modified Hummel–Dreyer method. These log nK values are listed in Table 18 of the Appendix (p. 315). The calculated results are shown

Table 9.2   Log k values of acidic and basic drugs. nrf represents a negative capacity ratio. Reproduced by permission of Elsevier, ref. 37.

Acidic drugs	log k[a]	log k[b]	Basic drugs	log k[a]	log k[c]
p-Aminohippuric acid	0.180	−0.202	Ajimaline	0.004	−0.394
Amoxicillin	0.208	−0.325	Alloprinol	−1.268	nrf
Barbital	0.179	−0.015	Atropine	−0.139	0.780
Benzoic acid	0.286	0.187	Caffeine	−0.638	−1.036
Chloramphenicol	0.607	0.453	Carbamazepine	0.384	−2.955
Diazepam	1.303	0.417	Dextromethorphan	0.735	0.819
Furosemide	0.424	1.239	Ethambutol	−0.905	nrf
p-Hydroxybenzoic acid	0.188	0.380	Homatropine	−0.312	0.806
Ibuprofen	1.150	0.409	Imipramine	1.265	0.540
Indomethacin	1.089	1.139	Lidocaine	0.923	−0.495
Iopanoic acid	1.184	1.362	Prazosin	0.122	−1.292
Isoproterenol	0.250	−0.045	Prednisolone	0.440	nrf
Mefenamic acid	1.184	1.391	Procaine	0.061	0.338
Nalidixic acid	0.416	−0.024	Quinine	0.769	0.433
Naproxen	0.645	0.798	Rifampin	1.226	nrf
Nicotinic acid	0.187	−0.343	Scopolamine	0.082	−0.411
Phenobarbital	0.438	0.319	Theobromine	−1.030	−1.129
Phenylbutazone	0.927	0.432	Theophylline	−0.762	nrf
Propenecid	0.471	0.349			
Salicylic acid	0.325	0.707			
Sulfamethoxazole	0.205	0.419			
Terbutaline	0.279	−0.281			
Tetracycline	0.476	0.484			
Tolazamide	0.742	0.080			
Tolbutamide	0.711	0.263			
Warfarin	0.554	0.681			

[a]Measured on a C5 phase.
[b]Measured on a guanidino phase.
[c]Measured on a carboxyl phase.

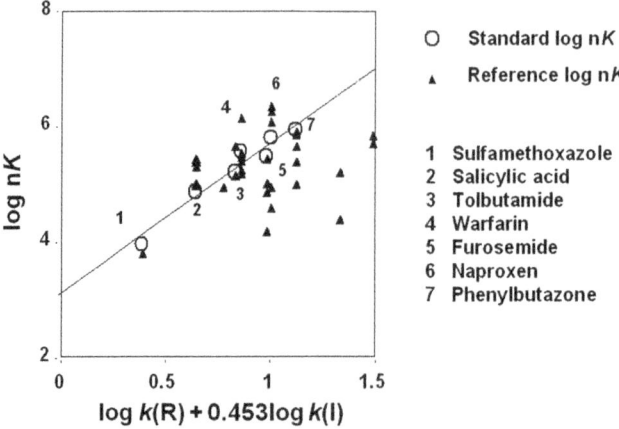

Figure 9.3   The relationship between log nK and log k values measured by reversed-phase and ion-exchange liquid chromatography for acidic drugs.

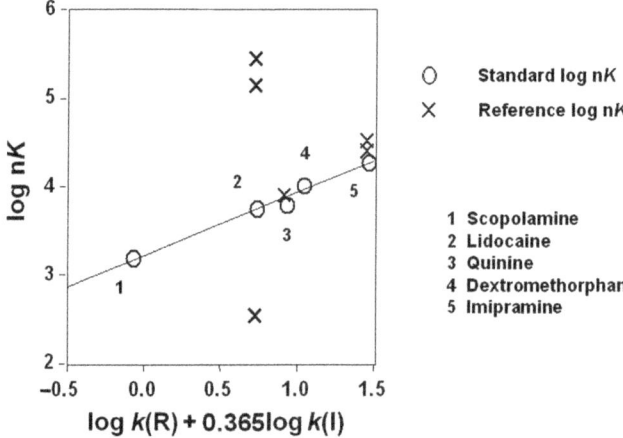

Figure 9.4   The relationship between log nK and log k values measured by reverse-phase and ion-exchange liquid chromatography for basic drugs.

in Figures 9.3 and 9.4, and the relationship is given by the following equation:

$$\log nK = 2.614 \times [\log k(R) + 0.453 \log k(I)] + 3.120, \quad r = 0.974, \quad n = 7,$$
$$\text{for acidic drugs}$$

$$\log nK = 0.708 \times [\log k(R) + 0.365 \log k(I)] + 3.211, \quad r = 0.991, \quad n = 5,$$
$$\text{for basic drugs}$$

where log k(R) is log k measured using reversed-phase liquid chromatography and log k(I) was determined using ion-exchange liquid chromatography. Log nK values measured by the modified Hummel–Dreyer method are summarized in Figures 9.3 and 9.4 with reference values.

The reference log nK values varied significantly, probably due to the different qualities of HSA and the different analytical systems used. The log nK values measured by the modified Hummel–Dreyer method using purified HSA correlated well with log k values measured by reversed-phase liquid chromatography using a pentyl-bonded silica gel column, and ion-exchange liquid chromatography using guanidyl-group-bonded silica gel and carboxyl-bonded silica gel columns. The pentyl-bonded silica gel was more stable than the butyl-bonded silica gel,[38] and simplified the estimation of the binding affinity compared to the previous study with a butyl-bonded silica gel column.[36] Such simple liquid chromatography may be useful for measuring the albumin–drug affinity without albumin. Therefore, the log nK values of 26 acidic and 18 basic drugs, listed in Table 9.2, were predicted using the above systems. The precision of these predicted log nK values will be studied in future for evaluation of this analytical method. The development of new packing materials for the modified Hummel–Dreyer method is necessary for drugs of low binding affinity.

The above guanidyl-group and carboxyl-bonded phases were first developed using a short ligand, therefore the measurement of hydrophobic interactions using a pentyl-bonded phase was required to obtain a good correlation with the HSA–drug binding affinity.[37] Later, a longer ligand was used to simplify the measurement process.[53] The latest ion exchangers have the combined properties of hydrophobicity and ion-exchange.

Log k values measured on a guanidyl-group-bonded silica gel column for acidic compounds, and on a carboxyl-bonded silica gel column for basic compounds correlated well with nK_3 and nK_4, respectively.

$$nK_1 = 0.926 \times (k_A) - 0.075, \quad r = 0.861, \quad n = 8,$$

$$nK_2 = 0.865 \times (k_A) - 0.172, \quad r = 0.869, \quad n = 8,$$

$$nK_3 = 1.346 \times (k_A) + 5.623, \quad r = 0.918, \quad n = 6,$$

$$nK_4 = 2.490 \times (k_A) + 5.716, \quad r = 0.934, \quad n = 12,$$

$$nK_6 = 1.287 \times (k_A) + 5.734, \quad r = 0.693, \quad n = 9,$$

$$nK_7 = 0.756 \times (k_A) + 5.398, \quad r = 0.593, \quad n = 9,$$

$$nK_1 = 0.596 \times (k_B) - 0.366, \quad r = 0.889, \quad n = 11,$$

$$nK_2 = 0.650 \times (k_B) + 0.491, \quad r = 0.912, \quad n = 11,$$

$$nK_4 = 0.542 \times (k_B) + 3.154, \quad r = 0.929, \quad n = 4,$$

$$nK_5 = 0.624 \times (k_B) + 3.034, \quad r = 0.842, \quad n = 23,$$

$$nK_6 = 0.263 \times (k_B) + 4.108, \quad r = 0.170, \quad n = 7.$$

This method required separate columns for acidic and basic compounds, but is feasible compared to the method using an immobilized-HSA column. The quality of these chemically synthesized guanidyl-group and carboxyl-bonded silica gels can be controlled in the same way as octadecyl-bonded silica gels. The correlation between log k (k_A and k_B) is reasonable, but that with nK_3 is poor, as described previously. The correlation with nK_4 and nK_5 is fine because nK_4 and nK_5 were derived from the relationship between nK_3 and k_A or k_B.

A simple approach was presented, that is, the measurement of the percentage concentration of bound drugs on an immobilized-HSA column. This one-point measurement method is simple even in free solution, and the tedious Scatchard plot is not necessary. This approach was applied for a limited number of drugs, and evaluated with reference values from ultra-filtration and dialysis methods, and good agreement was observed.[50] The data are summarized as PB_1 with reference values as PB_2 in Table 18 of the Appendix (p. 315). Good correlation was obtained between the percentage of solute bound to HSA and retention, expressed as $k/(k+1)$, and the extent of albumin binding for the benzodiazepines and coumarins studied, but not for triazole derivatives.[64]

9.6 Prediction of HSA–Drug Binding Affinity *In Silico*

Hummel–Dreyer and frontal analyses have been used to measure protein–drug binding affinity by liquid chromatography.[65,66] The protein binding affinity of drugs was determined using a physically protein-coated ODS column[67] and a chemically bonded bovine serum albumin column.[35,68–70] The immobilized protein column method is simple but the columns are not stable. The active sites are probably buried by the binding reaction used in the synthesis of packing materials for liquid chromatography.

The prediction of HSA–drug binding affinity is useful in the pharmaceutical industry to speed up the design of new compounds. The prediction of log K_{HSA} was studied using log k values measured with an immobilized-HSA column. After an extensive QSAR analysis using 53 descriptors, an acceptable correlation was obtained using 12 descriptors. The following equation was then obtained:

$$\log K_{HSA} = -0.607873 + 0.06784 \times (\text{HBondDon} - 3)^2 - (9 \times 10^{-6})(\text{JursTPSA})$$
$$- 0.028261 \times (E_{HOMO} + 7.4076)^2 + (0.005697) \times (\text{AM1dip}^2)$$
$$+ (0.182595)(\text{ClogP}) + (2.33529)(^6\chi\text{ring})$$

where dip represents dipole moment; χ are connectivity indices; HBondDon is the number of hydrogen bond donor atoms; JursTPSA is the Jurs descriptor, Total-Polar Surface Area; E_{homo} is the homo energy; and AM1dip is the dipole moment calculated using the AM1 program.

The statistics for this model are as follows: $r^2 = 0.83$ and cross-validated r^2, $q^2 = 0.79$.[49] Further QSAR analysis was carried out with E-state topological structure representations using the same log k values measured on the immobilized-HSA column. The following equation was obtained.

$$\log K_{HSA} = 0.0503 \times (\pm 0.0050) \times S^T(\text{arom}) + 0.0787 \times (\pm 0.0083) \times S^T(\text{CHsat})$$
$$+ 0.0291 \times (\pm 0.0059) \times S^T(-F, Cl) + 0.00871 \times (\pm 0.0032) \times S^T(-OH)$$
$$+ 2.38 \times (\pm 0.95) \times {}^6\chi^V\text{CH} - 2.96 \times (\pm 0.99) \times {}^5\chi^V\text{CH} - 1.18$$

where $S^T(\text{arom})$ is the sum of aromatic carbon atom types, $S^T(\text{CHsat})$ is the sum of the aliphatic carbons that bear hydrogens, and $S^T(-F, Cl)$ is the sum of the atom type E-state for fluorine and chlorine, $S^T(-OH)$ is the sum of the OH group, ${}^6\chi^V\text{CH}$ is the sixth order valence molecular connectivity χ chain index, and ${}^5\chi^V\text{CH}$ is the fifth order valence molecular connectivity χ chain index.

The statistics for this model as are follows: $r^2 = 0.77$ and $q^2 = 0.70$.[52] The correlation coefficient (r) for the latter log K_{HSA} and log k values measured was 0.88. The latter log K_{HSA} values, however did not correlate with the measured HSA–drug binding affinity using a modified Hummel–Dreyer method with the purified HSA.[55] The correlations between the topologically predicted log K values proposed based on log k values (nK_1) measured using an immobilized-HSA column liquid chromatography and the original measured nK values were:

$$nK_2 = 0.812 \times (nK_1) - 0.001, \quad r = 0.879, \quad n = 94,$$

$$nK_3 = -0.585 \times (nK_2) + 4.803, \quad r = 0.229, \quad n = 9,$$

$$nK_4 = 1.862 \times (nK_2) + 5.710, \quad r = 0.614, \quad n = 10,$$

$$nK_5 = 0.764 \times (nK_2) + 3.452, \quad r = 0.948, \quad n = 9,$$

$$nK_6 = 1.302 \times (nK_2) + 5.858, \quad r = 0.659, \quad n = 11,$$

$$nK_7 = -0.079 \times (nK_2) + 4.912, \quad r = 0.045, \quad n = 34.$$

These correlations indicate that the nK values of basic compounds (nK_5) derived from the log k values measured using the carboxyl-bonded silica gel correlated well with those measured using the immobilized-HSA column. The predicted log K_{HSA} (nK_2) values did not correlate with the nK values measured by free solution methods.

The mimic ion-exchange liquid chromatographic method was quantitatively analyzed *in silico*. Prediction of the binding affinity was possible. The molecular interaction energy values ($MIFS_A$ and $MIFS_B$) correlated well with log k values, and with binding affinity values.

For acidic compounds:

$$MIFS_A = 10.087 \times (k_A) + 34.795, \quad r = 0.844, \quad n = 14,$$

$$MIFS_A = 10.976 \times (nK_1) + 36.096, \quad r = 0.680, \quad n = 8.$$

$$MIFS_A = 11.682 \times (nK_2) + 37.284, \quad r = 0.670, \quad n = 8.$$

For basic compounds:

$$\text{MIFS}_B = 11.377 \times (k_B) + 18.468, \quad r = 0.932, \quad n = 16,$$

$$\text{MIFS}_B - 13.693 \times (nK_1) + 27.969, \quad r = 0.972, \quad n = 4,$$

$$\text{MIFS}_B = 14.326 \times (nK_2) + 25.957, \quad r = 0.986, \quad n = 5.$$

The correlation coefficients were similar to those obtained with log k values measured on a guanidino- or carboxyl-bonded columns. Molecular interaction energy values calculated using the MM2 program with the mimic guanidine phase carboxyl phase well correlated with the capacity ratios measured using the immobilized-HSA column. MIF_B values also correlated well with log K_{HSA} values.

The above results indicate that mimic ion-exchange liquid chromatography is feasible compared to using an immobilized-HSA column. The computational chemical analysis of molecular interactions using the mimic ion exchanger is practical for the rapid screening of drug candidates.

A computational chemical analysis using MM2 calculations was applied to analyze the retention mechanism of acidic drugs on a guanidino phase, and to estimate the albumin–drug binding affinity log nK values. The subtracted energy values, Δvalues, were considered to be molecular interaction energy values. The Δfinal (MIFS), Δhydrogen bonding (MIHB), Δelectrostatic (MIES) and Δvan der Waals (MIVW) values were used for the analyses. The maximum log k values (log k_{max}) measured both on the guanidyl-group-bonded polyvinylalcohol gel using 0.05 M sodium phosphate solution containing 50% methanol and on the guanidyl-group-bonded silica gel using 0.05 M sodium phosphate solution without methanol, did not show any meaningful correlation with these MI values for the retention of molecular-form analytes. The maximum retention was related to a combination of hydrophobic and ion–ion interaction forces, and could not be related to one type of interaction such as between a molecular-form compound and the guanidino phase or an ionic-form compound and the guanidino phase, because construction of a partially-ionized-form compound is difficult in computational chemical calculations. The poor correlation is due to the fact that the maximum capacity ratios, log k_{max}, did not correlate well with their VlogP values obtained from the literature, $\log k_{max} = 0.250 \times (\text{V log P}) + 0.654$, $r = 0.750$, $n = 17$, because the measurement of maximum retention time was difficult due to an ion-exclusion effect at low pH. This phenomenon differs from that in reversed-phase liquid chromatography.The dissociation constant was calculated from k values measured in a higher pH eluent than the pH used to obtain the highest k values. This phenomenon in ion-exchange liquid chromatography is different from that measured in reversed-phase liquid chromatography where k values in low-pH eluents are usually constant for acidic compounds. However, the k values were smaller in lower pH eluents in ion-exchange liquid chromatography due to an ion-exclusion effect. The pK_a values measured in ion-exchange liquid chromatography are relative pK_a

values of the analytes used, which differ from the pK_a values measured in reversed-phase liquid chromatography. The relative pK_a values were related to the ionic strength of the ion exchanger. The difference between the pK_a values measured by reversed-phase liquid chromatography and ion-exchange liquid chromatography is a property of the ion exchanger used. The relative pK_a values measured on the guanidyl-group-bonded polyvinylalcohol copolymer gel column were about one pK_a unit higher than their reference values. One reason will be the effect of the organic modifier concentration. The pK_a values were about 1.1 pK_a units higher in 50% acetonitrile[63] and methanol.[71] However, pH was measured before mixing with an organic modifier in this experiment. Therefore, the relatively high pH values should be a property of the ion exchanger used.[60] The relationship is given by the following equation:

$$pK_a \text{ measured} = 0.420 \times (pK_a \text{ reference}) + 4.089, \quad r = 0.878, \quad n = 17$$

where the pK_a values of barbituric acid, iopanoic acid, phenobarbital and probenecid were excluded due to the lack of experimental data.

The construction of a molecular form of Zwitter ion at pH 7.40 was difficult for computational chemical calculations, therefore *p*-aminohippuric acid and amoxicillin are eliminated from further discussion. The above results indicated that the calculated energy values were required for a pH effect because log k values were pH dependent.

Further analysis was performed to estimate the albumin–acidic drug binding affinity log nK values. The albumin–acidic drug binding affinity was measured at pH 7.40. The log k values measured at pH 7.40 correlated with the predicted log nK values from the literature. The following relationship was obtained, log $nK = 1.597 \times (\log k) + 5.808$, $r = 0.887$, $n = 13$.[55] The log nK values may be derived from log k values measured on a guanidino phase at pH 7.40. Therefore, the addition of k measured by reversed-phase liquid chromatography improved the correlation, as previously described.[55]

The calculated molecular interaction energy of iopanoic acid was very small. This may be due to the difficulty in calculating the energy level of iodine with high precision. Finally, the log nK values correlated with the molecular interaction energy values without these three compounds. The correlations are shown in the following equations:

$$MIVW = 0.764 \times (\log nK) + 4.981, \quad r = 0.261, \quad n = 16,$$

$$MIHB = -3.673 \times (\log nK) + 33.454, \quad r = 0.335, \quad n = 16,$$

$$MIES = 7.696 \times (\log nK) - 22.798, \quad r = 0.596, \quad n = 16,$$

$$MIFS = 4.702 \times (\log nK) + 14.314, \quad r = 0.932, \quad n = 16,$$

where these energy values were calculated from the following equation:

$$\text{MI energy} = \frac{\text{MI energy}_m + \text{MI energy}_i(K_a/[H^+])}{1 + (K_a/[H^+])}$$

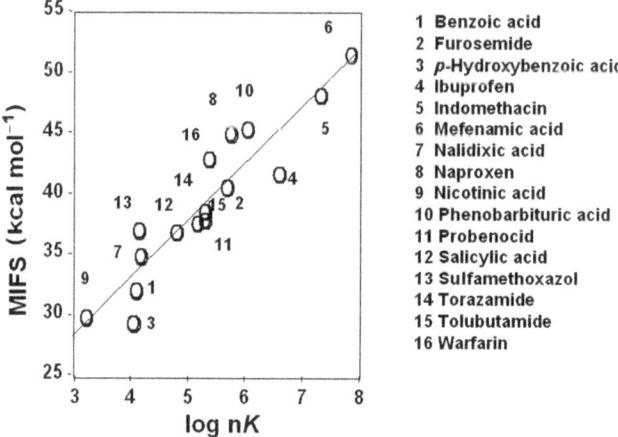

Figure 9.5   The relationship between log nK values and calculated molecular interaction energy values (MIFS) for acidic drugs.

where MI energy$_i$ is MI energy value of the ionic form of the analyte, MI energy$_m$ is the MI energy value of the molecular form of the analyte, and H$^+$ is the hydrogen ion concentration at pH 7.4. The dissociation constant pK_a was measured by ion-exchange liquid chromatography and the values are given in Table 18 of the Appendix (p. 315). The relationship between the MIFS and log nK values is shown in Figure 9.5. These results suggest that the albumin–acidic drug binding affinity, the log nK values, can be estimated by computational chemical calculations.

A simple model carboxyl-phase shown, as Figure 7.6, was used to investigate basic-drug–carboxyl phase interactions. The molecular interaction energy values (MI energy) were calculated and correlated with log nK values, however the correlation coefficient was poor ($r = 0.696$, $n = 17$). The electrostatic energy values were the major contributors for acidic drugs. The hydrophobic interaction, however, is important for basic drugs. Therefore, the three-dimensional model phase shown in Figure 7.10 was used to calculate the MI energy values. The relationships to log nK are given in the following equations:

$$\text{MIFS} = 14.934 \times (\log nK) - 23.390, \quad r = 0.791, \quad n = 17,$$

$$\text{MIHB} = 0.003 \times (\log nK) - 0.037, \quad r = 0.011, \quad n = 17,$$

$$\text{MIES} = -1.123 \times (\log nK) + 7.972, \quad r = 0.280, \quad n = 17,$$

$$\text{MIVW} = 20.284 \times (\log nK) - 46.055, \quad r = 0.821, \quad n = 17.$$

The contribution of the van der Waals energy values was important for the retention of basic compounds. The contribution of electrostatic energy was not important for basic drugs, compared to acidic drugs. The hydrogen-bonding energy values (hb) were almost constant before and after docking.

That is, the hydrogen-bonding energy values included in the energy values of the final structure did not contribute to the molecular interaction. This phenomenon was observed whilst analyzing the retention mechanisms of benzoic acid derivatives. Therefore, the hydrogen-bonding energy values of individual compounds were subtracted from the MIFS values and correlated with the log nK values. The following relationship was obtained:

$$\text{MIFS} + \text{hb} = 16.254 \times (\log \text{n}K) - 30.335, \quad r = 0.905, \quad n = 17.$$

A model experiment to analyze basic-drug–protein interactions demonstrated that the hydrophobic interaction was the major driving force, not the ion–ion interaction (the Coulombic force).

The pH effect on molecular interaction can be examined using liquid chromatography experimentally without proteins. The original pK_a values measured by titration were affected by an organic modifier in the eluent and an inductive effect of the ion-exchange groups of the bonded phase. However, the pK_a of the analytes was considered to be constant, and their log nK values not directly affected by the ion-exchange groups and organic modifiers. The MIFS energy was then recalculated using $\text{MIFS}_m + \text{hb}_m$, $\text{MIFS}_i + \text{hb}_i$ and pK_a values from the literature. The MI energy values are summarized in Table 9.3. The correlation coefficient was the same between pH 7.3 and 7.6. The correlation between MIFS and the new log nK is given in the following equations:

$$\text{MIFS(pH 7.50)} = 18.432 \times (\log \text{n}K) - 38.996, \quad r = 0.928, \quad n = 17.$$

This relationship is shown in Figure 9.6. Therefore, a small change in the eluent pH would not affect the predicted log nK values for these compounds. When the above calculation was performed using the pK_a values obtained

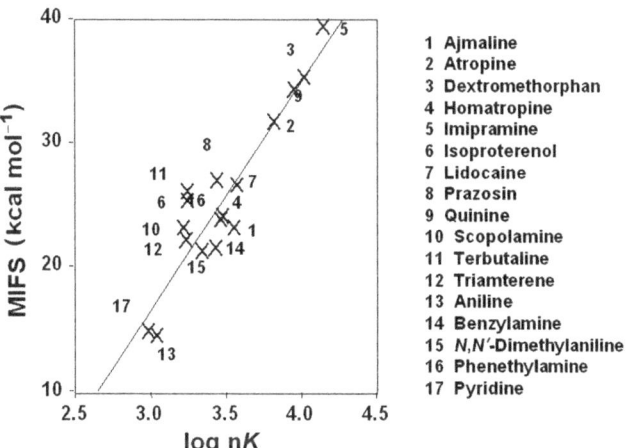

Figure 9.6 The relationship between log nK values and calculated molecular interaction energy values (MIFS) for basic drugs.

Table 9.3 Molecular properties of some basic compounds. nK_1 values were measured using a modified Hummel–Dreyer method, nK_2 values were measured by a two-column method, nK_3 values were measured by a one-column method, log k values were measured by a one-column method, MIFS values were calculated at pH 7.5. Reproduced by permission of Elsevier.

Analyte	log nK_1	log nK_2	log nK_3	log k	MIFS	hb_m	hb_i
Ajmaline	—	3.130	3.828	1.184	31.523	−2.317	−2.110
Allopurinol	—	2.373	2.701	−0.657	—	—	—
Atropine	—	3.509	3.569	0.679	22.991	−3.451	−3.387
Carbamazepine	—	3.356	3.328	0.360	—	—	—
Dextromethorphan	4.00	4.053	4.026	1.424	35.086	0.000	0.000
Homatropine	—	3.421	3.487	0.559	23.886	−0.606	−3.306
Imipramine	4.26	4.202	4.153	1.639	39.225	0.000	0.000
Isoproterenol	—	—	3.252	0.204	25.833	−11.250	−8.581
Lidocaine	3.74	3.677	3.586	0.757	26.459	−2.406	−2.226
Prazosin	—	3.200	3.443	0.581	26.741	−1.257	−1.099
Pyridine	—	—	2.994	−0.185	14.670	0.000	0.000
Quinine	3.78	3.843	3.959	1.370	34.034	−4.612	−4.621
Rifampicin	—	3.857	3.719	0.993	—	—	—
Scopolamine	3.18	3.177	3.227	1.573	23.013	−4.157	−4.146
Terbutaline	—	—	3.256	0.212	25.084	−6.111	−5.812
Theophylline	—	2.675	2.839	−0.433	—	—	—
Triamterene	—	—	3.244	0.246	21.994	−3.085	−2.844
Aniline	—	—	3.049	−0.172	14.270	−1.334	−1.334
Benzylamine	—	—	3.435	0.425	21.353	−0.973	−0.983
Phenylethylamine	—	—	3.473	0.427	23.581	−0.089	−0.089
N,N-Dimethylaniline	—	—	3.354	0.385	20.999	0.000	0.000

using ion-exchange liquid chromatography, the correlation coefficients were poor, 0.766–0.782, $n = 17$. The dissociation should occur based on the individual pK_a, but not at the artificial pK_a induced by the organic modifier concentration and the inductive effect of the ion exchanger.

9.7 Glycosylation effect on HSA–Warfarin Interactions

The influence of glycosylation on the drug binding of HSA was studied using HSA containing different amounts and degrees of GHSA. The drug used was warfarin. The drug–HSA affinity (log nK) was measured by a modified Hummel–Dreyer method using a mannose-bonded silica gel.[54] The modified Hummel–Dreyer method was the simplest method with the highest precision, and required the smallest amount of protein. Generally, the binding affinity of HSA decreased, perhaps due to a conformational change or steric hindrance, when further glycosylation occurred.

9.7.1 Purification of HSA Samples

HSA samples containing different amounts of GHSA were prepared from HSA and commercial HSA (Fraction V), purchased from Sigma (St. Louis,

Table 9.4 Contents of GHSA and the effect of HSA glycosylation on HSA–warfarin binding. mix 1 represents a mixture of Nos. 1 and 8 (2 : 1) and mix 2 represents a mixture of Nos. 1 and 8 (1 : 2). The GHSA concentration is the final concentration after purification and differs from the original concentration.

No.	Lot number	GHSA/%	n	$K/10^6\ M^{-1}$	$nK/10^6\ M^{-1}$	log nK
1	47F9338	3.2	0.893	11.126	9.940	6.997
2	26H9327	11.3	0.723	12.045	8.709	6.940
3	26H9327	15.1	0.892	5.158	4.600	6.663
4	mix 1	34.4	1.173	3.106	3.642	6.561
5	26H9327	36.9	1.080	1.424	1.538	6.187
6	mix 2	50.2	0.959	1.013	0.971	5.987
7	47F9338	63.0	1.229	1.134	1.431	6.156
8	26H9327	97.3	1.389	1.269	1.763	6.246

MO). Their molecular weights were assumed to be 66 000 Da. Commercial HSA was separated by affinity chromatography on a 3-aminophenylboronic acid monohydrate gel (Asahi Chemical Industry Co., Japan), and dialyzed against purified water using a Hollow Fiber Dialyzer ALF-210G Cuprophan · · (NIKKISO, Tokyo, Japan) and lyophilized. The percentage of GHSA was measured by an automated liquid chromatograph combined with size-exclusion and affinity chromatography.[72] The results are summarized in Table 9.4.

9.7.2 Glycosylation Effect

The drug–HSA binding affinity was measured using warfarin and the eight HSA samples listed in Table 9.4 using the modified Hummel–Dreyer method, and the binding affinity values were analyzed using Scatchard plots. Examples of the Scatchard plots obtained by the modified Hummel–Dreyer method are shown in Figure 9.2. The association constant (K) and the maximum number of drug molecules bound to albumin (n) were calculated, and the values are summarized in Table 9.4. However, only the results for the primary binding sites are presented because the values of the second sites were small and were not practical for discussing the glycosylation effect.

The GHSA content in original HSA of lot number 26H9327 was 59.9%. This value indicated that this HSA contained the most advanced glycosylated GHSA and advanced Maillard reaction products. The GHSA content in original (marketed) HSA of lot number 47F9338 was 35.5%, therefore the GHSA should have less advanced glycosylated GHSA. The contents of GHSA in normal and diabetic subjects were about 16 and 40%, respectively. The nK values did not linearly relate to the contents of GHSA, as shown in Figure 9.7. These results demonstrated that the HSA in the different lots contained different types of GHSA, as expected. In the early stage of a Maillard reaction, the principal glycosylation site of HSA is considered to be at Lys-525, and only about 33% of the glucose was covalently bound at this site.[18] However,

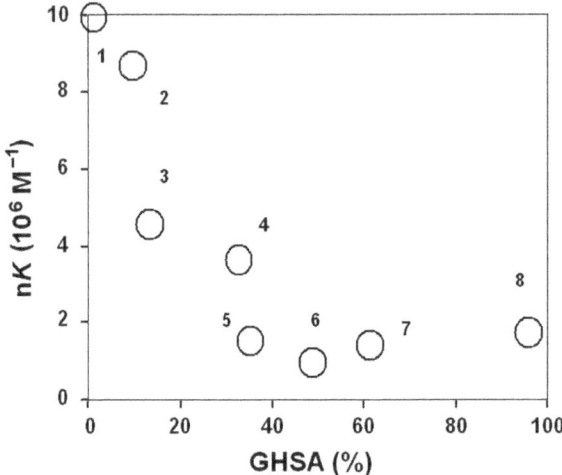

Figure 9.7   The effect of glycosylation on the HSA–warfarin binding affinity.
Reproduced by permission of Elsevier, ref. 37.

Lys-525 is located away from the warfarin binding site where many drugs
bind.[73] Therefore, the glycosylation of Lys-525 could not affect the drug
binding, since the binding of hemin was not inhibited by the glycosylation of
Lys-525.[18] Therefore, it was thought that if the glycosylation occurred only at
the Lys-525 site of GHSA contained in the HSA of lot number 47F9338, the
drug-binding capacity would not be altered. Further glycosylation of HSA in
the early stages occurs at the Lys-199, 281 and 439 sites, which are considered to be drug-binding sites. That is, the glycosylation of lot number
26H9327 HSA proceeded and inhibited the binding of drugs.

If the glycosylation proceeded further, that is, an advanced Maillard reaction took place, non-disulfide covalently linked proteins[74] and a sequence
of dehydrations and rearrangements would have occurred.[75,76] Evidence of
this was observed by the finding of two types of glucose-derived advanced
glycosylation products.[77–79] One type of compound is 2-furoyl-4(5)-(2-
furanyl)1-*H*-imidazole, resulting from a heterocyclic condensation of two
glucose molecules and two lysine-derived amino groups. The chemical
structure was revealed to be an advanced glycosylation end-product *in vivo*
and *in vitro*,[76,80] and is found in enzymatically hydrolyzed tissue.[77] Another
type of Amadori product is 3-deoxyglucasone.[78,79] The formation of such
compounds may enhance the hydrophobicity of HSA due to an increase of
hydrophobic regions, *i.e.*, the binding capacity of GHSA for hydrophobic
drugs should increase as seen in the results of Nos. 7 and 8 in Figure 9.7.

9.8   Modified Hummel–Dreyer Method

A modified Hummel–Dreyer method was carried out using an inner hydrophobic bonded silica gel, Wako WS GP-N6 from Wako Chemical (Osaka,

Japan) and a mannose-bonded silica gel packed in a 10 cm×2.1 mm i.d. column. The columns were thermostated at 37 °C. The flow rate was 0.2 mL min^{-1}. A fresh HSA sample solution was prepared for binding experiments, and the concentration was kept constant at 5×10^{-5} M in 50 mM phosphate buffer (pH 7.40). A 50 mM drug solution and the 50 mM phosphate buffer (pH 7.40) were mixed by high-pressure gradient programs to produce eluents containing various concentration of drug in 50 mM phosphate buffer (pH 7.40).

A blank control run was performed by injection of the above buffer. Then, at the same eluent composition, vacancy peaks were obtained by injection of the above sample solutions containing only HSA in the buffer. Blank and protein injections were run using the eluents containing different drug concentrations (0.4–50 μM) in an automated sequence, controlled by the workstation. The experimental data obtained by the modified Hummel–Dreyer method were analyzed using Scatchard plots. Binding affinity was calculated from a Langmuir-type equation.

References

1. T. Peters Jr., Serum albumin, *Adv. Protein Chem.*, 1985, **37**, 161–245.
2. A. Goldstein, The interaction of drugs and plasma proteins, *Pharmacol. Rev.*, 1949, **1**, 102–165.
3. M. C. Meyer and D. E. Guttman, The binding of drugs by plasma proteins, *J. Pharm. Sci.*, 1968, **57**, 895–917.
4. A. H. Anton and H. M. Solomon, Drug–protein binding, *Ann. N. Y. Acad. Sci.*, 1973, **226**, 1–362.
5. Y. Masubuchi, S. Narimatsu and T. Suzuki, Activation of propranolol and irreversible binding to rat liver microsomes: Strain differences and effects of inhibitors, *Biochem. Pharmacol.*, 1992, **43**, 635–637.
6. M. Brownlee and A. Cerami, The biochemistry of the complications of diabetes mellitus, *Annu. Rev. Biochem.*, 1981, **50**, 385–432.
7. H. F. Bunn, K. H. Gabbay and P. M. Gallop, The glycosylation of hemoglobin: Relevance to diabetes mellitus, *Science*, 1978, **200**, 21–27.
8. E. Schleicher, T. Deufel and O. H. Wieland, Non-enzymatic glycosylation of human serum lipoproteins, *FEBS Lett.*, 1981, **129**, 1–4.
9. A. E. Renold and G. F. Cahill Jr., Diabetes Mellitus, in *The Metabolic Basis of Inherited Disease*, ed. J. B. Stanbury, J. B. Wyngaarden and D. S. Fredrickson, McGraw–Hill, New York, 1978, pp. 69–108.
10. K. H. Gabbay, Hyperglycemia, polyol metabolism, and complications of diabetes mellitus, *Annu. Rev. Med.*, 1975, **26**, 521–536.
11. K. M. West, *Epidemiology of Diabetes and its Vascular Lesions*, Elsevier, New York, 1978, 231–247.
12. C. J. Fox, S. C. Darby, J. T. Ireland and P. H. Sonksen, Blood glucose control and glomerular capillary basement membrane thickening in experimental diabetes, *Br. Med. J.*, 1977, **2**, 605–607.

13. M. A. M. van Boekel, The role of glycation in aging and diabetes, *Mol. Biol. Rep.*, 1991, **15**, 57–64.

14. H. J. Kim and I. V. Kurup, Nonenzymatic glycosylation of human plasma low density lipoprotein. Evidence for in vitro and in vivo glucosylation, *Metabolism*, 1982, **31**, 348–353.

15. S. K. Williams, J. J. Devenny and M. W. Bitensky, Micropinocytic ingestion of glycosylated albumin by isolated microvessels: possible role in pathogenesis of diabetic microangiopathy, *Proc. Natl. Acad. Sci. U. S. A.*, 1981, **78**, 2393–2397.

16. R. Dolhofer and O. H. Wieland, Glycosylation of serum albumin in diabetes mellitus, *Diabetes*, 1980, **29**, 417–422.

17. M. Brownlee, H. Vlassara and A. Cerami, Advanced glycosylation end products in tissue and the biochemical basis of diabetic complications, *N. Engl. J. Med.*, 1988, **318**, 1315–1321.

18. N. Shaklai, R. L. Garlick and H. F. Bunn, Nonenzymatic glycosylation of human serum albumin alters its conformation and function, *J. Biol. Chem.*, 1984, **259**, 3812–3817.

19. S. Tsuchiya, S. Tamiko and S. Sekiguchi, Nonenzymatic glucosylation of human serum albumin and its influence on binding capacity of sulfonylureas, *Biochem. Pharmacol.*, 1984, **33**, 2967–2971.

20. G. L. Kearns, S. F. Kemp, C. P. Turley and D. L. Nelson, Protein binding of phenytoin and lidocaine in pediatric patients with type I diabetes mellitus, *Dev. Pharmacol. Ther.*, 1988, **11**, 14–23.

21. K. A. Mereish, H. Rosenberg and J. Cobby, Glucosylated albumin and its influence on salicylate binding, *J. Pharm. Sci.*, 1982, **71**, 235–238.

22. S. F. Kemp, G. L. Kearns and C. P. Turley, Alteration of phenytoin binding by glycosylation of albumin in IDDM, *Diabetes*, 1987, **36**, 505–509.

23. M. H. Murtiashaw and K. H. Winterhalter, Non-enzymatic glycation of human albumin does not alter its palmitate binding, *Diabetologia*, 1986, **29**, 366–370.

24. I. Sjšholm, The specificity of drug binding sites on human serum albumin, ed. M. M. Reidenberg and S. Erill, *Drug–Protein Binding*, Praeger, New York, 1986, pp. 36–45.

25. W. J. Jusko and M. Gretch, Plasma and tissue protein binding of drugs in pharmacokinetics, *Drug Metab. Rev.*, 1976, **5**, 43–140.

26. W. E. Müller and U. Wollert, Human serum albumin as a silent receptor for drugs and endogenous substances, *Pharmacology*, 1979, **19**, 59–67.

27. J. P. Tillement, G. Houin, R. Zini, S. Urien, E. Albengres, J. Barre, M. Lecomte, P. D'Athis and B. Sevbille, The binding of drugs to blood plasma macromolecules: recent advances and therapeutic significance, *Adv. Drug Res.*, 1984, **13**, 39–94.

28. J. Oravcová, B. Böhs and W. Lindner, Drug-protein binding studies, New trends in analytical and experimental methodology, *J. Chromatogr., B*, 1996, **677**, 1–28.

29. J. P. Hummel and W. J. Dreyer, Measurement of protein-binding phenomena by gel filtration, *Biochem. Biophys. Acta*, 1962, **63**, 530–532.

30. B. Sebille, N. Thuaud and J. P. Tillement, Study of low-molecular-weight ligand to biological macromolecules by high-performance liquid chromatography. Evaluation of binding parameters for two drugs bound to human serum albumin, *J. Chromatogr.*, 1978, **167**, 159–170.

31. P. F. Cooper and G. C. Wood, Protein-binding of small molecules: new gel filtration method, *J. Pharm. Pharmacol.*, 1968, **20**, 150S–156S.

32. L. Soltes, B. Sebille, J. P. Tillement and D. Berek, Study of bilirubin binding in human serum by high-performance liquid chromatography, *J. Clin. Chem. Clin. Biochem.*, 1989, **27**, 935–939.

33. B. Sebille, N. Thuaud and J. P. Tillement, Equilibrium saturation chromatographic method for studying the binding of ligands to human serum albumin by high-performance liquid chromatography. Influence of fatty acids and sodium dodecyl sulphate on warfarin–human serum albumin binding, *J. Chromatogr.*, 1979, **180**, 103–110.

34. B. Sebille, N. Thuaud and J. P. Tillement, Retention data methods for the determination of drug–protein binding parameters by high-performance liquid chromatography, *J. Chromatogr.*, 1981, **204**, 285–291.

35. D. S. Hage, T. A. G. Noctor and I. W. Wainer, Characterization of the protein binding of chiral drugs by high-performance affinity chromatography, interaction of *R*– and *S*–ibuprofen with human serum albumin, *J. Chromatogr., A*, 1995, **693**, 23–32.

36. T. Hanai, R. Miyazaki and T. Kinoshita, Quantitative analysis of human serum albumin–drug interactions using reversed-phase and ion-exchange liquid chromatography, *Anal. Chim. Acta*, 1999, **378**, 77–82.

37. T. Hanai, A. Koseki, R. Yoshikawa, M. Ueno, T. Kinoshita and H. Homma, Prediction of human serum albumin–drug binding affinity without albumin, *Anal. Chim. Acta*, 2002, **454**, 101–108.

38. T. Hanai, New developments in liquid chromatography stationary phases, ed. P. Brown and E. Grushka, *Adv. Chromatogr.*, 2000, 40 , 315–357.

39. A. Chen and M. J. Shapiro, NOE pumping as high-throughput method to determine binding affinity to macromolecules by NMR, *J. Am. Chem. Soc.*, 2000, **122**, 414–415.

40. L. H. Lucas, K. E. Price and C. K. Larive, Epitope mapping and competitive binding of HSA drug site II ligands by NMR diffusion, *J. Am. Chem. Soc.*, 2004, **126**, 14258–14266.

41. J. R. Simad, P. A. Zunszain, J. A. Hamilton and S. Curry, Location of high and low affinity fatty acid binding sites on human serum albumin reveled by NMR drug-competition analysis, *J. Mol. Biol.*, 2006, **361**, 336–351.

42. T. N. Parac-Vogt, K. Kimpe, S. Laurent, E. L. Vander, C. Burtea, F. Chen, R. N. Muller, Y. Ni, A. Verbruggen and K. Binnemans, Synthesis, characterization, and pharmacokinetic evaluation of a potential MR1 contrast agent containing two paramagnetic centers with albumin binding affinity, *Chem.–Eur. J.*, 2005, **11**, 3077–3086.

43. M.-X. Xie, X. Y. Xu and Y. D. Wang, Interaction between hesperetin and human serum albumin revealed by spectroscopic methods, *Biochim. Biophys. Acta*, 2005, **1724**, 215–224.

44. J. Sowell, K. A. Agnew-Heard, M. J. Christian, C. Mama, L. Strekowski and G. Patonay, Use of non–covalent labeling in illustrating ligand binding to human serum albumin via affinity capillary electrophoresis with near-infrared laser fluorescence detection, *J. Chromatogr., B*, 2001, **755**, 91–99.

45. C. Bertucci, V. Andrisano, R. Gotti and V. Cavrini, Development of a disulfiram-modified serum albumin-based HPLC column, *Chromatographia*, 2000, **52**, 319–324.

46. Z. D. Zhivkova and V. N. Russeva, Thermodynamic characterization of the binding process of sulindac to human serum albumin, *Arzneim. Forsch.*, 2003, **53**, 53–56.

47. J. Chen, C. Phnmacht and D. S. Hage, Studies of phenytoin binding to human serum albumin by high-performance affinity chromatography, *J. Chromatogr., A*, 2004, **809**, 137–145.

48. H. Xu, X.-D. Yu, X.-D. Li and H. Y. Chen, Determination of binding constants for basic drugs with serum albumin by affinity capillary electrophoresis with the partial filling technique, *Chromatographia*, 2005, **61**, 419–422.

49. G. Colmenarejo, A. Alvarez-Pedraglio and J.-L. Lavandera, Cheminformatic models to predict binding affinities to human serum albumin, *J. Med. Chem.*, 2001, **44**, 4370–4378.

50. Y. Cheng, E. Ho, B. Subramanyam and J.-L. Tseng, Measurement of drug–protein binding by using immobilized human serum albumin liquid chromatography-mass spectrometry, *J. Chromatogr., B*, 2004, **809**, 67–73.

51. W. A. Ritschel, *Handbook of Basic Pharmacokinetics*, Drug Intelligence Publications, Hamilton Press, Femandina Beach, FL, 1980.

52. L. M. Hall, L. H. Hall and L. B. Kier, Modeling drug albumin binding affinity with E-state topological structure representation, *J. Chem. Inf. Comput. Sci.*, 2003, **43**, 2120–2128.

53. T. Hanai, R. Miyazaki, E. Kamijima, H. Homma and T. Kinoshita, Computational prediction of drug–albumin binding affinity by modeling liquid chromatographic interaction, *Internet Electron. J. Mol. Des.*, 2003, **2**, 702–711.

54. T. C. Pinkerton and K. A. Koeplinger, Determination of warfarin–human serum albumin protein binding parameters by an improved Hummel–Dreyer high-performance liquid chromatographic method using internal surface reversed-phase columns, *Anal. Chem.*, 1990, **62**, 2114–2122.

55. K. Koizumi, C. Ikeda, M. Ito, J. Suzuki, T. Kinoshita, K. Yasukawa and T. Hanai, Influence of glycosylation on the drug binding of human serum albumin, *Biomed. Chromatogr.*, 1998, **12**, 203–210.

56. T. Hanai, Y. Masuda and H. Homma, Chromatography in silico, retention of basic compounds on a carboxyl ion-exchanger, *J. Liq. Chromatogr. Relat. Technol.*, 2005, **28**, 3087–3097.

57. U. Kragh-Hansen, V. T. G. Chuang and M. Otagiri, Practical aspects of the ligand-binding and enzymatic properties of human serum albumin, *Biol. Pharm. Bull.*, 2002, **25**, 695–704.

58. Personal Collection of Y. Matsushita, Kitasato University, School of Pharmaceutical Sciences, Tokyo.

59. T. Hanai, R. Miyazaki, J. Suzuki and T. Kinoshita, Computational chemical analysis of newly developed guanidino-phase for quantitative analysis of saccharides in liquid chromatography, *J. Liq. Chromatogr. Relat. Technol.*, 1997, **20**, 2941–2948.

60. T. Hanai and J. Hubert, Chromatography of aromatic acids on ion-exchangers, *J. Chromatogr.*, 1984, **316**, 261–265.

61. R. Miyazaki, T. Hanai, J. Suzuki and T. Kinoshita, Study of ion–ion interaction for protein–drug binding using a newly developed guanidino-bonded phase in liquid chromatography, *J. Liq. Chromatogr. Relat. Technol.*, 1998, **21**, 2887–2895.

62. T. Hanai, Structure–retention correlation in liquid chromatography, *J. Chromatogr.*, 1991, **550**, 313–324.

63. T. Hanai, K. C. Tran and J. Hubert, Prediction of retention times for aromatic acids in liquid chromatography, *J. Chromatogr.*, 1982, **239**, 385–395.

64. I. Fitos, J. Visy, A. Magyar, J. Kajtar and M. Simonyi, Inverse stereoselectivity in the binding of acenocoumarol to human serum albumin and to alpha 1-acid glycoprotein, *Biochem. Pharmacol.*, 1989, **38**, 2259–2262.

65. T. Hanai, Selection of chromatographic methods for biological materials, in *Advanced Chromatographic and Electromigration Methods in BioScience*, ed. Z. Deyl, I. Miksik, F. Tagliaro and E. Tesarova, Elsevier, Amsterdam, 1998.

66. T. Cserhati and K. Valko, *Chromatographic Determination of Molecular Interactions*, CRC Press, Boca Raton, 1993, pp. 173–204.

67. H. Yoshida, I. Morita, G. Tamai, T. Masujima, T. Tsuru, N. Takai and H. Imai, Some characteristics of a protein-coated ODS column and its use for the determination of drugs by the direct injection analysis of plasma samples, *Chromatographia*, 1984, **19**, 466–472.

68. N. Lammers, H. D. Bree, C. P. Groen, H. M. Ruijten and B. J. D. Jong, Determination of drug protein–binding by high performance liquid chromatography using a chemically bonded bovine albumin stationary phase, *J. Chromatogr.*, 1989, **496**, 291–300.

69. C. Lagercrantz, T. Laisson and H. Karisson, Binding of same fatty acids and drugs to immobilized bovine serum albumin studied by column affinity chromatography, *Anal. Biochem.*, 1979, **99**, 352–364.

70. J. Yang and D. S. Hage, Characterization of the binding and chiral separation of D- and L- tryptophan on a high-performance immobilized HSA column, *J. Chromatogr.*, 1993, **645**, 241–250.

71. T. Hanai, unpublished data measured from methanol concentration effect of standard buffer (pH 4.01 and 6.87).

72. K. Yasukawa, F. Abe, N. Shida, Y. Koizumi, T. Uchida, K. Noguchi and K. Shima, High-performance affinity chromatography system for the

rapid, efficient assay of glycated albumin, *J. Chromatogr.*, 1992, **597**, 271–275.

73. K. J. Fehske, W. E. Müller and U. Wollert, The location of drug binding sites in human serum albumin, *Biochem. Pharmacol.*, 1981, **30**, 687–692.

74. M. S. Swamy, C. Tsai, A. Abraham and E. C. Abraham, Glycate mediated lens crystallin aggregation and cross-linking by various sugars and sugar phosphates *in vitro*, *Exp. Eye Res.*, 1993, **56**, 177–185.

75. M. Brownlee, H. Vlassara and A. Cerami, Nonenzymatic glycosylation and the pathogenesis of diabetic complications, *Ann. Intern. Med.*, 1984, **101**, 527–537.

76. F. G. Njoroge, L. M. Sayre and V. M. Monnier, Detection of D-glucose-derived pyrrole compounds during Maillard reaction under physiological conditions, *Carbohydr. Res.*, 1987, **167**, 211–220.

77. S. Pongor, P. C. Ulrich, F. A. Bencsath and A. Cerami, Aging of proteins: isolation and identification of a fluorescent chromophore from the reaction of polypeptides with glucose, *Proc. Natl. Acad. Sci. U. S. A.*, 1984, **81**, 2684–2688.

78. F. G. Njoroge, A. A. Fernandes and V. M. Monnier, Mechanism of formation of the putative advanced glycosylation end product and protein cross-link 2-(2-furoyl)-4(5)-(2-furanyl)-1H-imidazole, *J. Biol. Chem.*, 1988, **263**, 10646–10652.

79. J. C. F. Chang, P. C. Ulrich, R. Bucala and R. A. Cerami, Detection of an advanced glycosylation product bound to protein in situ, *J. Biol. Chem.*, 1985, **260**, 7970–7974.

80. J. Farmar, P. C. Ulrich and A. Cerami, Novel pyrrole from sulfite-inhibited Maillard reaction: insight into the mechanism of inhibition, *J. Org. Chem.*, 1988, **53**, 2346–2749.

CHAPTER 10

Quantitative Analyses of Protein Affinity Chromatography

10.1 Introduction

Protein affinity is defined as the selective binding of proteins to a certain number of compounds. The selective molecular recognition is called the *affinity*.[1] However, there is no numerical value to explain the degree of affinity. The chromatographic retention time difference and enzyme reactivity indicate the relative degree of affinity. Proteins differ in their degree of affinity and their steric hindrance properties. These protein characteristics can be used for highly selective separation and purification of compounds from a complex mixture. The quantitative analysis of protein molecular recognition is of considerable research interest. The complexity and flexibility of protein structures make analysis of selective molecular recognition particularly challenging.

Molecular interaction forces are based on solubility factors. The Coulombic force is the strongest, followed by the Lewis acid–base interaction, including hydrogen bonding and charge-transfer effects, and van der Waals forces. Steric hindrance also affects the molecular interaction. Individual interaction forces can be studied using chromatography. The main interaction force in reversed-phase liquid chromatography is the van der Waals force. The main interaction force in normal-phase liquid chromatography is the Lewis acid–base interaction, including hydrogen bonding, and that in ion-exchange liquid chromatography is the Coulombic force. Enantioseparation is achieved by a combination of these molecular interaction forces, with steric hindrance. The retention mechanism in reversed-phase liquid chromatography was demonstrated using quantitative *in silico* analysis of the chromatographic data of various compounds, such as phenolic

RSC Chromatography Monographs No. 19
Quantitative *In Silico* Chromatography: Computational Modelling of Molecular Interactions
By Toshihiko Hanai
© Toshihiko Hanai, 2014
Published by the Royal Society of Chemistry, www.rsc.org

compounds, aromatic acids, and acidic and basic drugs in Chapter 6. The retention mechanism of ion-exchange liquid chromatography was then quantitatively analyzed *in silico*, in Chapter 7. Carboxyl and guanidyl phases were selected for studying the basic molecular recognition mechanism of proteins. Affinity liquid chromatography can be used as a mimic tool for protein molecular recognition even when the flexibility of immobilized proteins is limited.

There are numerous reports on the three-dimensional structures of proteins determined by X-ray crystallography and/or NMR, and by computational chemical calculation from the results of amino acid sequencing. An empirical approach to identifying catalytic sites, the location of metal ion- and carbohydrate-binding sites, and folding and unfolding, has been used with molecular dynamics simulations. A variety of different search and optimization methods have been developed for protein–ligand docking applications.[2–5] Once the binding site of a protein, the structure of a protein, and a small molecule interaction are determined, the fitted small molecule is further modified, and a variety of drug candidates can be designed. One question is, how does a small molecule reach the center of the protein? Agre and MacKinnon clearly demonstrated how ions and water cross cell membranes, in which the existence of ion channels in a protein is important. Ions and water are driven through by electrostatic forces. The electronic movement is Arrhenius theory of electrolytic dissociation.[6–11]

The active sites of some enzymes are nearly identical, but their catalytic reactions are different. One enzyme can have a different chemistry at the same active site to another enzyme. These observations indicate that a protein will bind different compounds in the same manner.[12–15] The binding mechanism should be basically the same, and it is not with 100% affinity. The basic docking mechanism can be explained using a simple model experiment.

The guanidyl groups of arginine should function as anion-exchange groups, and the carboxyl groups of aspartic and glutamic acids should function as cation-exchange groups. Determining how to design a new bonded phase to measure protein–drug binding affinity using a single chromatography method is important for developing a quick screening method for drug candidates. Several model phases were constructed and a suitable model phase was screened to demonstrate the importance of ion–ion interactions, and to explain how a small molecule reaches the center of a protein.[16]

The method has been applied to study the enantiomer recognition of the protein monoamine oxidase (MAO), which has been used for affinity liquid chromatography. A mutant preparation method has been described using D-amino acid oxidase (DAO). The optimized structure of the docked complex between a protein and a substrate is obtained by performing molecular mechanics calculations, and the data indicates the degree of complex tightness.

10.2   Model Study of Protein–Drug Docking

The model support consisted of 730 carbons, 988 hydrogens, and 11 275 connectors as shown in Figure 3.12. The center hydrogen was replaced by a methyl group or a guanidino group. The surrounding 6 hydrogens were re-placed by methyl groups to make room for the analyte to contact the center substituent, and 54 hydrogens of the 2nd, 3rd, and 4th circles were replaced by dodecyl groups. The bonded phase was like a tulip, and a small pocket remained where a small molecule could reach the methyl or ion-exchange group at the center of the bonded phase, like a bee entering the center of a tulip.

The molecular or ionized form of benzoic acid was placed outside a hole and faced toward the center methyl or guanidyl group at the bottom of the pocket before starting docking, then the complex structure was optimized to measure the direct hydrophobic and ion–ion interactions. The center sub-stituent in the bottom of the model phase was a methyl or guanidyl group. The methyl and guanidyl groups represent hydrophobic and ion-exchange phases, respectively. The pocket size of the phase was 8.8 Å internal diameter at the entrance, 4.4 Å at the bottom, and 12.5 Å deep. According to these pocket sizes, benzoic acid might reach the bottom without encountering a physical barrier, as described in Section 3.4, but a larger molecule could not reach the bottom without a strong interaction to push aside the alkyl groups.

The molecular interaction energy (MIES) between benzoic acid and a guanidino phase demonstrated the existence of an ion–ion interaction be-tween a carboxyl ion and a guanidyl group. No such energy value changes were observed for the hydrophobic phases. Even when the phenyl group was placed toward the guanidyl group, a significant energy change was not ob-tained because benzoic acid was trapped by hydrophobic interactions before it reached the bottom of the phase. The molecular form of benzoic acid underwent hydrogen bonding with the guanidyl group, but the energy change was less than that of the electrostatic energy of the ionized form. A hydrogen-bonding energy change was also observed on the hydrophobic phase, whose pocket size was large enough for the free access of a small molecule to the bottom.

Basic studies were performed using larger model molecules ('A'), shown in Figure 10.1. The MI energy values between the authentic molecule and guanidyl group (GUA) were obtained after a computational chemical analysis using a molecular mechanics calculation program (MM2). The optimized energy value was less than 0.00001 kcal mol^{-1} by MM2 optimization. The calculated energy values are listed in Table 10.1. After subtraction of the energy value of the complex from the sum of the individual energies of the analytes and the model bonded phase, the remaining value was considered to be the interaction energy value.

The order of MIFS strength is Phase–ACOO$^-$ > Phase–ACOOH > Phase–AMe. MIES contributed to the strong interactions of the ionized authentic molecule, and MIHB contributed to the interactions of the molecular form of the authentic molecule. AMe interacted with GUA by van der Waals forces. After docking, there was little change in the atomic partial charge (apc) of the authentic molecule, and the effect was as a weak as the MIHB of the Phase–ACOOH complex. These complex forms are shown in Figure 10.2.

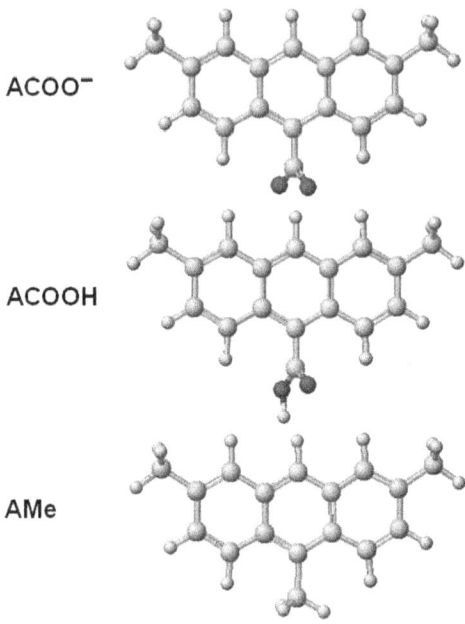

Figure 10.1 Structure of authentic molecules. White, light-gray, and black balls represent hydrogen, carbon, and oxygen, respectively.
Reproduced by permission of Taylor and Francis, ref. 26.

Table 10.1 Molecular properties of the model phase, analytes, and molecular interaction energy values. FS, HB, ES, and VW represent the energy of the final (optimized) structure, the hydrogen-bonding energy, the electrostatic energy, and the van der Waals energy (kcal mol^{-1}), respectively. GUA represents a guanidine phase.

Analyte	FS	HB	ES	VW
ACOO$^-$	−23.4730	0.000	−0.596	14.664
ACOOH	−35.3459	−3.591	−7.233	14.394
AMe	−31.2530	0.000	−0.189	15.005
Phase (GUA)	3681.1654	−0.809	−21.050	835.088
Complex of phase–ACOO$^-$	3619.1131	−1.014	−26.931	818.510
Complex of phase–ACOOH	3609.0592	−8.041	−29.342	820.137
Complex of phase–AMe	3614.9854	−1.002	−21.060	816.435
MI of phase–ACOO$^-$	38.5800	0.205	5.285	31.216
MI of phase–ACOOH	36.7610	3.641	1.059	29.319
MI of phase–AMe	34.9277	0.193	−0.179	33.632

The above results demonstrated that the Coulombic force (the ion–ion interaction) is strong enough to move a distance of more than 6 Å and push aside an alkyl wall whose hydrophobicity is approximately log $P = 5$. This means there might be a hole where a large molecule can enter the protein using ion–ion interactions.

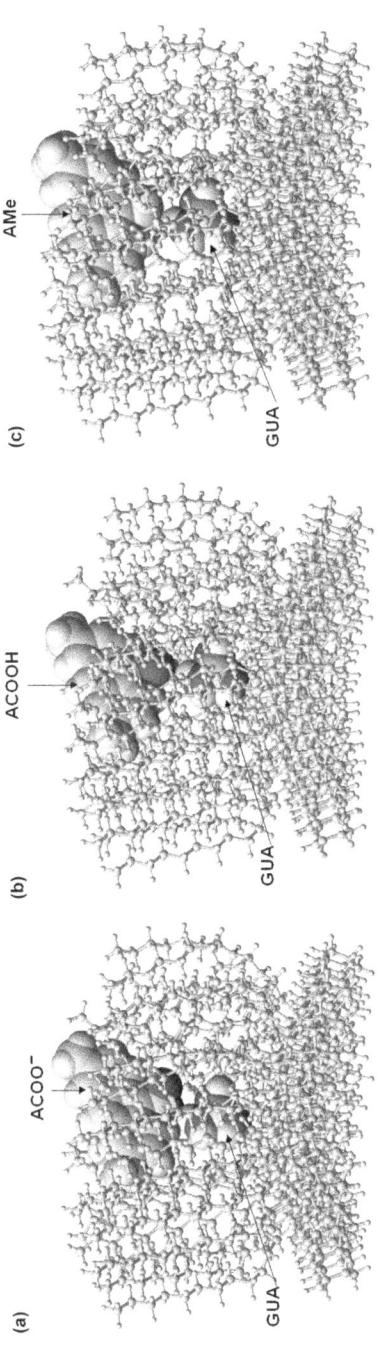

Figure 10.2 Structures of complexes between the model phase and (a) $ACOO^-$, (b) ACOOH, and (c) AMe. White, light-gray, dark-gray, and black balls represent hydrogen, carbon, nitrogen, and oxygen, respectively. The atomic size of the authentic molecule and part of guanidyl group is five times that of the other atoms. Reproduced by permission of Taylor and Francis, ref. 26.

10.3   Selectivity of Monoamine Oxidase

10.3.1   Introduction

DAO and MAO are both families of flavoenzymes. DAO and related modified (engineered) enzymes selectively oxidize various D-amino acids.[17] The molecular interaction center is clear; the positively charged guanidyl group of the enzyme attaches to the ionized carboxyl group of the amino acid by Coulombic forces, and the carbonyl group of the flavine ring interacts with the ionized amino group of the amino acid by a Lewis acid–base interaction. The two possible points of interaction allow for molecular modeling studies of DAO enzyme activity. MAO, however, does not have a strong interaction site, and only the carbonyl group of the flavine ring interacts with the ionized amino group of the analyte by a Lewis acid–base interaction. The one-point interaction makes it difficult to model the complex form. The analyte may also interact with a carbonyl group on the protein. In this experiment, the hydrophobic interaction is not considered to be the main interaction force.

The enzyme activity of MAO is inhibited by amphetamine and ephedrine, and the enantiomers are chirality selective.[18] The selectivity was studied by comparing it to that of the complex form of DAO. The downloaded structure of MAO was carefully fixed after rejection by the MM2 program. The substrate was replaced with optimized (R)- or (S)-amphetamine, and the complex conformation was further optimized using the MM2 program. The complexes with the lowest energy values were selected as the ideal conformations.

10.3.2   Conformational Analysis

The inhibition of amphetamine and ephedrine optical isomers was further studied in an enzyme reaction of MAO. There are two types of MAO, and the enzyme reaction was selectively inhibited.[18] The stereostructures of amphetamine–MAO-A and –MAO-A are shown in Figures 10.3 and 10.4. MAO-A is 2BXR, and MAO-B is 1GOS in the Protein Data Bank (PDB). The irregularity of their structures was corrected using MM2 calculations. The structure of MAO-A with the flavine adenine dinucleotide (FAD) binding site contains 7102 atoms, 7196 bonds, and 59 727 connectors. That of MAO-B with the FAD binding site contains 8008 atoms, 8110 bonds, and 67 289 connectors. The structures were optimized using the augmented MM2 function of the CAChe™ program. The minimum energy was 10^{-7} kcal mol^{-1}. The properties of the MAO structures and their inhibitors are shown in Table 10.2.

No arginine residue exists in the reaction cavity, and no carboxyl group exists in the inhibitors, compared to DAO. This means that monoamine does not dock strongly using Coulombic forces, rather, it is docked by a Lewis acid–base interaction between the amino group of the ligand monoamine and the carbonyl group of the flavine ring. The docking required several trials due to the uncertain conformation, compared to that of docking an

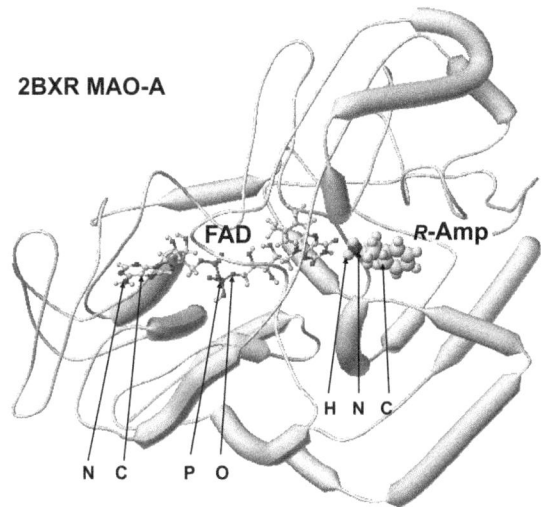

Figure 10.3 Stereostructure of the complex between (R)-amphetamine and 2BXR MAO-A. For the ligand (R-Asp) and coenzyme (FAD), small white, large gray, black, and dark gray balls represent hydrogen, carbon, oxygen, and nitrogen. Phosphorus is separately indicated.
Reproduced by permission of American Laboratory, ref. 27.

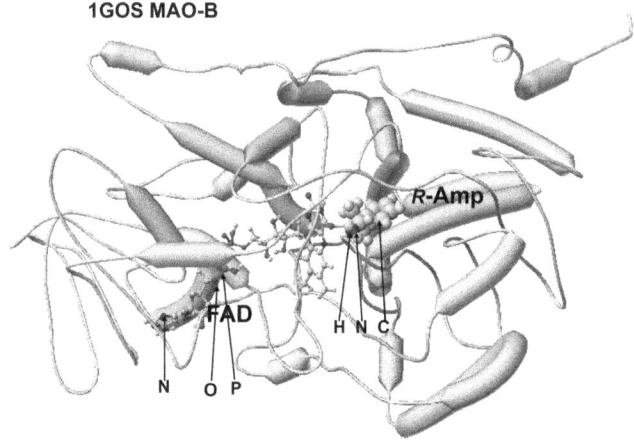

Figure 10.4 Stereostructure of the complex between (R)-amphetamine and 1GOS MAO-B. For the ligand (R-Asp) and coenzyme (FAD), small white, large gray, black, and dark gray balls represent hydrogen, carbon, oxygen, and nitrogen. Phosphorus is separately indicated.
Reproduced by permission of American Laboratory, ref. 27.

amino acid with DAO. Other inhibitors used were (1S,2S)-(+)-ψ-ephedrine and (1R,2R)-(−)-ψ-ephedrine. The optimized molecular interaction energy and apc values are summarized in Tables 10.3 and 10.4.

Table 10.2 Molecular properties of MAO-A and MAO-B inhibitors and their complexes. Reproduced by permission of American Laboratory, ref. 27.

Enzyme substrate	FS	HB	ES	VW
(R)-Amphetamine	−7.5275	−1.184	−0.815	4.286
(R),(S)-Amphetamine	−12.2431	−5.772	−1.628	4.672
(1R,2R)-(−)-ψ-Ephedrine	−12.1699	−3.390	−7.274	3.998
(1S,2S)-(+)-ψ-Ephedrine	−32.5772	−19.151	−14.754	8.052
MAO-A	−10 608.7570	−6782.575	−5325.214	−532.712
MAO-B	−11 557.6431	−7729.357	−5822.221	−436.481
MAO-A-(R)-amphetamine	−10 742.0528	−6871.464	−5390.366	−531.390
MAO-A-(R),(S)-amphetamine	−10 762.6740	−6876.673	−5408.041	−525.431
MAO-B-(R)-amphetamine	−11 841.2832	−7898.868	−6009.620	−405.129
MAO-B-(R),(S)-amphetamine	−11 955.5046	−7975.872	−6047.347	−423.915
MAO-A-(1R,2R)-(−)-ψ-ephedrine	−10 815.2317	−6865.672	−5459.998	−553.150
MAO-A-(1S,2S)-(+)-ψ-ephedrine	−10 899.2051	−6930.607	−5532.184	−522.696
MAO-B-(1R,2R)-(−)-ψ-ephedrine	−11 959.7410	−7882.375	−6118.227	−445.863
MAO-B-(1S,2S)-(+)-ψ-ephedrine	−11 926.5853	−7970.062	−6037.787	−416.383

The molecular interaction energy values were calculated using the following equation:

ΔMI = energy value of enzyme including FAD + energy value of substrate − energy value of complex.

(S)-Amphetamine more effectively inhibits the enzyme reaction of MAO-A than that of MAO-B. On the other hand, (1S,2S)-(+)-ψ-ephedrine inhibits the enzyme reaction of MAO-A, but (1R,2R)-(−)-ψ-ephedrine inhibits the enzyme reaction of MAO-B. The molecular interaction energy values indicate the strength of the molecular interaction. The larger the change in the final structure energy, the stronger the molecular interaction. The Δapc values also indicate the strength of the contact between the monoamine ligand and the carbonyl group of flavine. The absolute value of Δapc is small compared to that of the DAO mutant M215R. The enantiomer selectivity of MAO-A and -B against (R)- or (S)-amphetamine was clear based on the atom distance between the carbonyl group of the flavine and the ionized amino group of (R)- or (S)-amphetamine and the apc of the key atoms. The nitrogen of ephedrine is not a primary amine. The direct interaction with the flavine carbonyl group was minimal, leading to weak inhibition.

10.4 Selectivity of D-amino acid oxidase

10.4.1 Introduction

Proteins naturally recognize enantiomers. The study of protein recognition of enantiomers was applied to analyze the reactivity of DAO, which selectively oxidizes (R)-amino acids [(R)-AAs]. DAO was the second flavoenzyme to

Table 10.3 Molecular interaction energy values for MAO-A and MAO-B inhibitors and their complexes. MIFS, MIHB, MIES, MIVW represent the molecular interaction energy of the final (optimized) structure, the hydrogen-bonding energy, the electrostatic energy, and the van der Waals energy (kcal mol^{-1}), respectively. Reproduced by permission of American Laboratory, ref. 27.

Enzyme substrate	$IC_{50}/\mu M^a$	MIFS	MIHB	MIES	MIVW
Complexes with MAO-A					
(R)-Amphetamine	31.1	125.768	−87.705	64.337	2.964
(R,S)-Amphetamine	3.1	141.674	−88.326	81.199	−2.609
Complexes with MAO-B					
(R)-Amphetamine	246	276.113	168.327	186.584	−27.066
(R,S)-Amphetamine	62.5	385.618	240.743	223.498	−7.894
Complexes with MAO-A					
(1R,2R)-(−)-ψ-Ephedrine	5350	194.305	79.707	127.510	24.436
(1S,2S)-(+)-ψ-ephedrine	880	257.871	128.881	192.216	−1.964
Complexes with MAO-B					
(1R,2R)-(−)-ψ-Ephedrine	5030	389.928	149.628	288.732	13.380
(1S,2S)-(+)-ψ-Ephedrine	10 000	336.365	221.554	200.812	−12.046

^{a}Values from ref. 30.

Table 10.4 Atomic partial charge (apc) of target atoms. C is the α-carbon of the amino acid, N the nitrogen of the amino acid, and O the oxygen of the carbonyl group on the flavine ring, respectively. Reproduced by permission of American Laboratory, ref. 27.

Enzyme	Substrate (inhibitor)	Δapc (C)	Δapc (N)	Δapc (O)
MAO-A	(R)-amphetamine	−0.031	0.030	0.139
	(R,S)-amphetamine	−0.046	0.010	0.047
MAO-B	(R)-amphetamine	−0.056	0.031	0.101
	(R,S)-amphetamine	−0.069	0.014	0.097
MAO-A	(1R,2R)-(−)-ψ-ephedrine	0.014	0.039	0.035
	(1S,2S)-(+)-ψ-ephedrine	0.036	0.036	0.093
MAO-B	(1R,2R)-(−)-ψ-ephedrine	0.041	0.040	−0.024
	(1S,2S)-(+)-ψ-ephedrine	0.056	0.029	0.047

be discovered. Krebs first detected DAO activity in tissue specimens in 1935.[19] This enzyme is present in a variety of organisms, such as bacteria, yeast, fungi, mollusks, insects, fish, amphibians, reptiles, birds, and mammals. Although there is a small difference in the total number of amino acid residues, the amino acid sequences are highly conserved. The physiological function of DAO has been reviewed,[20] and the catalytic activity varies for (R)-AAs. DAO has a broad substrate specificity with a preference for (R)-AAs bearing hydrophobic side chains of up to four carbon atoms, followed by those carrying polar and aromatic groups.[21] The physiological role, stereostructure, and reaction mechanism of DAO were reviewed, including the mutants.[22]

Many crystallographic structures of DAO are readily available from the Protein Data Bank, and several stereostructures of DAO were downloaded as PDB files.[23] The substrates of the DAO complexes were varied, and were replaced with an amino acid. The conformation of the DAO–amino acid complex was then optimized using MM2 calculations to analyze which amino acid residue of DAO was directly involved in the binding, based on various atomic distances, and on electron transfer which was determined from the apc of neighboring atoms. Furthermore, the stereostructure of human DAO was estimated from the sequence data NP001908 (ref. 24) and the stereostructure of pig kidney from DAO 1VE9 (ref. 21) because of the similarity of these sequences. The selectivity of the estimated human DAO was analyzed. The selectivity of DAO mutants was also analyzed after docking with various amino acids and drugs.

10.4.2 Preparation and Evaluation of Mutants

Naturally occurring proteins are generally unstable, and the purification is a tedious process. Tailor-made proteins can be synthesized by using biology, chemistry or biotechnology methods. Biotechnology methods are preferred for synthesizing a large quantity of proteins. Typically, a suitable protein is designed before performing DNA preparation. After the protein is designed, the messenger RNA, complimentary DNA, and DNA are prepared for initiating protein synthesis.

First, a database search is performed by using PDB to identify a suitable stereostructure for a naturally occurring protein. Next, the protein structure is downloaded from the database and the amino acid sequence is obtained. Target amino acids are replaced in the amino acid sequence table, while the protein stereostructure is preserved and the protein is optimized by MM2 calculations. One example was preparation of the yeast DAO mutants M215R and L120H. Briefly, the mutants were biologically synthesized to develop a selective analyzer for (R)-amino acid such as (R)-Ala, (R)-Asp, and (R)-Glu, that are formed during food processing, and that also originate from microbial-rich food contaminants such as water, and soil. The content of (R)-amino acids has been shown to be a reliable molecular marker for ripening and also an index of food product quality.[17] The M215R mutant oxidized both the neutral and acidic (R)-Ala.[25] The L120H DAO mutant response had limited dependence on the mixture composition.[17] Yeast DAO mutants M215R and L120H were prepared from the stereostructure of 1COP by replacing one amino acid using the sequence table, followed by optimization of whole structure using MM2 calculations. The optimized conformation of the mutant was identical to the yeast DAO 1COP structure. The structure of M215R is shown in Figure 10.5. The feasibility of each mutant was analyzed by performing conformational analysis of their amino acid complexes.

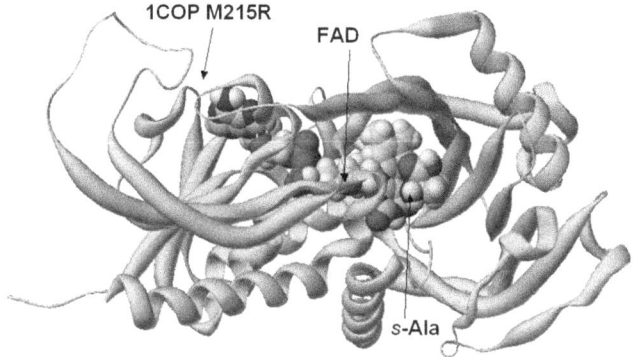

Figure 10.5 Stereostructure of the complex between 1COP M215R (yeast DAO mutant) and (*S*)-alanine. White, light-gray, dark-gray, and black balls represent hydrogen, carbon, nitrogen, and oxygen, respectively.

The significant difference in the adsorption of (*R*)-Ala was the location of Tyr240, which moved toward Arg287 and increased the size of the entrance in both M215R and L120H mutants. The difference in selectivity between M215R and L120H mutants was due to the atomic distances N(1)–O(5) and H(1)–N(2) in their D-Pro and (*R*)-Asp complexes. N(1)–O(5) was short in M215R and H(1)–N(2) was short in L120H. The ion–ion interaction conformation between the guanidyl group of the arginine and the carboxyl group of the amino acid varied depending on the complex. The parallel interaction form was observed for the M215R–(*R*)-Ala, L120H–(*R*)-Ala, and L120H–(*R*)-Asp complexes, and the perpendicular form was observed for M215R–(*R*)-Pro, M215–(*R*)-Asp, and L120H–(*R*)-Pro complexes.

When we had to replace many amino acids, we needed a little training. One example was the preparation of homo DAO. The stereostructure of human (*Homo sapiens*) DAO was constructed as a mutant of pig DAO based on the sequence.[24] The amino acid identity was 25.8 and 84.7% of yeast and pig kidney DAO, respectively. Therefore, pig kidney DAO was selected. The stereostructures of pig kidney DAO and human DAO were considered to be basically the same, due to having the same reaction mechanisms.

The arginine attracts (*R*)-alanine by Coulombic forces like the molecular interaction between arginine and benzoic acid demonstrated in Section 3.4. The optimized energy values calculated using MM2 calculations are summarized in Table 10.5.

The difference in energy values of the complexes with (*R*)-alanine and (*S*)-alanine is due to the optimized conformation of the initial protein conformation. Molecular interaction energy values favor complex formation with (*R*)-alanine. That is, the feasibility of mutants can be analyzed from their conformation and reactivity. However, mutants for chromatographic separation should be prepared with co-enzyme or a co-enzyme-like compound to maintain their stereostructure. The structure of homo DAO is shown in Figure 10.6. The optimized structure is very similar to the 1COP M215R mutant shown in Figure 10.5.

Table 10.5 Molecular interaction energy values for homo DAO and its complexes with alanine. Reproduced by permission of Taylor and Francis, ref. 26.

	MIFS	*MIHB*	*MIES*	*MIVW*
Homo DAO + FAD	−6790.7798	−5147.518	−3511.330	−343.590
(*R*)-Alanine	18.9948	0.000	16.751	1.318
(*S*)-Alanine	19.5358	0.000	17.393	1.358
Homo DAO + FAD + (*R*)-Ala complex	−6989.0151	−5229.354	−3650.106	−340.359
Homo DAO + FAD + (*S*)-Ala complex	−6973.0917	−5198.819	−3651.600	−344.330
MI (*R*)-Ala	179.2405	−81.836	−155.527	−1.913
MI (*S*)-Ala	162.7761	−51.301	−157.663	2.098

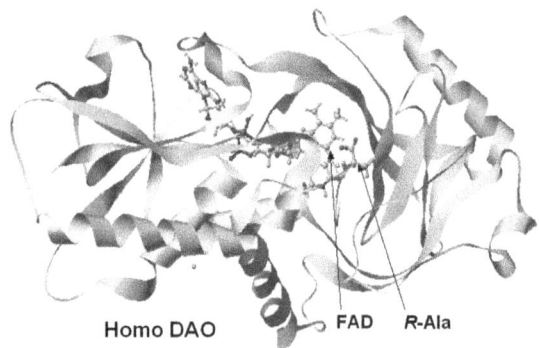

Homo DAO   FAD   R-Ala

Figure 10.6 Stereostructure of the complex between homo DAO and (*R*)-alanine.

Further studies were performed using a DAO mutant as an enantiomer separator. The original enzyme can recognize small amino acids because of its narrow cavity entrance. Figure 10.7 shows that mandelic acid is sandwiched by tyrosine (Y228), methionine (M215), and serine (S337). Figure 10.7(a) shows the extracted side view of the oxidase mutant conformation and indicates that the atom size of serine is 20% of that of the other amino acids. Figure 10.7(b) shows the top view. Tyrosine (Y240) blocks the entrance of large molecules into the cavity, and binds to key amino acids. Y240 was replaced with alanine to widen the cavity entrance. The mutant Y240A demonstrated enantiomer recognition for ibuprofen, ketoprofen, mandelic acid, and naproxen, but it lost enantiomer recognition of alanine. The molecular interaction energy ratios are summarized in Table 10.6.

Thus, this enzyme cannot be used for amino acid enantiomer purification because it oxidizes amino acids. If the oxygen of the flavine ring is replaced, the enzyme may be able to separate amino acid enantiomers. The tentative replacement of oxygen with fluorine yielded an enzyme that recognized alanine enantiomers, and the MI energy difference between the mutant and (*R*)- or (*S*)-alanine complexes was 3 kcal mol^{-1}. The MI energy difference

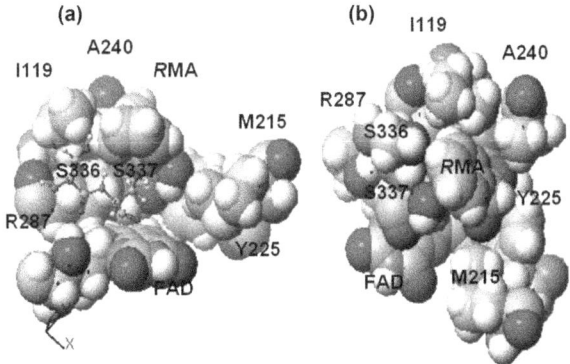

Figure 10.7 Stereostructure of the complex between Y240A and mandelic acid. (a) Side view where the atom size of serine is 20% of that of the other amino acids. (b) Top view. White, light-gray, dark-gray, and black balls represent hydrogen, carbon, nitrogen, and oxygen, respectively.

Table 10.6 Molecular interaction energy ratios $(R)/(S)$ for complexes of 1COL Y240A. Reproduced by permission of Taylor and Francis, ref. 26.

Analyte	MIFS	MIVW	MIFS	MIVW
Alanine	1.08	1.18	−2.48	1.02
Ibuprofen	—	—	1.42	0.59
Ketoprofen	—	—	1.13	1.17
Mandelic acid	0.62	1.88	1.12	0.49
Naproxen	0.70	2.13	1.63	1.80
Phenoxypropionic acid	—	—	0.83	1.63

between the original enzyme (1COP) and (R)- or (S)-alanine complexes was also 3 kcal mol^{-1}. However, FAD removal decreased enantiomer recognition.

Taken together, these results indicate that new enantioseparation proteins can be designed by using a computational chemical method. The feasibility of this approach was supported by the results of the quantitative analysis of enzyme reactions.

References

1. P. Cuatrecasas, M. Willcheck and C. B. Anfimeen, Selective enzyme purification by affinity chromatography, *Proc. Nat. Acad. Sci. U. S. A.*, 1968, **61**, 636–643.
2. R. L. Mancera, P. Källblad and N. P. Todorov, Ligand-protein docking using a quantum stochastic tunneling optimization method, *J. Comput. Chem.*, 2004, **25**, 858–864.

3. R. Tatsumi, Y. Fukunishi and H. Nakamura, A hybrid method of molecular dynamics and harmonic dynamics for docking of flexible ligand to flexible receptor, *J. Comput. Chem.*, 2004, **25**, 1995–2005.

4. P. Källblad, R. L. Mancera and N. P. Todorov, Assessment of multiple binding modes in ligand–protein docking, *J. Med. Chem.*, 2004, **47**, 3334–3337.

5. M. Kontoyianni, L. M. McClellan and G. S. Sokol, Evaluation of docking performance: comparative data on docking algorithms, *J. Med. Chem.*, 2004, **47**, 558–565.

6. G. M. Preston, Y. P. Carroll, W. B. Guggino and P. Agre, Appearance of water channels in Xenopus oocytes expressing red cell CHIP28 protein, *Science*, 1992, **256**, 385–387.

7. T. Walz, T. Hirai, K. Murata, J. B. Heymann, K. Mitusoka, Y. Fujiyoshi, B. L. Smith, P. Agre and A. Engel, The three-dimensional structure of aquaporin-1, *Nature*, 1997, **387**, 624–627.

8. D. A. Doyle, C. J. Morais, P. A. Pfuetzner, A. Kuo, J. M. Gulbis, S. L. Cohen, B. T. Chait and R. MacKinnon, The structure of the potassium channel: molecular basis of K+ conduction and selectivity, *Science*, 1998, **280**, 69–77.

9. D. Fu, A. Libson, L. J. Miercke, C. Weitzman, P. Nollert, J. Krucinski and R. M. Stroud, Structure of a glycerol-conducting channel and the basis for its selectivity, *Science*, 2000, **290**, 481–486.

10. H. Sui, B.-G. Han, J. K. Lee, P. Walian and B. K. Jap, Structural basis of water-specific transport through the AQP-1 water channel, *Nature*, 2001, **414**, 872–878.

11. A. Yarnell, Renaissance men, *Chem. Eng. News*, 2003, **81**, 35–38.

12. L. C. James and D. S. Tawfik, Catalytic and binding poly-reactivities shared by two unrelated proteins: the potential roll of promiscuity in enzyme evolution, *Protein Sci.*, 2001, **10**, 2600–2607.

13. J. L. Seffernick and L. P. Wackett, Rapid evolution of bacterial catabolic enzymes: a case study with atrazine chlorohydrolase, *Biochemistry*, 2001, **40**, 12747–12753.

14. D. M. Z. Schmidt, E. C. Mundorff, M. Dojka, E. Bermudez, J. E. Ness, S. Govindarajan, P. C. Babbitt, J. Minshull and J. A. Gerlt, Evolutionary potential of (β/α)8-barrels: Functional promiscuity produced by single substituents in the enalase superfamily, *Biochemistry*, 2003, **42**, 8387–8793.

15. A. Yarnell, The power of promiscuity, *Chem. Eng. News*, 2003, **81**, 33–35.

16. T. Hanai, Molecular modeling for quantitative analysis of molecular interaction, *Lett. Drug Des. Discovery*, 2005, **2**, 232–238.

17. S. Sacchi, E. Rosini, G. Molla, M. S. Pilone and L. Pollegioni, Modulating D-amino acid oxidase substrate specificity: production of an enzyme for analytical determination of all D-amino acids by directed evolution, *Prot. Eng., Des. Sel.*, 2004, **17**, 517–525.

18. N. Markoglou, R. Hsuesh and I. Wainer, Immobilized enzyme reactions based upon the flavoenzymes monoamine oxidase A and B, *J. Chromatogr., B*, 2004, **804**, 295–302.

19. H. A. Krebs, Metabolism of amino acids III. Determination of amino acids, *Biochem. J.*, 1935, **29**, 1620–1644.

20. R. Konno and Y. Yasumura, D-Amino-acid oxidase and its physiological function, *Int. J. Biochem.*, 1992, **24**, 519–524.

21. D. Malcolm and K. Kjell, D-Amino acid oxidase II. Specificity, competitive inhibition and reaction sequence, *Biochem. Biophys. Acta*, 1965, **96**, 368–382.

22. M. S. Pilone, D-Amino acid oxidase: new findings, *Cell. Mol. Life Sci.*, 2000, **57**, 1732–1747.

23. RCSB Protein Data Bank, www.rcsb.org/pdb/.

24. NP 001908, D-amino acid oxidase [Homo sapiens], www.ncbi.nlm.nih. gov/entrez/query.fcgi?CMD.

25. S. Sacchi, S. Lorenzi, G. Molla, M. S. Pilone, C. Rossetti and L. Pollegioni, Engineering the substrate specificity of D-amino acid oxidase, *J. Biol. Chem.*, 2002, **277**, 27510–27516.

26. T. Hanai, Quantitative *in silico* analysis of ion exchange from chromatography to protein, *J. Liq. Chromatogr. Rel. Technol.*, 2007, **30**, 1251–1275.

27. T. Hanai, Quantitative *in silico* analysis of enzyme reactions: Comparison of D-amino acid oxidase and monoamine oxidase, *American Biotechnology Laboratory*, 2007, **25**, 8–13.

Mechanisms of Highly Sensitive Detection

11.1 Detection of Bromate in Ion Liquid Chromatography

11.1.1 Introduction

Bromate is considered a carcinogen, and the World Health Organization (WHO) has recommended a provisional bromate guideline value of 25 mg L^{-1}, which is associated with an excess lifetime cancer risk of 7×10^{-5}, because of limitations in the available analytical and treatment methods.[1] A highly sensitive analytical method was therefore developed. Bromate in ozonized water was detected with very high sensitivity by ion liquid chromatography with post-column reaction detection using ultra-violet absorption. With the addition of nitrite for the post-column reaction, the sensitivity was improved 738-fold. The detection limit was 0.35 mg L^{-1}, and the linear range was over four orders of magnitude, from 0.5 to 10 mg L^{-1} (ref. 2). The addition of ClO$^-$ improved the sensitivity 327-fold.[2]

Chiu and Eubanks examined bromide spectrophotometrically; they proposed a reaction mechanism and suggested that the end product is tribromide (Br$_3^-$).[3] The proposed reactions are as follows:

$$Br^- + 3ClO^- \rightarrow BrO_3^- + 3Cl^-$$

$$BrO_3^- + 5Br^- + 6H^+ \rightarrow 3Br_2 + 3H_2O$$

$$Br_2 + Br^- \rightarrow Br_3^-$$

In addition, bromate and chlorate were determined by potentiometric titration after reduction with sodium nitrite.[4] Sodium nitrite was added to

RSC Chromatography Monographs No. 19
Quantitative *In Silico* Chromatography: Computational Modelling of Molecular Interactions
By Toshihiko Hanai
© Toshihiko Hanai, 2014
Published by the Royal Society of Chemistry, www.rsc.org

sodium bromide for online hydrobromic acid generation in this system, and highly sensitive detection was achieved.[2] However, the reaction mechanism and the final product were not determined. Tuchler *et al.* studied bimolecular interactions, and directly detected the internal conversion involving $Br(^2P_{1/2}) + I_2$ initiated from a van der Waals dimer.[5] The reaction complex was formed from a van der Waals dimer precursor, $HBr \cdot I_2$. The resulting product, the highly vibrationally excited molecular I_2, was monitored by resonance-enhanced multiphoton ionization combined with time-of-flight mass spectroscopy. The HBr constituent of the precursor, $HBr \cdot I_2$, photodissociated at 220 nm. The H atom departed instantaneously, allowing the remaining electronically excited $Br(^2P_{1/2})$ to form a collision complex, $(BrI_2)^*$, in a restricted region along with the $Br + I_2$ reaction coordinate determined by the precursor geometry. Sims *et al.* reported the femtosecond real-time probing of the bimolecular reaction $Br + I_2$, and summarized a number of trihalogen intermediates observed in matrix isolation studies.[6] Computational chemistry can predict the electronic spectra of a variety of compounds which cannot be obtained in their pure form. This tool was applied to study the highly sensitive detection of bromate in ion chromatography. Several possible ions and molecules and their complexes were constructed by a molecular editor, and optimized by molecular mechanics (MM2) and MOPAC (PM3 and AM1) calculations. Their possible electronic spectra were then obtained the ZINDO (INDO/1)-Vizualyzer in the CAChe™ program. The λ_{max} of the spectra and the transition dipole were calculated using the ProjectLeader program. Further study was carried out using the version-up programs PM6 and RM1. The properties used for the molecular mechanics calculations were: bond stretch, bond angle, dihedral angle, improper torsion, van der Waals, electrostatic (MM2 bond dipole), hydrogen bond, and the cut-off distance for van der Waals interactions: 9.00 Å, updating van der Waals interactions every 50 interactions. The parameters for the MOPAC calculation were geometry search options (precise, minimized by NLLSQ, optimized geometry by BFGS), and properties [Mulliken population, energy partitioning, polarizabilities, localize, thermo, rotational symmetry (C1)] in the CAChe™ program. The predicted data were compared with those obtained experimentally.

11.1.2 Theory

According to the Beer–Lambert law, the ratio of the intensity of the light at the inlet site, $I_0(v)$, and the outlet site, $I(v)$ is given by the following equation:

$$\frac{I(v)}{I_0(v)} = 10^{-\kappa(v)Dx} = e^{-\ln 10 \kappa(v)Dx}$$

That is, the absorbance is given by:

$$A = \log 10 \frac{I(v)}{I_0(v)} = \kappa(v)Dx$$

where $I(v) = I_0(v) \times 10^{\kappa(v)Dx}$, and $\kappa(v)$ is the molar extinction coefficient *i.e.*, the molar absorptivity.

The following equation gives the relationship between the experimentally measured and theoretical absorption intensities:[7]

$$\frac{10^3 \ln 10 c}{N_A h} \int \frac{\kappa(v)}{v} dv = \frac{8\pi^3}{h^2} |\langle j | \hat{k} \cdot (\text{mean } x) er | i \rangle|^2$$

The intensity of the spectrum is given by the following equation:

$$f(\text{theoretical}) = \frac{8\pi^2 m v}{3h} |\langle j | \hat{k} \cdot (\text{mean } x) er | i \rangle|^2$$

where $|\langle j | \hat{k} \cdot (\text{mean } x) er | i \rangle|^2$ is the transition dipole, D is the concentration of the analyte, x is the pass length of light, c is the speed of light, N_A is Avogadro's constant, h is Plank's constant, v is the frequency, j is the excited state, i is the ground state, κ is Boltzmann's constant, er is the transition dipole moment, $\hat{k}$ is the polarized light vector and m is the mass of the atom.

$$\kappa(v) = \frac{1}{Dx} \log 10 \frac{I}{I_0} \propto |\langle j | \hat{k} er | i \rangle|^2$$

That is, the molar absorptivity, $\kappa(v)$, is related to the transition dipole.

11.1.3   *In Silico* Analysis of the Reaction Mechanisms

The computational chemical calculation was performed using the CAChe™/ Scigress programs. The molar absorptivity of several ions, molecules and complexes was directly measured on spectra obtained by ZINDO-Visualization after their conformations were optimized using the MM2 and MOPAC (PM3, AM1, PM6 and RM1) programs. The spectra were also obtained using the ZINDO-Visualizer at INDO/1 geometry. Their transition dipoles were calculated by the ProjectLeader program using the MM2 and MOPAC (PM3, AM1, PM6 and RM1) programs. The values of molar absorptivity and the transition dipoles are summarized in Table 11.1; the values of their complexes with nitrite and chlorite are included. The relationship between the transition dipole and the molar absorptivity was:

$$Y = 1.57.422 X^2 + 3017.582 X - 2368.256, \quad r^2 = 0.993, \quad n = 14$$

where Y is the molar absorptivity $\kappa(v)$ (mol cm^{-1}) and X is the transition dipole (Debye).

Chromatographic sensitivity is directly related to the molar absorptivity of the analytes. The molar absorptivities of Br_3^-, and the $Br_2 + Br^-$ complex were very high, about 190 000 mol cm^{-1}. The measurements of molar absorptivity and λ_{max} were not easily obtained, but these values can be automatically calculated using the ProjectLeader program. Br_3^- and the $Br_2 + Br^-$ complex have similar structures, as shown in Figure 11.1.

Table 11.1 Analyte properties. $\Delta_f H$ represents heat of formation (from PM3 calculations), $\lambda_{max} > 200$ nm, td represents the transition dipole, ma represents the molar absorptivity. Reproduced by permission of American Chemical Society, ref. 22.

Analyte	PM3				PM6[b]		RM1[c]		INDO/1[d]	
	$\Delta_f H$/kcal mol^{-1}	λ_{max}/ nm	td/ Debye	$\kappa(v)$/mol cm^{-1}	λ_{max}/ nm	td/ Debye	λ_{max}/ nm	td/ Debye	λ_{max}/ nm	td/ Debye
Br$^-$	−56.00	—	—	^{a}	—	—	—	—	—	—
Br$_2$	−4.92	602	0.277	81	518	0.332	220	6.645	519	0.330
Br$_3^-$	−105.69	258	12.300	188 200	245	11.887	242	11.801	242	11.801
BrO$_3^-$	−39.59	462	0.927	595	330	0.466	424	1.325	625	0.549
NO$_2^-$	−42.93	208	4.005	24 660	206	4.067	204	4.021	207	4.038
ClO$^-$	−32.97	234	0.409	458	212	0.459	223	0.432	237	0.404
Br$_2$ + NO$_2^-$ /A	−98.49	224	7.227	74 550	219	6.550	218	6.994	231	6.474
Br$_2$ + NO$_2^-$ /B	−104.80	239	8.183	91 440	243	8.610	271	7.438	271	7.428
Br$_2$ + NO$_2^-$ /C	−99.93	230	5.550	43 370	222	6.371	221	6.645	253	5.263
Br$_2$ + Br$^-$	−105.69	258	12.327	188 250	245	11.901	242	11.801	242	11.804
Br$_2$ + ClO$^-$/A	−113.02	228	4.758	30 670	226	5.742	208	3.274	222	5.166
Br$_2$ + ClO$^-$/B	−74.52	228	10.385	148 400	209	9.823	441	4.149	209	10.014
Cl$^-$	−51.22	—	—	^{a}	—	—	—	—	—	—
Cl$_2$	−11.57	410	0.464	336	380	0.499	425	0.447	407	0.468
Cl$_3^-$	−91.06	214	10.615	168 200	206	10.330	201	10.125	204	10.233
Cl$_2$Br$^-$	−95.30	247	10.727	148 760	223	10.730	214	10.472	224	10.566
Cl$_2$ + ClO$^-$	−87.51	243	5.220	34	446	4.360	416	4.413	233	4.098
BrO$_3^-$ + NO$_2^-$	−51.11	209	4.117	30 166	199	3.958	205	4.370	198	4.336
BrO$_3^-$ + ClO$^-$	−70.28	219	2.185	15 888	292	2.739	201	2.551	202	2.551
I$_3$	−85.58	221	12.738	236 800	218	12.606	219	12.659	240	14.108
I$_2$Br	−87.59	229	12.276	209 360	226	11.992	225	12.208	243	13.292

^{a}Molecule lacks electronic state information.
^{b}PM6 is the version-up program of PM3, values from refs. 43 and 54.
^{c}RM1 is the version-up program of AM1, values from refs. 43, 54 and 55.
^{d}INDO/1 values from refs. 45 and 47–49.

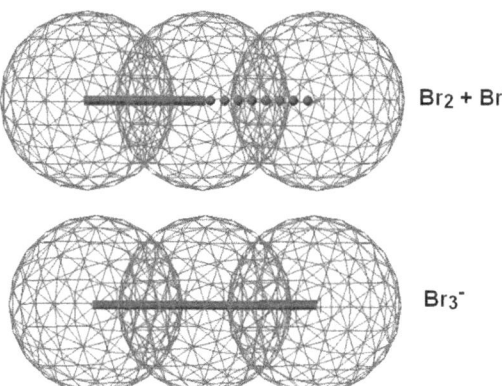

Figure 11.1 Conformation of Br$_3^-$, and the Br$_2$ + Br$^-$ complex.

The complex between Br$_2$ and Br$^-$ was automatically formed after optimization of the structure, and the heat of formation $(\Delta_f H)$ value was the lowest among the analytes listed in Table 11.1; the values were about

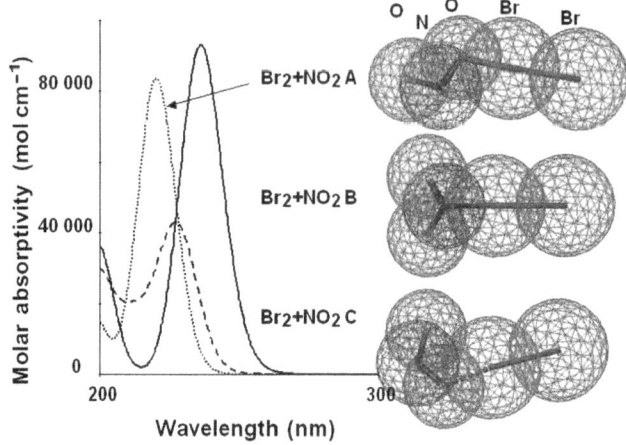

Figure 11.2 Possible Br_2 and NO_2^- complexes A–C and their spectra.

−106 kcal mol^{-1}. The $(\Delta_f H)$ value of the complex was the same as that for Br_3^-. This result indicated that Br_3^- can be formed where Br_2 and Br^- co-exist as the BrI_2 collision complex.[5,6] The question arises as to how NO_2^- and ClO^- act in the reaction; do these ions form different compounds or complexes with bromide or bromine during the highly sensitive detection of bromate? The $Br_2 + NO_2^-$ and $Br_2 + ClO^-$ complexes were thus constructed, and their structures was optimized by MM2 and PM3 calculations. Br_2 and NO_2^- formed three types of conformation, as shown in Figure 11.2. The structures A and B were obtained as molecules and the structure C was obtained as a transition state. Their $\Delta_f H$ values are given in Table 11.1 as $Br_2 + NO_2^-$/A, $Br_2 + NO_2^-$/B and $Br_2 + NO_2^-$/C, respectively. The $\Delta_f H$ values were low; the lowest energy value was −105 kcal mol^{-1}, about the same as that of the $Br_2 + Br^-$ complex. The structure with the lowest energy value was structure C in Figure 11.2.

However, its molar absorptivity was less than half that of the $Br_2 + Br^-$ complex. This result suggested that NO_2^- may form a complex with Br_2. However, such a complex may not be the final product, due to its low sensitivity. The λ_{max} values of structures A, B and C in Figure 11.2 were 224, 239 and 230 nm, and differed from those of the $Br_2 + Br^-$ complex and Br_3^-, whose λ_{max} values were 258 nm, as shown in Figure 11.3. The λ_{max} of 258 nm was the closest wavelength to that observed experimentally (265 nm). This result also suggested that such a complex might not be the final product. The formation of these complexes was supported by the negative values of their van der Waals energies calculated using the MM2 program (Table 11.1). $Br_2 + ClO^-$ formed two types of conformation, as shown in Figure 11.3. Their λ_{max} values were 228 nm, and structure B showed the highest transition dipole value. However, λ_{max} was lower than that of the $Br_2 + Br^-$ conformation. Bromide did not form a complex with NO_2^-. Bromide, bromine,

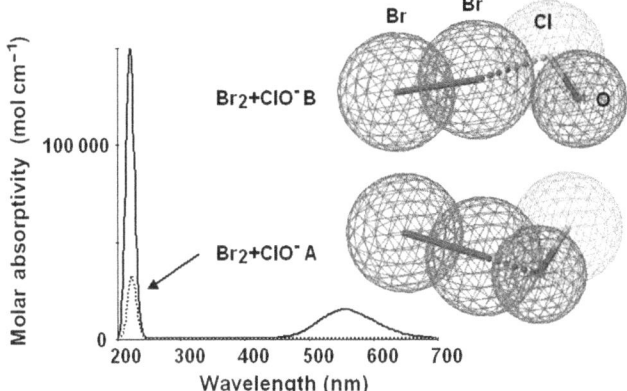

Figure 11.3 Possible Br_2 and ClO^- complexes and their spectra.

bromate, and nitrite were not highly sensitive analytes, due to their low transition dipole values and λ_{max} values.

Another question was, why the sensitivity measured in the presence of ClO^- about the half of that measured in the presence of NO_2^-? The reaction processes were estimated according to the proposal by Chiu and Eubanks.[3]

$$2BrO_3^- + 4NO_2^- + 4H^+ \rightarrow Br_2 + 4HNO_3 + 2H_2O$$

$$Br_2 + Br^- \rightarrow Br_3^-$$

$$2BrO_3^- + 4ClO^- + 6H^+ \rightarrow Br_2 + Cl_2 + 2HClO_3 + 3H_2O$$

$$Br_2 + Br^- \rightarrow Br_3^- \text{ and } Cl_2 + Br^- \rightarrow Cl_2Br^-$$

The value of the molar absorptivity of Cl_2Br^- (148 760) was lower than that of Br_3^- (188 200), and the λ_{max} of Cl_2Br^- (247 nm) was also lower than that of Br_3^- (258 nm) as shown in Figure 11.4. Therefore, the final sensitivity using ClO^- as the reaction reagent was less than that using NO_2^-.

Bromate formed a complex with nitrite. However, the complex may be unstable, due to the high value of $\Delta_f H$. The electronic spectra of the complexes are shown in Figure 11.5. This complex is not a candidate for the highly sensitive detection of bromate due to its low transition dipole value and λ_{max}. Bromine can form a complex with ClO^-. However, the $\Delta_f H$ was low for a complex with a higher transition dipole. This means that the Br_2-$+ ClO^-$ complex may not be a candidate for the highly sensitive detection of bromate.

The above results indicated that the highly sensitive detection of chlorate and iodinate can be achieved by using the techniques employed for bromate analysis. The sensitivity of chlorate and iodinate will be 90 and 111% of bromate. However, the λ_{max} values of Cl_2Br^- and I_2Br^- are 10 and 30 nm lower than that of Br_2Br^-. If Cl_3^- and I_3^- are the final products, a specific

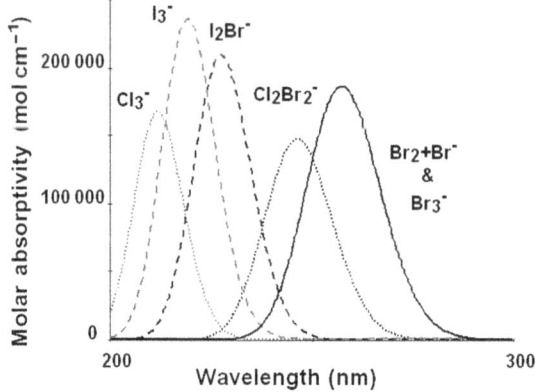

Figure 11.4 Electronic spectra of Br_3^-, $Br_2 + Br^-$, Cl_2Br^-, I_2Br^-, I_3^-, and Cl_3^-.

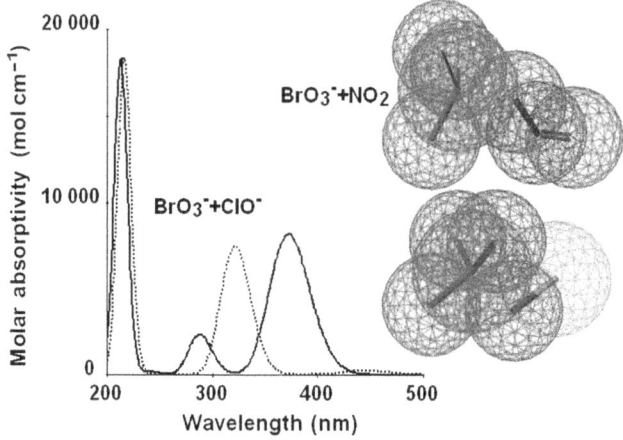

Figure 11.5 Possible bromate complexes and their spectra.

ion generator should be constructed. However, the detection wavelengths of Cl_3^- and I_3^- are lower than those of Cl_2Br^- and I_2Br^-, and selective detection may not be easy, as shown in Figure 11.4. The computational chemical analysis of fluorinate could not performed due to the lack of stable electronic information for fluorinate. An AM1 calculation can be used to optimize these structures. However, the present AM1 calculation did not give complex forms due to the fixed atomic distances. The λ_{max} values were usually shorter than those obtained by PM3, and the values of molar absorptivity were smaller. For example, the maximum atomic distances of Br_3^- calculated by PM3 and AM1 were 5.065 and 4.575 Å, respectively. Their λ_{max} and molar absorptivity values, calculated by PM3 and AM1 were 258 nm/188 200 mol cm^{-1} and 217 nm/175 150 mol cm^{-1}, respectively. Furthermore, the version-up programs PM6 and RM1 did not demonstrate better results than those obtained by using PM3 program, based on the λ_{max} of Br_3^-. These up-graded-version programs produced the blue-shift in λ_{max}.

The above results demonstrated the feasibility of computational chemistry to predict the spectra of a variety of compounds that cannot be obtained as pure compounds, and it was used to study the highly sensitive detection of bromate in ion chromatography. Comparison of the experimental and predicted λ_{max} of the spectra and the transition dipole indicated that Br_3^- was the probable reaction product, and that NO_2^- and ClO^- accelerated the reaction.

11.2 Chemiluminescence Detection

11.2.1 Introduction

Fluorescence detection is commonly used for highly sensitive detection in chromatography. The limiting factor in fluorescence detection is the stray light that increases the baseline noise, and in order to avoid this problem, chemical excitation (chemiluminescence) methods are applied. The efficiency of a chemiluminescence reaction can be expressed as the number of light-emitting molecules related to the number of all exited molecules (luminescence efficiency). Peroxyoxalate luminescence can also be used for assaying hydrogen peroxide or the number of fluorophores. However, most of the compounds assayed by peroxyoxalate chemiluminescence do not posses luminescence, and a suitable fluorescence tagging operation must precede the actual assay.[8] On the other hand, reducing agents can be analyzed without pre-derivatization. Phenacylalcohol derivatization has been used for the highly sensitive analysis of carboxylic acids by chemiluminescence. The chemiluminescence detection mechanisms of phenacylesters and steroids have been suggested to be the same, based on their chemical structures, and the sensitivity difference has been suggested to be due to the stability of the compounds in alkaline solution.[9]

Chemiluminescence analysis has been applied to a variety of atoms and compounds in different techniques, such as gas chromatography, liquid chromatography, supercritical fluid chromatography, capillary electrophoresis, chip analysis, flow-injection analysis, and immunoassay. The high reactivity makes it suitable for post-column reaction detection for chromatography and flow-injection analysis. A targeted glycosylated albumin has been analyzed[10] using chemiluminescence detection without separation from glycosylated albumin mixtures in Maillard reaction products,[11] where fluorescence detection would have required chromatographic separation of such complex mixtures.

Organic reducing compounds, reducing sugars, ascorbic acid, uric acid, *etc.*, have been detected by the chemiluminescence method using lucigenin and luminol.[12–15]

11.2.2 Theory

A variety of steroids have been detected by chemiluminescence methods using lucigenin and luminol, in highly alkaline solution with different sensitivities. The important effect of the 1,2-enediol-type structure on the chemiluminescence reaction has also been reported. The reaction processes are shown below:

It is known that corticosteroids and phenacylesters of carboxylic acids produce intense chemiluminescence in basic solution,[9] and a chemiluminescence method has been applied for the selective detection of glycosylated human serum albumin.[10] Glycosylated albumin changes to 1,2-enaminol form after Amadori rearrangement during the Maillard reaction. 1,2-Enaminol produced superoxide during conversion to glucosone,[11,16,17] and the reaction center of steroids is considered to be carbon (17) in Figure 11.6.

The reaction process is considered to be the same for similar compounds, but the sensitivity of chemiluminescence has been suggested to be structure-dependent.[8,18] The reactivity of a compound contributes to its sensitivity. Phenacylalcohol derivatives have been detected with different sensitivities in chemiluminescence analysis.[9,19] In radical reactions, the reaction proceeds as follows: in buffered solutions, a compound such as phenacylalcohol is easily attacked by oxidation if traces of a copper or iron salt are present,[20] and the superoxide reacts with luminol or lucigenin to produce chemiluminescence[21] as shown in Figure 11.7.

The chemiluminescence reaction of lucigenin with reducing sugars was proposed as the formation of the 1,2-enediol tautomer. The intermediate

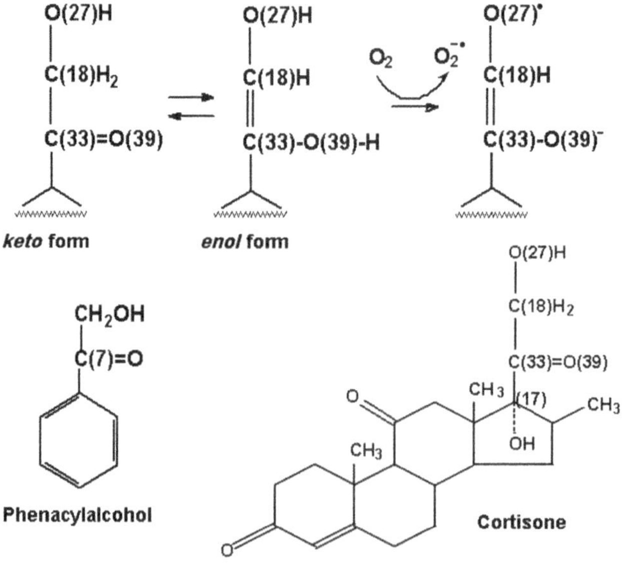

Figure 11.6   Reaction processes of steroids.

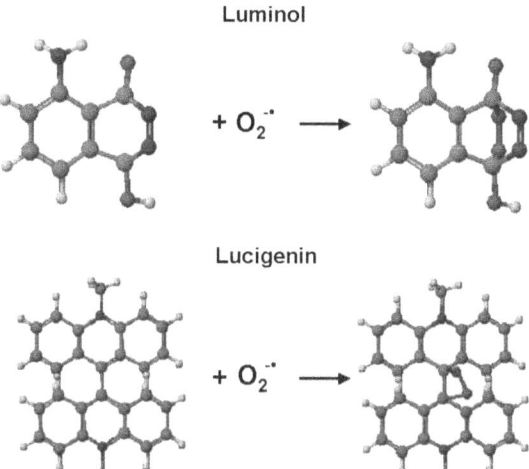

Figure 11.7 The reaction of luminol and lucigenin with the superoxide radical. Reproduced by permission of Society of Computer Chemistry, ref. 59.

enediol is oxidized by lucigenin in subsequent reaction steps.[22] The most important reaction process was, therefore, considered to be the *keto–enol* rearrangement. The partial charge of the carbon atom of the carbonyl group changed significantly. The different chemiluminescence detection mechanisms of phenacylesters and steroids have been suggested to be due to differences in the stability of these compounds in alkaline solution. Further computational chemical analyses of the relative sensitivity of other compounds, measured by liquid chromatography and flow-injection analysis[9,16] were performed quantitatively to determine the reactivity of the analytes.

11.2.3 Computational Chemical Analysis of Reaction Processes

Since the chemiluminescence reaction detects superoxide, the sensitivity depends on the amount of superoxide present. The reaction from the *keto* to the *enol* form was considered to be an important process, similar to that observed for glycosylated albumin;[10,11,16,17] the reaction process was the same, but the reactivity was different, and the amount of superoxide present may have been related to the amount of *enol*-form compound produced. According to the Maillard reaction, the transition state should be the Amadori rearrangement process for glycosylated albumin.[10] The *keto–enol* rearrangement of phenacyl alcohol and steroids was considered to be the key point for analyzing sensitivity. The change in properties of key atoms during the reaction process is important in predicting the sensitivity. A variety of molecules were therefore constructed using the Molecular Editor of the CAChe™ program, and their properties were calculated using the MOPAC program after optimizing their structures using molecular mechanics.

Table 11.2 Properties of phenacylalcohol derivatives and steroids. rs represents relative sensitivity,[20] C7 and C33 represent the 7th and 33rd carbons of the phenacylalcohol derivatives and steroids, A represents the *enol* form, and Δapc represents the difference between apcC7/apcC7A and apcC33/apcC33A. The units for apc values are au. Reproduced by permission of Society of Computer Chemistry, ref. 59.

Phenacylalcohol derivative	rs^a	apcC7	apcC7A	ΔapcC7
Phenacylalcohol	1.00	0.2631	0.0641	0.1990
2-Acetoxy-	1.09	0.2646	0.0660	0.1986
2-Acetoxy-4-bromo-	2.07	0.2630	0.0585	0.2045
2-Acetoxy-4-nitro-	3.61	0.2589	0.0465	0.2124
2-Acetoxy-4-phenyl-	1.11	0.2651	0.0684	0.1967
Steroids	rs	apcC33	apcC33A	ΔapcC33
Betamethasone	0.16	0.1804	−0.1343	0.3147
Cortisone	2.62	0.2181	−0.1544	0.3765
Corticosterone	0.65	0.2082	−0.1083	0.3165
Dexamethasone	1.38	0.1724	−0.1661	0.3385
Deoxycorticostereone	0.20	0.2059	−0.0969	0.3028
Hydrocortisone	1.00	0.1755	−0.1528	0.3293
Methylprednisolone	0.99	0.1791	−0.1499	0.3290
Tetrahydrocortisol	1.45	0.1767	−0.1683	0.3450

[a]Values from ref. 1.

The change in the atomic partial charge (Δapc) of key atoms before and after their *keto–enol* form rearrangement was used to analyze their reactivity.The HOMO density, frontier density, superdelocalizability, LUMO density, and apc (units au) of key elements were calculated. The reaction processes are shown in Figure 11.6 where only the important atoms are numbered. The sensitivity of phenacylalcohol derivatives and steroids and the calculated apc values are summarized in Table 11.2.

Generally, the electron density of the LUMO is a key property for studying reactivity, but the values did not explain the sensitivity difference, and the electron density of the HOMO was also not useful for explaining the sensitivity, thus, the electron density values of the LUMO and HOMO are not given. The frontier density and the superdelocalizability also should not be used for this purpose, and their values are also not given.

The apc of atoms was correlated with the relative sensitivity of phenacyl alcohol derivatives and steroids. The correlation coefficients for several *keto*-form steroids varied from 0.221 to 0.842, and for *enol* forms from 0.163 to 0.800, as summarized in Table 11.3.

The apc values of other atoms did not change significantly and were not used for further study. This change in apc of key atoms is useful information for studying the feasibility of the *keto–enol* form rearrangement, since the Δapc values of these atoms correlated with their sensitivity. The best correlation coefficient was obtained between the relative sensitivity and Δapc of C7 of the phenacylalcohol derivatives, and that of C33 of the steroids.

Table 11.3 Correlation coefficients between the apc values of key atoms for eight steroids. Reproduced by permission of Society of Computer Chemistry, ref. 59.

Atom	keto form	enol form
18th Carbon	0.842	0.163
33rd Carbon	0.221	0.655
27th Oxygen	0.684	0.800
39th Oxygen	0.784	0.502

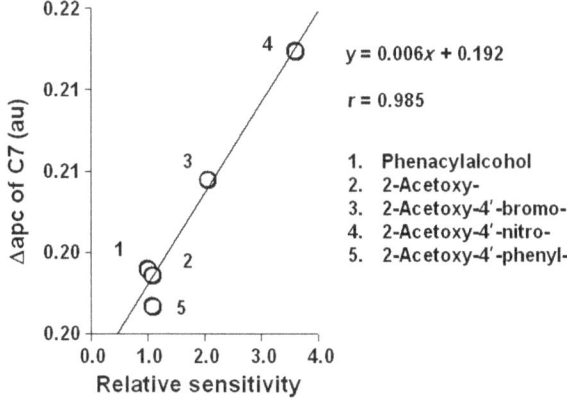

Figure 11.8 The relationship between relative sensitivity and the difference in atomic partial charge of the C7 atom of steroids.
Reproduced by permission of Society of Computer Chemistry, ref. 59.

The correlation coefficients were 0.985 ($n = 5$) and 0.964 ($n = 8$) for phenacylalcohol derivatives and steroids, respectively. The relationship between phenacylesters and steroids is shown in Figures 11.8 and 11.9. The correlation coefficients for other atoms of steroids were 0.066 for C18, 0.342 for O27, and 0.240 for O39.

The analytical method described here can predict the relative sensitivity detected by chemiluminescence reactions using luminol, and computational chemical analysis can help to predict sensitive detection in liquid chromatography. The reaction mechanisms of other compounds under similar conditions should be the same as those described for the above compounds. Further computational chemical study will clarify the reaction mechanisms of chemiluminescence and their sensitivity differences.

The reaction from the *keto* to the *enol* form was considered to be an important process, similar to that observed for phenacylalcohols and steroids.[10] The reaction process was the same, but the reactivity was different, and the amount of superoxide present may have been related to the amount of *enol*-form compound produced. The relative sensitivities of derivatized phenacylalcohols (from the literature) and the calculated apc values of carbonyl carbons are summarized in Table 11.4.

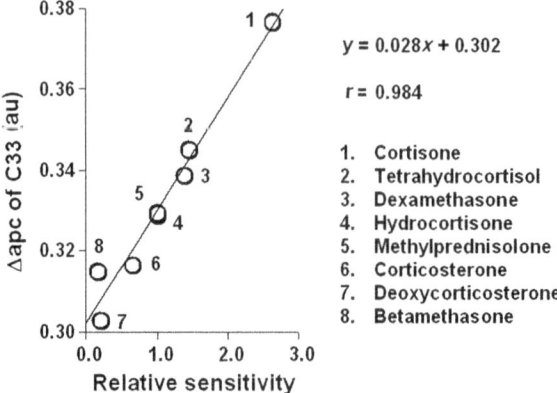

$$y = 0.028x + 0.302$$

$$r = 0.984$$

1. Cortisone
2. Tetrahydrocortisol
3. Dexamethasone
4. Hydrocortisone
5. Methylprednisolone
6. Corticosterone
7. Deoxycorticosterone
8. Betamethasone

Figure 11.9 The relationship between relative sensitivity and the difference in atomic partial charge of the C33 atom of phenacylesters. Reproduced by permission of Society of Computer Chemistry, ref. 59.

Table 11.4 Relative sensitivity and Δapc values for key carbon atoms of phenacylalcohols, $R_1-C_6H_3-CO-CH_2-R_2$. Reproduced by permission of Taylor and Francis, ref. 60.

Phenacylalcohol derivative			*Relative sensitivity*		
R_1	R_2	$Δapc/au$	Benzyl[a]	Acetyl[a]	Acetyl[b]
Hydrogen	Benzoyl	0.1918	0.65	—	—
Bromo	Benzoyl	0.2029	0.13	—	—
Nitro	Benzoyl	0.2170	1.25	—	—
Phenyl	Benzoyl	0.1908	0.39	—	—
Hydrogen	Hydroxyl	0.1990	1.00	1.00	1.00
Hydrogen	Acetyl	0.1986	—	0.83	1.09
Bromo	Acetyl	0.2045	—	1.54	2.07
Nitro	Acetyl	0.2124	—	1.51	3.61
Phenyl	Acetyl	0.1967	—	0.39	1.11
r	—	—	0.558	0.842	0.985

[a]Values taken from ref. 9.
[b]Values taken from ref. 19.

The relative sensitivities of acetylated phenacylalcohols, measured by different systems were related to the Δapc values of the carbonyl carbon between the original and Amadori rearranged forms. The r value was 0.8 using lucigenin[9] and 0.985 ($n = 5$) using luminol.[20] These results indicated that the relative sensitivity can be predicted using the apc calculated by a computational chemical method. However, the relationship for benzoylesters was poor. The r value was 0.558 ($n = 5$). This previous result was based on experimental data in which the data point for 4′-bromophenacyl-benzoylesters was 0.13. This value may actually be 1.3, and r should be 0.826 ($n = 5$) based on the results obtained for acetylated compounds. Unfortunately, this old result could not be reconfirmed.

Atomic partial charges were subjected to analysis of the detection sensitivity of a variety of compounds measured by the same system. The detection limits and apc values are shown in Table 11.5. The apc values of saccharides were calculated as the aldehyde form, but these sugars exist as hexoses and the concentration of the aldehyde form was not known under these experimental conditions. Therefore, these values were eliminated from the correlation coefficient calculations. The sensitivity of creatinine was very low, therefore creatinine was also eliminated from the calculations. The correlation coefficients were 0.704 ($n = 7$) for Group 1, and 0.960 ($n = 8$) for Group 2. In the above calculation, uric and ascorbic acid have two carbonyl groups and, therefore, two apc values were combined.

The HOMO density, frontier density, superdelocalizability, LUMO density, and apc of all elements were calculated using the MOPAC program after optimizing their structures with molecular mechanics. The apc of the *keto* and *enol* forms of phenacylalcohol were constructed using the CAChe™ system and are shown in Figure 11.10. The electron density of the carbonyl carbon was significantly reduced from the *keto* to the *enol* form.

The Δapc value of key atoms is useful for studying the feasibility of the *keto-enol* form rearrangement, as the balance of apc values of these atoms can be correlated with their sensitivity, and the Δapc values of a series of compounds such as phenacylalcohol derivatives.

Table 11.5 Chemiluminescence intensity (CLI) and Δapc values. Reproduced by permission of Taylor and Francis, ref. 60.

	Analyte	Δapc/au	CLI[a] Group[b]	Group[c]
1	Uric acid	0.4421	900.4	—
2	Phenacylalcohol	0.3459	534.7	267
3	Cortisone	0.3725	384.8	42
4	Ascorbic acid	0.4428	250.4	—
5	Corticosterone	0.3165	95	—
6	Glutathione	0.1368	27.6	—
7	Cysteine	0.0477	24.7	—
8	Fructose	0.3605	2.6	—
9	Glucose	0.1463	1	1
10	Creatinine	0.1962	0.1	—
11	Galactose	0.1463	—	1
12	Mannose	0.1463	—	1
13	Glucosamine	0.1789	—	1.9
14	Gluceroaldehyde	0.2837	—	138
15	Glycoaldehyde	0.2792	—	181
16	Cortisol	0.328	—	41
17	Tetrahydrocortisol	0.345	—	47
18	Dihydroxyacetone	0.3245	—	212
19	Benzoin	0.246	—	173

[a]CLI values are relative to glucose as a standard.
[b]Values from ref. 16.
[c]Values from ref. 9.

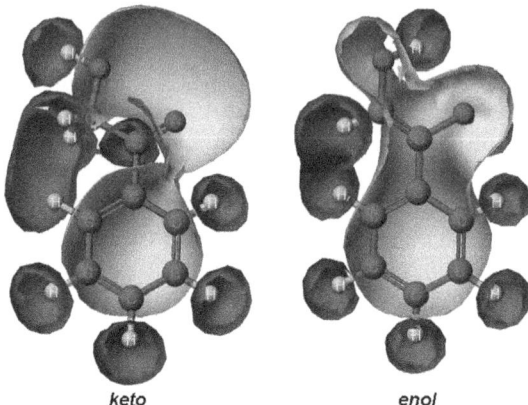

keto   enol

Figure 11.10   Electron density of the *keto* and *enol* forms of phenacylalcohol.

The analytical method described here can predict the relative sensitivity detected by the chemiluminescence reactions of luminol and lucigenin, and computational chemical analysis can help to predict sensitive detection in liquid chromatography and ion chromatography.[23] In the latter, reaction products are not easily obtained. Moreover, the reaction mechanisms of other compounds under similar conditions should be the same as those described for the above compounds. Further computational chemical study will elucidate the reaction mechanisms of chemiluminescence and the sensitivity differences, and facilitate further improvement of sensitivity.

11.2.4   Chemiluminescence Intensity Related to Toxicity

The computational chemical analysis above, targeted the productivity of superoxide from a *keto–enol* rearrangement to study chemiluminescence intensity in analytical chemistry. Superoxide is toxic *in vivo*. The apc was therefore related to biological activities, such as toxicity (rat oral LD_{50}), the efficacy of the steroids as an endermic liniment, and the contraction index of blood vessel by steroids.

The toxicity lethal dose 50% (LD_{50}, mg kg^{-1}) was calculated using the TOPKAT program from Fujitsu. The values for the phenacylalcohol derivatives with apc are summarized in Table 11.6. The relationship between the chemiluminescence intensity or Δapc, and rat oral LD_{50} for phenacylalcohol derivatives is summarized in Figures 11.11 and 11.12.

The high correlation coefficient indicated that the measurement of chemiluminescence intensity provides a quantitative measurement of the toxicity of an analyte. Furthermore, the calculation of Δapc by computational chemical methods can be used to estimate rat oral LD_{50}.

These experimental and computational chemical methods will facilitate the rapid screening of drug candidates by chemiluminescence assay. Many steroid drugs are used for the treatment of skin diseases. Superoxide

Table 11.6 The molecular properties of some phenacylalcohol derivatives. Reproduced by permission of World Scientific, ref. 25.

Phenacylalcohol derivative	$\Delta apc/au^a$	CLI^b	$LD_{50}/\text{mg kg}^{-1c}$
Phenacylalcohol	0.1990	1.00	2.083
2-Acetyl-	0.1986	1.09	1.942
2-Acetyl-4-bromo-	0.2045	2.07	1.937
2-Acetyl-4-nitro-	0.2124	3.61	1.513
2-Acetyl-4-phenyl-	0.1967	1.11	2.003

[a]Values from ref. 16.
[b]Values from ref. 11.
[c]Calculated using the TOPKAT program.

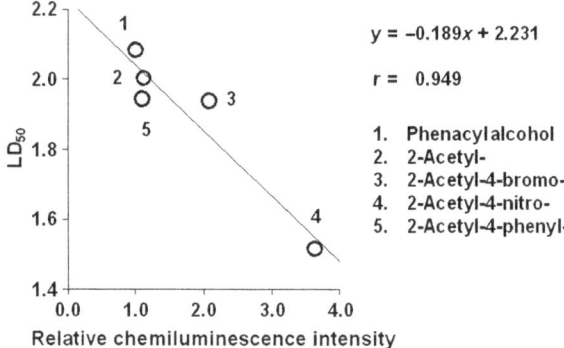

Figure 11.11 The relationship between LD_{50} and relative chemiluminescence intensity for phenacylalcohol derivatives.

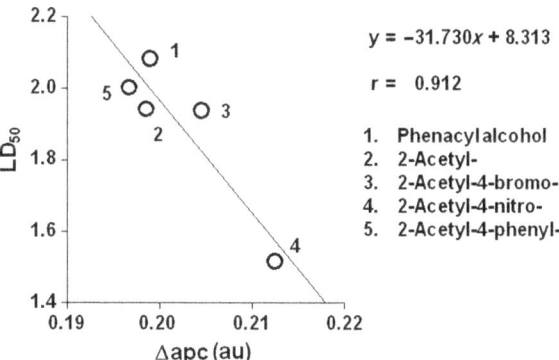

Figure 11.12 The relationship between LD_{50} and the difference in atomic partial charge for phenacylalcohol derivatives.

produced from steroids should also produce chemiluminescence by the same mechanism. Therefore, the above approaches were applied to study the efficacy, *i.e.*, the toxicity, of steroid drugs. The properties are summarized in Table 11.7. These properties were not related to their log P values, indicating

Table 11.7   Molecular properties of some steroids. EEL represents the efficacy index of steroids as an endermic liniment, and CIBV represents the logarithmic contraction index of blood vessels. Reproduced by permission of World Scientific, ref. 25.

Steroid	log P	EEL	LD_{50}	CIBV	
1	Alclomethasone dipropionate	3.352	—	3.811	—
2	Amcinonide	3.581	2.71	2.601	360
3	Beclomethasone 17,21-dipropionate	3.683	3.77	2.153	500
4	Betamethasone	1.657	4.26	2.643	—
5	Betamethasone butyrate propionate	4.460	—	2.208	—
6	Betamethasone 17,21-dipropinate	3.559	1.99	—	1660
7	Betamethasone 17-valerate	3.572	3.10	1.942	360
8	Deoxycorticosterone	2.663	—	2.622	—
9	Dexamethasone	1.657	5.26	2.643	—
10	Dexamethasone acetate	2.145	5.21	2.085	43
11	Dexamethasone 17,21-dipropionate	3.559	—	2.149	1700
12	Dexamethasone 17-valerate	3.572	3.02	1.942	—
13	Diflorasone diacetate	2.825	1.87	3.222	1600
14	Diflucortolone 21-valerate	4.830	2.93	3.479	500
15	Difluprednate	4.310	2.19	3.081	1600
16	Fludroxycortide	1.151	3.45	4.139	—
17	Flumethasone pivarate	3.625	4.79	4.052	361
18	Fluocinonide	2.966	2.44	3.069	600
19	Fluocinolone acetonide	2.497	3.73	2.859	100
20	Hydrocortisone	1.596	5.96	3.493	0.1
21	Hydrocortisone acetate	2.106	5.79	3.704	—
22	Hydrocortisone 17-butyrate	3.103	4.93	2.858	50
23	Hydrocortisone 17-butyrate 21-propionate	4.150	3.42	3.215	360
24	Methylprednisolone acetate	2.231	5.65	3.797	—
25	Predonisolone	1.930	5.49	3.841	0.5
26	Predonisolone 17-valerate 21-acetate	4.401	3.88	3.716	360
27	Triamcinolone acetonide	2.293	4.35	2.258	75

that efficacy does not depend on molecular mass or solubility due to diffusion. The analysis of chemical reactivity was the important factor, and apc contributes to the activity.

The correlation between Δapc and the efficacy index of steroids as an endermic liniment (EEL)[24] has been found to be 0.873 ($n = 23$) as shown in Figure 11.13. The computational chemical calculation allowed us to estimate the efficacy of these compounds. The linear relationship between Δapc and the logarithmic contraction index of blood vessels (CIBV)[24] was obtained with a correlation coefficient of 0.724 ($n = 19$) as shown in Figure 11.14.

The Δapc values did not exhibit a good linear relationship with LD_{50}. Dexamethasone valerate, betamethasone valerate, dexamethasone acetate and triamcinolone acetonide were approximately one-fold more toxic and difluoro-substituted steroids were approximately one-fold less toxic than that estimated from the apc Specifically, steroids with a fluorine at the C6 position were less toxic. The correlation coefficient between LD_{50} and Δapc was calculated without these positively and negatively affected compounds. The correlation coefficient was not satisfactory. The analysis of a molecule

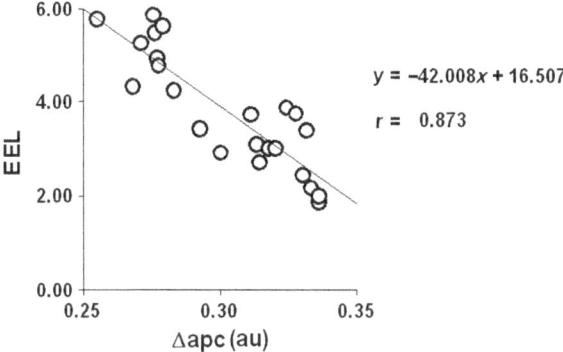

Figure 11.13 The relationship between the efficacy of steroids as an endermic liniment and the difference in atomic partial charge for phenacylalcohol derivatives.

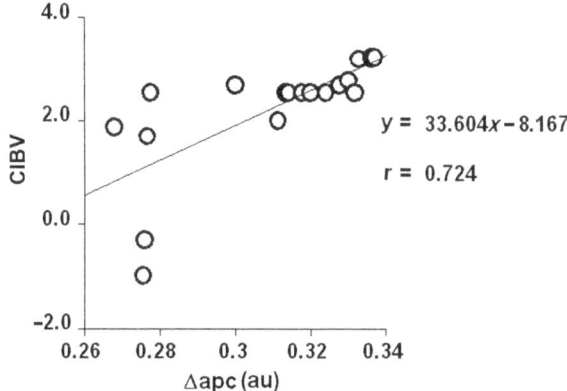

Figure 11.14 The relationship between the logarithmic contraction index of blood vessels and the difference in atomic partial charge for steroids.

with one site of action is a good indicator for the LD_{50}, such as in phenacylalcohol derivatives. Steroids, however, are complex molecules, and their metabolites contribute to the experimentally measured LD_{50} values. If the toxicity can be related to superoxide, the measurement of chemiluminescence intensity can be used for drug candidate screening. Further studies of the substituent effect of steroids are required before LD_{50} values can be estimated from computational chemical calculations.

11.3 Derivatization Reagents for the Highly Sensitive Analysis of Amino Acids

11.3.1 Introduction

Derivatization is performed to enable highly sensitive analyses. Selection of the derivatizing reagent depends on the desired sensitivity, selectivity, and reactivity. Post-column derivatization is generally preferred because of its reproducibility. However, this operation is costly and less sensitive than pre-column derivatization. A variety of derivatizing reagents have been developed for trace amino acid analysis in biological samples. In particular, the analysis of trace amounts of (R)-amino acids in high-quantity (S)-amino acids is a subject of interest because low levels of (R)-amino acids have been discovered in mammals. Fluorescent derivatives are preferred because of their higher sensitivity and selectivity, and the elimination of the solvent background. Mass spectrophotometric detection is theoretically the most sensitive method, but it is practically not sensitive and is very expensive.

In Sections 11.2 and 11.3, highly sensitive detection mechanisms involving bromate formation[23] and chemiluminescence[25] were quantitatively analyzed using computational chemical methods. In the case of bromate, the MOPAC/ZINDO in CAChe™ programs were used to calculate spectra, while in the case of chemiluminescence, a MOPAC/PM5 program was used to calculate apc values as indicators of electron localization.

The MOPAC/ZINDO and MO-S programs can also be used to construct electronic (absorption) spectra. Although the absolute spectrum and intensity cannot be obtained, their relative values indicate the wavelength and intensity differences between compounds. At present, fluorescence spectra cannot be derived by computational chemical methods, but the absorption wavelength and intensity can be related to excitation wavelength and fluorescence intensity, respectively. Therefore, the calculated absorption spectra were used to study sensitivity based on molar absorptivity values. However, a major limitation of this method is that solvent effects cannot be included in the calculation, and no program can produce a reliable spectrum or an absolute intensity. The relative sensitivities of derivatized amino acids are investigated by computational chemical techniques to understand reagent properties, and to design new reagents.

11.3.2   *In Silico* Prediction of Derivatized Amino Acid Spectra

Structures of various amino acids derivatized using different reagents were obtained. Their conformations were optimized using molecular mechanics, and then further optimized using semi-empirical methods. The spectra and HOMO and LUMO electron density maps were visualized using Tabulator and Visualize in the CAChe™ and SciGress programs.[26] The absorption wavelengths (nm) of (S)-alanine derivatives, calculated using ZINDO (INDO/S and CI combination) and MO-S (MNDO using CI at the MNDO geometry) programs, are summarized in Table 11.8, along with their reference values. Some possible excitation wavelengths have been identified, and these are listed as proposed wavelengths with their corresponding molar absorptivities. The optimized chemical structures and spectra calculated using MO-S are summarized in Figure 11.15.

Table 11.8 Properties of some derivatized (*S*)-alanine compounds. ma represents molar absorptivity ($\kappa(v)$) (mol cm^{-1}), λ represents the wavelength of fluorescence detection (nm). The sensitivity depended upon the chromatographic conditions. Chemical structures were from refs. 27 and 28.

Derivatizing agent	ZINDO λ	ZINDO $\kappa(v)$	MO-S λ	MO-S $\kappa(v)$	Reference value λ	Sensitivity
(a) DNB	205	57 500	227	15 167	254	0.2 μmol
	247	22 630	—	—	—	—
(b) DNS-Cl	208	77 907	246	13 350	335–340/525–530	10 pmol
	236	48 290	333	2220	—	—
	299	15 890	—	—	—	—
(c) Fluorescein	197	68 260	244	14 892	490/518	0.5 pmol
	374	64 047	431	22 838	—	—
(d) Fluorescein-ITC	241	81 223	245	22 838	—	—
	372	72 085	429	4900	—	—
(e) MNS-Cl	238	29 533	210	17 140	321/450	100 pmol
	292	28 300	247	9983	—	—
	—	—	324	6170	—	—
	—	—	397	1400	—	—
(f) NBD-Cl (F)	197	47 510	177	8687	448–482/530–540	10 fmol
	217	42 338	237	7550	—	—
	366	38 700	395	3380	—	—
(g) NDA (CN)^a	256	142 800	272	23 070	246/490	—
(h) NDA (imidine)	234	117 250	253	20 847	270/432	2 fmol
	272	8800	388	240	—	—
(i) OPA^a	223	90 304	240	12 606	340/450–455	200 fmol
	331	21 742	404	2239	—	—
(j) OPA (imidine)	198	79 180	227	14 776	260/311	—
	225	23 230	341	114	—	—
	274	2360	—	—	—	—
(k) PITC (PTH)	195	31 300	216	14 367	254	1 pmol
	279	29 180	—	—	—	—
(l) OPA-NAC	221	105 206	244	12 118	340/450–455	2 pmol
	327	23 553	378	2450	—	—
(m) OPA-BTCC	324	49 894	243	16 427	335–340/420–450	50 fmol
	328	25 440	378	2450	—	—
	390	18 710	—	—	—	—
(n) OPA-NaTC	226	160 560	244	32 325	—	—
	235	85 630	376	2875	—	—
	326	28 787	—	—	—	—
(o) OPA-NPh	245	78 187	236	24 822	—	—
	320	51 594	378	4925	—	—
(p) NDA-NAC	252	145 435	270	21 174	—	—
	402	24 150	435	2380	—	—

^{a}Calculated using MO-S MNDO using CI at MM2 geometry.

The selection of a reagent depends on the required reactivity. Fluorescein [Figure 11.15(c)] and 7-nitro-2,1,3-benzoxadiazole (NBD) [Figure 11.15(f)] were used when highly sensitive reagents were required. Fluorescein is also used for fluorometric analysis of proteins. *Ortho*-phthaldialdehyde (OPA) is a

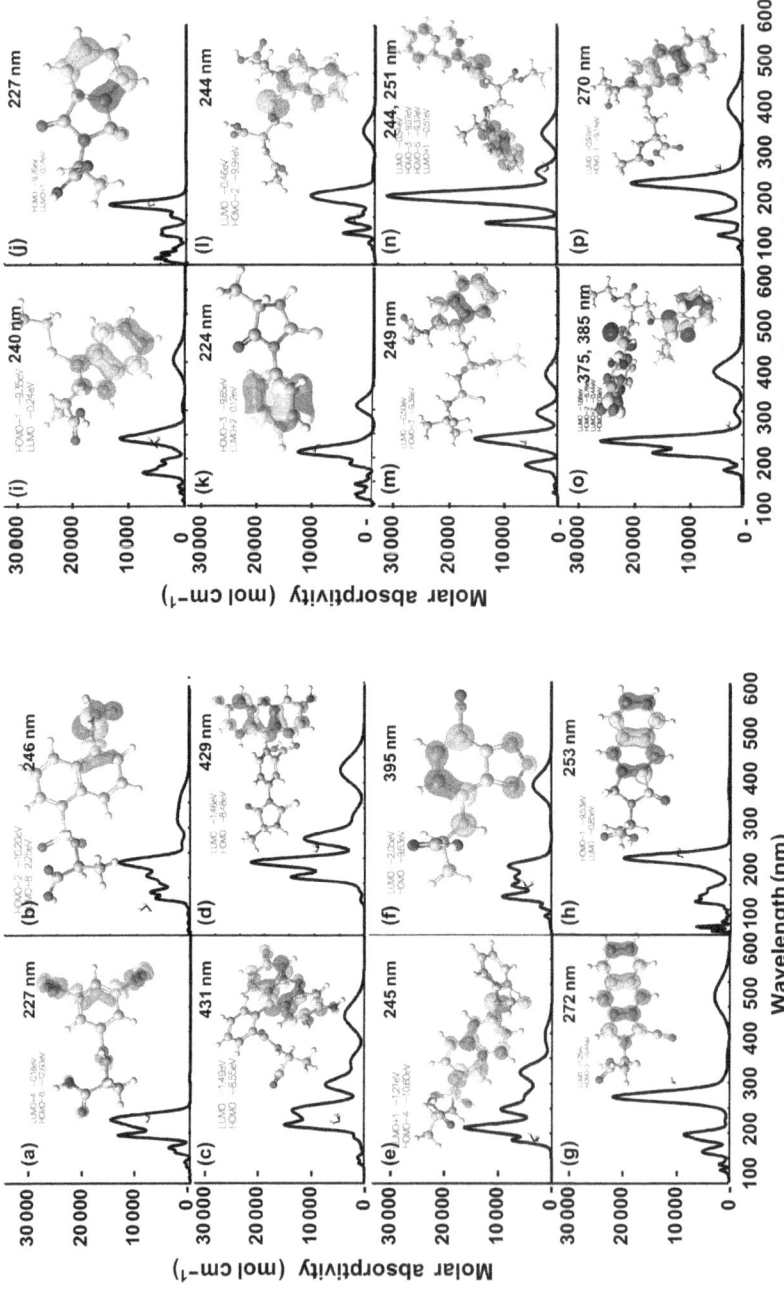

Figure 11.15   Absorption spectra and HOMO and LUMO density maps of (S)-alanine derivatives (a)–(p) at different wavelengths. Small white, large white, light-gray, dark-gray, and black balls represent hydrogen, sulfur, carbon, nitrogen, and oxygen, respectively.

popular reagent because of its reactivity, but the reaction products are not stable. Naphthalene-2,3-dialdehyde (NDA) [Figure 11.15(d)] is a family of OPA reagents and is more sensitive, and its reaction products are also stable. However, the catalytic reagent for NDA is very toxic and not easily handled within law. This problem was overcome by synthesizing imidine [Figure 11.15(h)][29] and using it at a low-absorption wavelength for highly sensitive detection. The results obtained were supported by computational chemical analyses, which indicated that the selection of excitation wavelength should improve the sensitivity and the possibility of designing a new reagent. The molar absorptivity values are quite large at UV wavelengths, where even OPA-derivatized compounds [Figure 11.15(i), (l) and (m)] may demonstrate higher sensitivities.

Additionally, the separation of diastereomers by reversed-phase liquid chromatography is preferred over normal-phase liquid chromatography, because the former method does not require the need to switch to chirally selective columns. The diastereomers used in reversed-phase liquid chromatography are relatively large molecules whose optimized energies vary widely. Their molecular shapes can be used to derive the differences in their contact surface areas using a model phase.

Dansyl (Dns)-amino acids [Figure 11.15(b)] were separated using reversed-phase liquid chromatography with either metal chelates[30,31] or (S)-histidine additives.[32] The separation efficiency of derivatized diastereomers has been studied using many selective reagents.[33–37,39,40]

Diastereomers are easily synthesized using a combination of OPA, a derivatized cysteine, and a target amino acid. The cysteine provides enantiomer selectivity, and several modifications of cysteine have been performed, such as N-acetyl-(S)-cysteine [Figure 11.15(l)].[33,34,38–41] However, these derivatives are not sufficiently sensitive to be able to analyze trace quantities of (R)-amino acids; therefore, a more highly sensitive diastereomer derivative is required. One candidate is N-naphthalenethiocarbonyl-(S)-cysteine ethyl ester (NaTC) in place of N-tert-butylthiocarbonyl-(S)-cysteine ethyl ester (BTCC) [Figure 11.15(m)]. NaTC [Figure 11.15(n)] is related to BTCC and has a naphtyl group instead of the tert-butyl group. NaTC was synthesized from 1-naphthylisothiocyanate, (S)-cysteine ethyl ester, and diethylamine in N,N-dimethylformamide, and the structure, shown in Figure 11.15(n), was determined by NMR. The reactivity with amino acids was, however, poor and required high reaction temperatures that resulted in numerous impurities. NaTC is, therefore, not suitable for diastereomer separation at present. NBD derivatives [Figure 11.15(f)] have been demonstrated to be highly sensitive compounds at relatively high excitation wavelengths. Therefore, p-nitrophenylisothiocyanate [Figure 11.15(o)] may be used in place of naphthyl or tert-butyl. Another possibility is a combination of NDA imidine and (S)-cysteine [Figure 11.15(p)]. Further study is necessary using these derivatized amino acids.

Above results demonstrate that a powerful application of computational chemistry is the estimation of absorption spectra, even though solvent effects cannot be included (handled) in the calculation. Computational chemical studies can be used to design a highly sensitive reagent, but its synthesis and practical reactivity remain to be overcome. The reactivity, with amino acids in particular, needs to be managed using a catalyst because of the low-temperature reaction conditions required to avoid racemization.

Abbreviations

BFGS	Broyden Fletcher Goldfarb Shanno algorithm, an iterative metho for solving unconstrained nonlinear optimization problems[57]
BTCC	*N-tert*-Butylthiocarbonyl-L-cysteine ethyl ester
DNB	3,5-Dinitrobenzene
DNB-Cl	3,5-Dinitrobenzylchloride
DNS-Cl	5-Dimethylaminonaphthalene-1-sulphonylchloride (Dansylchloride)
Fluorescein-ITC	Fluorescein isothiocyanate
MNS-Cl	6-*N*-methylanilinonaphthalene-2-sulphonylchloride
NAC	*N*-Acetyl-L-cysteine ethyl ester
NBD-Cl (F)	4-Chloro(fluoro)-7-nitro-2,1,3-benzoxadiazole
NDA (CN)	Naphthalene 2,3-dialdehyde
NDA	Isoindolin-1-one naphthalimidine
NPh	*N-p*-Nitrophenylthiocarbonyl-L-cysteine ethyl ester
OPA	*ortho*-Phthaldialdehyde
PITC (PTH)	Phenyl isothiocyanate (phenylthiohydantoin)
AM1	Austin Model 1, a semi-empirical method for the quantum calculation of molecular electronic structure in computational chemistry based on NDDO.[42–44]
CI	Configuration Interaction.[45–47]
INDO	Intermediate Neglect of Differential Overlap.[45,47–49]
INDO/S	INDO for spectroscopy.[46,48]
MNDO	Modified NDDO; does not take into account delocalization effect, uses only s and p orbital sets.[43,50–53]
MO-S	Molecular orbital package used to calculate the spectroscopic properties of a molecule.[26]
NDDO	Neglect of Diffential Diatomic Overlap[42,43,47,58]
NLLSQ	Non-Linear Least Squares gradient minimization[56]
PM3	Parametric Model 3; reparametrization of MNDO with a core–core repulsion term similar to those of AM1.[42,43,52]
PM5	Parametric Model 5; version-up program of PM3.[43]
PM6	Version-up program of PM5, however, PM6 can handle more elements compared to PM5 according to J. Stewart.[43,54]
RM1	Recife Model 1, semi-empirical methods; reparameterization of AM1 for H, C, N, O, P, S, F, Cl, Br, and I.[43,54,55]
ZINDO	Zerner INDO.[45]

References

1. Guidelines for drinking-water quality, 2nd edn, vol. 1, World Health Organization, Geneva, 1993.
2. Y. Inoue, T. Sakai, H. Kumagai and Y. Hanaoka, High selective determination of bromate in ozonized water by using ion chromatography with postcolumn derivatization equipped with reagent preparation device, *Anal. Chim. Acta*, 1997, **346**, 299–305.
3. G. Chiu and R. D. Eubanks, Spectrophotometric determination of bromide, *Mikrochim. Acta*, 1989, **11**, 145–148.
4. R. C. Duty and J. S. Ward, Determination of bromate and chlorate via reduction with sodium nitrite, *Analyst*, 1994, **119**, 2141–2143.
5. M. F. Tuchler, S. Wright and J. D. McDonald, Real time study of bimolecular interactions: Direct determination of internal conversion involving $Br(^2P_{1/2}) + I_2$ initiated from a van der Waals dimer, *J. Chem. Phys*, 1997, **106**, 2634–2645.
6. I. R. Sims, M. Gruebele, E. D. Potter and A. H. Zewail, Femtosecond real-time probing of reactions. VIII. The bimolecular reaction $Br + I_2$, *J. Chem. Phys.*, 1992, **97**, 4127–4148.
7. T. Yonezawa, T. Nagata, H. Kato, S. Imamura and I. Murokuma, Ryoshikagaku Nyumon (Introduction to Quantum Chemistry), Kagakudojin, Kyoto, 1983, p. 480.
8. Z. Deyl, I. Miksik and E. Tesarova, Selected Derivatization Reactions, in *Advanced Chromatographic and Electromigration Methods in Biosciences*, ed. D. Deyl, I. Miksik, F. Tagliano and E. Tesarova, Elsevier, Amsterdam, 1998, pp. 166–169.
9. M. Maeda and A. Tsuji, Chemiluminescence with lucigenin as post-column reagent in high-performance liquid chromatography of corticosteroids and p-nitrophenacyl esters, *J. Chromatogr*, 1986, **352**, 213–229.
10. T. Hanai, M. Uchida, M. Amao, C. Ikeda, K. Koizumi and T. Kinoshita, Selective chemiluminescence analysis of Amodori form of glycated human serum albumin, *J. Liq. Chromatogr. Relat. Technol*, 2000, **23**, 3119–3131.
11. S. P. Wolf, Free radicals and glycation theory, in *The Maillard Reaction: Consequences for the Chemical and Life Sciences*, ed. R. Ikan, John Wiley & Sons, New York, 1996, pp. 77–88.
12. R. L. Veazey and T. A. Nieman, Chemiluminescence high-performance liquid chromatographic detector applied to ascorbic acid determinations, *J. Chromatogr.*, 1980, **200**, 153–162.
13. L. L. Klopf and T. A. Nieman, Determination of conjugated glucuronic acid by combining enzymatic hydrolysis with lucigenin chemiluminescence, *Anal. Chem*, 1985, **57**, 46–51.
14. R. L. Veazey and T. A. Nieman, Chemiluminescence determination of clinically important organic reductants, *Anal. Chem.*, 1979, **51**, 2092–2096.

15. J. W. Baynes, S. R. Thorpe and M. H. Mutiashaw, Nonenzymatic glucosylation of lysine residues in albumin, *Methods Enzymol.*, 1984, **106**, 88–98.

16. V. M. Monnier and A. Cerame, Non-enzymatic glycosylation of proteins in diabetes, *Clin. Endocrinol. Metab*, 1982, **11**, 431–452.

17. B. C. Baguley, W. A. Denny, G. J. Atwell and B. F. Cain, Potential anti-tumor agents. 35. Quantitative relationships between antitumor (L210) potency and DNA binding for 4′-(9-acridinylamino)methanesulfon-m-anisidine analogs, *J. Med. Chem.*, 1981, **24**, 520–525.

18. K. Nakashima and K. Imai, LC-Chemiluminescence detection, in *Advances in Liquid Chromatography*, ed. T. Hanai and H. Hatano, World Scientific, Singapore, 1996, pp. 99–122.

19. A. Toriba and H. Kubo, Chemiluminescence high performance liquid chromatography of corticosteroids and p-nitrophenacylesters based on the luminol reacton, *J. Liq. Chromatogr. Relat. Technol*, 1997, **20**, 2965–2977.

20. W. A. Waters, *Mechanism of Oxidation of Organic Compounds*, Methuen, London, 1965, p. 97.

21. K. Faulkner and I. Fridovich, Luminol and lucigenin as detectors for $O2^{\bullet-}$, *Free Radical Biol. Med*, 1993, **25**, 447–451.

22. T. Hanai, Y. Inoue, T. Sakai and H. Kumagai, Computational chemical analysis of the highly sensitive detection of bromate in ion chromatography, *J. Chem. Inf. Comput. Sci.*, 1998, **38**, 885–888.

23. R. L. Veazey, H. Nekimen and T. A. Nieman, Chemiluminescent reaction of lucigenin with reducing sugars, *Talanta*, 1984, **31**, 603–606.

24. H. Nakayama, K. Masubuchi and M. Sugawara, Saishinno Hifugaiyo-kuzai, Namzando, Tokyo, 1991.

25. T. Hanai and T. Tachikawa, Quantitative analysis of chemiluminescence intensity and toxicity *in silico*, presented at the 13th International Symposium on Bioluminescence and Chemiluminescence, Yokohama, 2004.

26. CAChe™ and SciGress programs from Fujitsu, Japan.

27. F. Li and C. K. Lim, Coloured and UV-absorbing derivatives, in *Handbook of Derivatives for Chromatography*, ed. K. Blau and J. M. Halket, John Wiley & Sons, New York, 1993, pp. 157–174.

28. N. Seiler, Fluorescent derivatives, in *Handbook of Derivatives for Chromatography*, ed. K. Blau and J. M. Halket, John Wiley & Sons, New York, 1993, pp. 175–213.

29. K. Gyimesi-Forras, A. Leitner, K. Akasaka and W. Lindner, Comparative study on the use of ortho-phthalaldehyde, naphthalene-2,3-dicarboaldehyde and anthracene-2,3-dicarboaldehyde reagents for α-amino acids followed by the enantiomer separation of the formed isoindolin-1-one derivatives using quinine-type chiral stationary phases, *J. Chromatogr., A*, 2005, **1083**, 80–88.

30. W. Lindner, J. N. LePage, G. Davies and B. L. Karger, Reversed-phase separation of optical isomers of Dns-amino acids and peptides using chiral metal chelate additives, *J. Chromatogr*, 1979, **185**, 323–344.

31. Y. Tapuhi, N. Miller and B. L. Karger, Practical consideration in the chiral separation of Dns-amino acids by reversed-phase liquid chromatography using metal chelate additives, *J. Chromatogr*, 1981, **205**, 325–337.

32. S. Lam and A. Karmen, Resolution of optical isomers of Dns-amino acids by high-performance liquid chromatography with L-histidine and its derivatives in the mobile phase, *J. Chromatogr*, 1982, **239**, 451–462.

33. H. Bruckner, R. Wittner and H. Godel, Automated enantioseparation of amino acids by derivatization with o-phthalaldehyde and N-acetylated cysteines, *J. Chromatogr*, 1989, **476**, 73–82.

34. N. Nimura and T. Kinoshita, o-Phthalaldehyde-N-acetyl-L-cysteine as a chiral derivatization reagent for liquid chromatographic resolution of amino acid enantiomers and its application to conventional amino acid analysis, *J. Chromatogr., A*, 1986, **352**, 169–177.

35. M. R. Euerby, L. Z. Partridge and W. A. Gibbons, Study of the chromatographic behaviour and resolution of α-amino acid enantiomers by high-performance liquid chromatography utilizing pre-column derivatization with o-phthaldialdehyde and new chiral thiols, *J. Chromatogr*, 1989, **483**, 239–252.

36. H. Burckner, M. Langer, M. Lupke, T. Westhauser and H. Godel, Liquid chromatographic determination of amino acid enantiomers by derivatization with o-phthaldialdehyde and chiral thiols Applications with reference to food science, *J. Chromatogr., A*, 1995, **697**, 229–245.

37. N. Nimura, T. Fujiwara, A. Watanabe, M. Sekine, T. Furuchi, M. Yohda, A. Yamagishi, T. Ohshima and H. Homma, A novel chiral thiol reagent for automated pre-column derivatization and high-performance liquid chromatographic enantioseparation of amino acids and its application to the aspartate racemase assay, *Anal. Biochem*, 2003, **315**, 262–269.

38. D. W Aswad, Determination of D- and L-aspartate in amino acid mixtures by high-performance liquid chromatography after derivatization with a chiral adduct of o-phthaldialdehyde, *Anal. Biochem*, 1984, **137**, 405–409.

39. R. H. Buck and K. Krummen, Resolution of amino acid enantiomers by high-performance liquid chromatography using automated pre-column derivatization with a chiral reagent, *J. Chromatogr., A*, 1984, **315**, 279–285.

40. R. H. Buck and K. Krummen, High-performance liquid chromatographic determination of enantiomeric amino acids and amino alcohols after derivatization with o-phthaldialdehyde and various chiral mercaptans: Application to peptide hydrolysates, *J. Chromatogr., A*, 1987, **357**, 255–265.

41. T. Takeuchi, T. Niwa and D. Ishii, Application of micro HPLC to the analysis of amino acids after precolumn derivatization with o-phthalaldehyde and N-acetyl-L-cysteine, *J. High Res. Chromatogr*, 1988, **11**, 343–346.

42. K. Kahn, Semi-empirical quantum chemistry, http://web.chem.ucsb.edu/~kalju/chem226/public/semiemp_intro.html.

43. J. J. P. Stewart, Optimization of parameters for semi-empirical methods V: Modification of NDDO approximations and application to 70 elements, *J. Mol. Model.*, 2007, **13**, 1173–1213.

44. T. D. Bouman, A. E. Hansen, K. L. Bak, T. B. Pedersen and R. A. Kirby, Random phase application/semi-empirical computations of electronic structure of extended organic molecules, *Recent Res. Dev. Phys. Chem.*, 1996, **5**, 1–24.

45. W. P. Anderson, W. D. Edwards and M. C. Zerner, Calculated spectra of hydrated ions of the first transition-metal series, *Inorg. Chem.*, 1986, **25**, 2728–2732.

46. S. Tretiak, V. Chernyak and S. Mukamel, Two-dimensional real-space analysis of optical excitations in acceptor-substituted carotenoids, *J. Am. Chem. Soc.*, 1997, **119**, 11408–11419.

47. R. C. Bingham, M. J. S. Dewar and D. H. Lo, Ground states of molecules. XXV. MINDO/3. An improved version of the MINDO semi-empirical SCF-MO method, *J. Am. Chem. Soc.*, 1975, **97**, 1285–1293.

48. S. Tretiak, A. Saxena, R. L. Martin and A. R. Bishop, CEO/semi-empirical calculations of UV–visible spectra in conjugated molecules, *Chem. Phys. Lett.*, 2000, **331**, 561–568.

49. M. C. Zerner, G. H. Loew, R. K. Kirchner and U. T. Mueller-Westerhoff, An intermediate neglect of differential overlap technique for spectroscopy of transition-metal complexes. Ferrocene, *J. Am. Chem. Soc.*, 1980, **102**, 589–599.

50. M. J. S. Dewar, E. G. Zoebisch, E. F. Healy and J. J. P. Stewart, AM1: A new general purpose quantum mechanical molecular model, *J. Am. Chem. Soc.*, 1985, **107**, 3902–3909.

51. M. J. S. Dewar and W. Thiel, Ground states of molecules. 38. The MNDO method. Approximations and parameters, *J. Am Chem. Soc.*, 1977, **99**, 4899–4907.

52. M. P. Repasky, J. Chandrasekhar and W. L. Jorgensen, PDDG/PM3 and PDDG/MNDO: improved semi-empirical methods, *J. Comput. Chem.*, 2002, **16**, 1601–1622.

53. M. R. Silva-Junior and W. Thiel, Benchmark of electronically excited states for semiempirical methods: MNDO, AM1, PM3, OM1, OM2, OM3, INDO/S, and INDO/S2, *J. Chem. Theory Comput.*, 2010, **6**, 1546–1564.

54. A. da Silva Goncalves, T. C. C. Franca, J. D. Figueroa-Villar and P. G. Pascutti, Conformational analysis of toxobonine, TMB-4 and HI-6 using PM6 and RM1 methods, *J. Braz. Chem. Soc*, 2010, **21**, 179–184.

55. G. B. Rocha, R. O. Freire, A. M. Simas and J. J. P. Stewart, RM1: A reparameterization of AM1 for H, C, N, O, P, S, F, Cl, Br, and I, *J. Comput. Chem.*, 2006, **27**, 1101–1111.

56. R. H. Bartels, Report CNA-44, University of Texas and Center for Numerical Analysis, 1972.

57. D. GoldfarbA Family of variable metric updates derived by variational means, *Mathematics of Computation*, 1970, **24**, 23–26.

58. J. Pople and D. Bevenridge, *Approximate Molecular Orbital Theory*, McGraw-Hill, New York, 1970.
59. T. Hanai, Computational chemical analysis of the sensitivity of phenacylesters and steroids in chemiluminescence detection, *Japan Chemistry Program Exchange Journal* 2001, **12**, 123–128.
60. T. Hanai, Quantitative computational chemical analysis of the sensitivity of chemiluminescence detection, *J. Liq. Chromatogr. Rel. Technol.*, 2002, **25**, 2425–2431.

Appendix

Table 1 Molecular properties and molecular interaction energy values calculated using a model methylsilicone phase, and measured log k values on CPSil5 and DB1. *fsp*, *hbp*, *esp*, and *vwp* represent the energy of the calculated final (optimized) structure, the hydrogen-bonding energy, the electrostatic energy, and the van der Waals energies for each pair of compounds, respectively. MIFS, MIVW, MIHB, and MIES represent the same molecular interaction energies (kcal mol^{-1}). Reproduced by permission of Institute for Chromatography, ref. 13.

Vwv, ΔH^a, ΔH^b, $\Delta vapH$, *Tb* represent van der Waals volume (cm^3 mol^{-1}), enthalpy measured on DB1 column, enthalpy measured on CPSil5 column, enthalpy of vaporization corrected using vwv, Boiling point (°C) from reference 39 in Chapter 4.

No.	Analyte	vwv	ΔH^a	ΔH^b	$\Delta vapH$	Tb
1	Heptane	78.49	—	—	36.62	114.008
2	Octane	88.72	—	—	41.39	133.486
3	Nonane	98.95	—	—	46.16	152.964
4	Decane	109.18	8.62	8.48	50.90	172.466
5	Undecane	119.41	9.19	9.16	55.70	191.944
6	Dodecane	129.64	10.11	9.86	60.40	211.422
7	Tridecance	139.87	10.51	10.56	65.20	230.899
8	Tetradecane	150.10	11.33	11.12	70.00	250.377
9	Pentadecane	160.33	11.55	11.85	74.70	269.855
10	Hexadecane	170.56	12.32	12.57	79.50	289.811
11	Heptadecane	180.79	13.10	13.31	84.30	308.811
12	Octadecane	191.02	13.87	14.04	89.00	328.289
13	Nonadecane	201.25	—	14.60	93.80	347.767
14	Eicosane	211.48	—	15.22	98.60	367.245

RSC Chromatography Monographs No. 19
Quantitative *In Silico* Chromatography: Computational Modelling of Molecular Interactions
By Toshihiko Hanai
© Toshihiko Hanai, 2014
Published by the Royal Society of Chemistry, www.rsc.org

Table 1 (*Continued*)

No.	Analyte	vwv	ΔH^a	ΔH^b	$\Delta vapH$	Tb
15	Heneicosane	221.71	—	—	103.36	386.699
16	Docosane	231.94	—	—	108.13	406.177
17	Tricosane	242.17	—	—	112.90	425.655
18	Tetracosane	252.40	—	—	117.66	445.133
19	Pentacosane	262.63	—	—	122.43	464.611
20	Hexacosane	272.86	—	—	127.20	484.088
21	1-Pentanol	62.63	—	—	42.90	142.483
22	1-Hexanol	72.86	—	—	46.81	159.250
23	1-Heptanol	83.09	8.22	7.84	50.72	176.017
24	1-Octanol	93.32	8.97	8.37	54.63	192.783
25	1-Nonylalcohol	103.55	—	—	58.54	209.550
26	1-Decylalcohol	113.78	10.23	9.86	62.44	226.317
27	1-Undecylalcohol	124.01	—	—	66.35	243.084
28	1-Dodecylalcohol	134.24	11.17	10.93	70.26	259.851
29	1-Tridecylalcohol	144.49	11.90	11.76	74.17	276.651
30	1-Tetradecylalcohol	154.70	12.65	12.62	78.07	293.385
31	1-Pentadecylalcohol	164.93	13.41	13.42	81.98	310.152
32	1-Hexadecylalcohol	175.16	—	13.79	85.89	326.919
33	1-Heptadecylalcohol	185.93	—	14.34	90.00	344.571
34	1-Octadecylalcohol	195.62	—	15.08	93.71	360.453
35	1-Nonadecylalcohol	205.85	—	—	97.61	377.220
36	1-Eicosanol	216.08	—	—	101.52	393.987
37	1-Docosanol	236.54	—	—	109.34	427.521
38	1-Tricosanol	246.77	—	—	113.25	444.288
39	Benzene	48.36	4.20	4.86	33.80	80.287
40	Naphthalene	73.97	8.65	8.46	62.00	204.240
41	Biphenyl	90.08	10.13	9.89	—	—
42	Fluorene	93.22	10.83	11.00	75.05	297.410
43	Phenanthrene	99.56	11.87	12.02	81.38	328.095
44	Anthracene	99.56	11.90	12.08	81.70	328.095
45	Pyrene	109.04	13.07	13.37	90.97	373.979
46	Chrycene	125.16	—	—	103.32	451.999
47	Tetracene	—	—	—	—	—
48	Benzpyrene	—	—	—	—	—
49	Pentacene	—	—	—	—	—
50	1-Heptacene	75.00	—	—	35.71	106.482
51	1-Nonene	95.46	7.48	7.62	45.30	146.974
52	1-Decene	105.69	8.57	8.19	50.00	167.220
53	1-Tridecene	136.38	10.28	10.26	64.40	227.955
54	1-Tetradecene	146.61	11.14	11.10	69.10	248.200
55	1-Hexadecene	167.07	12.09	12.17	78.70	288.691
56	1-Heptadecene	177.30	12.91	12.98	83.40	308.936
57	1-Octadecene	187.53	13.70	13.33	88.20	329.191
58	1-Nonadecene	197.76	—	14.18	93.00	349.425
59	1-Eicocene	207.90	—	14.71	97.70	369.493
60	Methylpentanoate	73.47	—	—	51.78	129.630
61	Methylhexanoate	83.66	—	—	54.75	149.384
62	Methyloctanoate	104.12	9.11	9.61	61.35	188.892
63	Methylnonanoate	114.35	—	—	64.97	208.646
64	Methyldecanoate	124.58	10.11	9.12	67.16	228.400
65	Methylundecanoate	134.81	11.31	11.01	71.50	248.154

Table 1 (*Continued*)

No.	Analyte	vwv	ΔH^a	ΔH^b	$\Delta vapH$	Tb
66	Methyldodecanoate	145.04	11.65	11.83	76.28	267.908
67	Methyltridecanoate	155.27	12.43	12.31	81.06	287.662
68	Methyltetradocanoate	165.50	13.15	13.11	85.84	307.416
69	Methylpentadocanote	175.73	—	13.52	90.62	327.171
70	Methylhexadecanoate	185.96	—	14.17	95.40	346.925
71	Methylheptadocanoate	196.19	—	15.03	100.18	366.679
72	Methyloctadocanoate	206.42	—	15.79	104.96	386.433
73	Methylnonadecanoate	216.65	—	—	112.61	406.187
74	Methyleicosanoate	226.88	—	—	118.52	425.941
75	Methylhenicosanoate	237.11	—	—	124.65	445.695
76	Methyldocosanoate	247.34	—	—	130/98	465.449
77	Methyltricosanoate	257.57	—	—	137.52	485.203
78	Methyltetracosanoate	267.80	—	—	144.27	504.957
79–109	no data					

No.	Analyte	fsp	hbp	esp	vwp	MIFS	MIHB
1	Heptane	1.9218	0.000	0.000	−0.197	11.5612	0.000
2	Octane	2.2065	0.000	0.000	−0.278	13.5452	0.000
3	Nonane	2.5030	0.000	0.000	−0.335	15.3358	0.000
4	Decane	2.7874	0.000	0.000	−0.403	16.5148	0.000
5	Undecane	3.0924	0.000	0.000	−0.437	18.1734	0.000
6	Dodecane	3.3718	0.000	0.000	−0.539	20.0847	0.000
7	Tridecance	3.6846	0.000	0.000	−0.575	21.1949	0.000
8	Tetradecane	3.9531	0.000	0.000	−0.675	22.8340	0.000
9	Pentadecane	4.2462	0.000	0.000	−0.747	24.6200	0.000
10	Hexadecane	4.5375	0.000	0.000	−0.807	25.9265	0.000
11	Heptadecane	4.8088	0.000	0.000	−0.877	27.8450	0.000
12	Octadecane	5.1196	0.000	0.000	−0.940	28.9292	0.000
13	Nonadecane	5.4176	0.000	0.000	−1.006	30.5334	0.000
14	Eicosane	5.7045	0.000	0.000	−1.071	32.3747	0.000
15	Heneicosane	5.9943	0.000	0.000	−1.136	33.6032	0.000
16	Docosane	6.2869	0.000	0.000	−1.205	35.2645	0.000
17	Tricosane	6.5729	0.000	0.000	−1.271	36.8734	0.000
18	Tetracosane	6.8666	0.000	0.000	−1.336	38.4414	0.000
19	Pentacosane	7.1545	0.000	0.000	−1.409	39.9656	0.000
20	Hexacosane	7.4570	0.000	0.000	−1.472	41.7582	0.000
21	1-Pentylalcohol	0.9210	−0.079	−0.066	−0.479	9.7935	0.000
22	1-Hexylalchol	1.2063	−0.083	−0.057	−0.562	11.4458	0.000
23	1-Heptylalcohol	1.4933	−0.077	−0.061	−0.628	12.7760	0.000
24	1-Octylalcohol	1.7802	−0.083	−0.059	−0.701	14.1010	0.000
25	1-Nonylalcohol	2.0721	−0.080	−0.061	−0.764	—	—
26	1-Decylalcohol	2.3693	−0.081	−0.060	−0.833	17.6850	0.000
27	1-Undecylalcohol	2.6678	−0.086	−0.059	−0.883	—	—
28	1-Dodecylalcohol	2.9418	−0.081	−0.057	−0.973	20.7350	0.000
29	1-Tridecylalcohol	3.2485	−0.081	−0.059	−1.023	21.8230	0.000
30	1-Tetradecylalcohol	3.5447	−0.081	−0.061	−1.096	23.3810	0.000
31	1-Pentadecylalcohol	3.8276	−0.092	−0.060	−1.154	25.2860	0.000
32	1-Hexadecylalcohol	4.1157	−0.083	−0.060	−1.210	26.5180	0.000
33	1-Heptadecylalcohol	4.4004	−0.087	−0.059	−1.288	27.3170	0.000
34	1-Octadecylalcohol	4.7121	−0.094	−0.060	−1.386	29.2570	0.000
35	1-Nonadecylalcohol	4.9819	−0.081	−0.062	−1.425	31.1398	0.000

Table 1 (*Continued*)

No.	Analyte	fsp	hbp	esp	vwp	MIFS	MIHB
36	1-Eicosanol	5.2751	−0.082	−0.061	−1.497	33.0836	0.000
37	1-Docosanol	5.8543	−0.087	−0.059	−1.624	35.9133	0.000
38	1-Tricosanol	6.1514	−0.086	−0.060	−1.697	37.4116	0.000
39	Benzene	−19.6550	0.000	0.000	2.511	8.5140	0.000
40	Naphthalene	−43.7744	0.000	0.000	5.151	12.9300	0.000
41	Biphenyl	−42.1636	0.000	0.000	6.883	17.4750	0.000
42	Fluorene	−35.8723	0.000	−0.003	−1.493	16.6700	0.000
43	Phenanthrene	−58.7912	0.000	0.000	10.503	17.9390	0.000
44	Anthracene	−68.0926	0.000	0.000	7.639	17.8920	0.000
45	Pyrene	−77.3877	0.000	0.000	9.374	19.4960	0.000
46	Chrycene	−72.6256	0.000	0.000	16.693	22.2880	0.000
47	Tetracene	−92.4331	0.000	0.000	10.097	—	—
48	Benzpyrene	−92.3571	0.000	0.000	14.778	—	—
49	Pentacene	−116.7744	0.000	0.000	12.564	—	—
50	1-Heptacene	0.9522	0.000	0.005	−0.064	11.7012	0.000
51	1-Nonene	1.5297	0.000	0.005	−0.187	15.9730	0.000
52	1-Decene	1.8176	0.000	0.005	−0.248	17.3490	0.000
53	1-Tridecene	2.6873	0.000	0.005	−0.461	21.7710	0.000
54	1-Tetradecene	2.9782	0.000	0.005	−0.521	23.7420	0.000
55	1-Hexadecene	3.5613	0.000	0.005	−0.645	26.3390	0.000
56	1-Heptadecene	3.8544	0.000	0.005	−0.725	27.6730	0.000
57	1-Octadecene	4.1467	0.000	0.005	−0.791	29.3490	0.000
58	1-Nonadecene	4.4338	0.000	0.005	−0.856	30.9840	0.000
59	1-Eicocene	4.7350	0.000	0.005	−0.907	33.1180	0.000
60	Methylpentanote	−3.9406	0.000	−5.675	1.987	11.3533	0.000
61	Methylhexanoate	−3.5518	0.000	−5.609	1.947	13.2176	0.000
62	Methyloctanoate	−3.0185	0.000	−5.711	1.833	16.2340	0.000
63	Methylnonanoate	−2.7649	0.000	−5.713	1.738	17.6130	0.000
64	Methyldecanoate	−2.4539	0.000	−5.679	1.658	19.2710	0.000
65	Methylundecanoate	−2.1841	0.000	−5.748	1.632	21.1630	0.000
66	Methyldodecanoate	−1.8738	0.000	−5.520	1.589	22.3070	0.000
67	Methyltridecanoate	−1.6012	0.000	−5.735	1.515	24.1560	0.000
68	Methyltetradecanoate	−1.3050	0.000	−5.752	1.435	25.9120	0.000
69	Methylpentadecanoate	−1.0143	0.000	−5.756	1.372	27.0240	0.000
70	Methylhexadecanoate	−0.7154	0.000	−5.735	1.296	28.7810	0.000
71	Methylheptadecanoate	−0.4336	0.000	−5.789	1.268	30.1740	0.000
72	Methyloctadecanoate	−0.1298	0.000	−5.757	1.178	31.7090	0.000
73	Methylnonadecanoate	0.1444	0.000	−5.709	1.071	33.4139	0.000
74	Methyleicosanoate	0.4368	0.000	−5.717	1.013	34.7206	0.000
75	Methylhenicosanoate	0.7264	0.000	−5.723	0.949	36.4853	0.000
76	Methyldocosanoate	1.0219	0.000	−0.789	0.939	37.8673	0.000
77	Methyltricosanoate	1.3187	0.000	−5.869	0.909	39.3747	0.000
78	Methyltetracosanoate	1.6069	0.000	−5.731	0.755	41.1900	0.000
79	Toluene	−22.0960	0.000	0.000	1.177	—	—
80	Ethylbenzene	−19.2850	0.000	0.000	2.362	—	—
81	Propylbenzene	−18.7720	0.000	0.000	2.461	—	—
82	Butylbenzene	−18.5830	0.000	0.000	2.292	—	—
83	Hexylbenzene	−18.0058	0.000	0.000	2.142	—	—
84	Heptylbenzene	−17.7167	0.000	0.000	2.053	—	—
85	Octylbenzene	−17.4332	0.000	0.000	1.976	—	—
86	Nonylbenzene	−17.1515	0.000	0.000	1.910	—	—

Table 1 (*Continued*)

No.	Analyte	fsp	hbp	esp	vwp	MIFS	MIHB
87	Decylbenzene	−16.8498	0.000	0.000	1.855	—	—
88	1,2-Dimethylbenzene	−22.6282	0.000	0.269	1.153	—	—
89	1,3Dimethylbenzene	−24.8721	0.000	−0.227	0.155	—	—
90	1,4-Dimethylbenzene	−24.1888	0.000	0.072	0.251	—	—
91	1,2,3-Trimethylbenzene	−22.5910	0.000	0.320	1.398	—	—
92	1,2,4-Trimethylbenzene	−25.7387	0.000	0.112	−0.589	—	—
93	1,3,5-Trimethylbenzene	−27.9339	0.000	−0.688	−1.543	—	—
94	1,2,3,4-Tetramethylbenzene	−21.2391	0.000	0.380	2.766	—	—
95	1,2,3,5-Tetramethylbenzene	−25.7257	0.000	−0.101	−0.049	—	—
96	1,2,4,5-Tetramethylbenzene	−27.2349	0.000	0.244	−1.402	—	—
97	Pentamethylbenzene	−20.4707	0.000	0.224	3.850	—	—
98	Chlorobenzene	−20.0143	0.000	0.104	1.921	—	—
99	1,2-Dichlorobenzene	−13.5793	0.000	4.957	2.807	—	—
100	1,3-Dichlorobenzene	−25.2582	0.000	−4.430	1.129	—	—
101	1,4-Dichlorobenzene	−18.8957	0.000	1.820	1.164	—	—
102	1,2,3-Trichlorobenzene	−11.0233	0.000	5.910	3.996	—	—
103	1,2,4-Trichlorobenzene	−16.9752	0.000	2.336	1.946	—	—
104	1,3,5-Trichlorobenzene	−35.2035	0.000	−13.706	0.405	—	—
105	1,2,3,4-Tetrachlorobenzene	−7.4736	0.000	5.995	7.066	—	—
106	1,2,3,5-Tetrachlorobenzene	−20.0373	0.000	−4.310	5.174	—	—
107	1,2,4,5-Tetrachlorobenzene	−13.3040	0.000	3.411	3.901	—	—
108	Pentachlorobenzene	−8.4290	0.000	2.323	9.524	—	—
109	Hexachlorobenzene	−0.7447	0.000	4.147	15.397	—	—

No.	MIES	MIVW	280^a	260^a	240^a	220^a	200^a
1	0.000	11.658	—	—	—	—	—
2	0.000	13.717	—	—	—	−1.189	−0.856
3	0.000	15.549	—	—	−1.111	−1.026	−0.734
4	0.000	16.725	−1.253	−1.110	−0.974	−0.869	−0.606
5	0.000	18.431	−1.121	−0.981	−0.839	−0.571	−0.469
6	0.000	20.329	−1.045	−0.863	−0.708	−0.423	−0.322
7	0.000	21.470	−0.897	−0.742	−0.575	−0.277	−0.173
8	0.000	23.108	−0.790	−0.622	−0.440	−0.133	−0.020
9	0.000	24.854	−0.672	−0.503	−0.309	−0.016	0.132
10	0.000	26.308	−0.555	−0.384	−0.176	0.156	0.288
11	0.000	28.230	−0.441	−0.266	−0.045	0.296	0.442
12	0.000	29.363	−0.329	−0.149	0.084	0.433	0.597
13	0.000	30.533	−0.220	−0.033	0.211	0.578	0.753
14	0.000	32.712	−0.112	0.082	0.339	0.716	0.905
15	0.000	34.062	−0.004	0.196	0.469	0.855	1.060
16	0.000	35.810	0.102	0.311	0.596	0.992	—
17	0.000	37.188	0.205	0.424	0.722	1.125	—
18	0.000	39.007	0.317	0.538	0.850	—	—
19	0.000	40.511	0.420	0.649	0.945	—	—
20	0.000	42.330	0.523	0.760	1.099	—	—

Table 1 (*Continued*)

No.	MIES	MIVW	280[a]	260[a]	240[a]	220[a]	200[a]
21	0.008	9.952	—	—	—	−1.220	−0.992
22	0.037	11.566	—	—	—	−1.066	−0.853
23	0.037	12.802	−1.185	1.115	−1.152	−0.914	−0.714
24	0.054	14.248	−1.071	−1.011	−1.000	−0.787	−0.563
25	—	—	—	—	—	—	—
26	−0.020	17.963	−0.870	−0.763	−0.864	−0.469	−0.257
27	—	—	—	—	—	—	—
28	0.032	21.003	−0.655	−0.529	−0.601	−0.176	0.051
29	0.031	22.168	−0.548	−0.415	−0.339	−0.032	0.206
30	0.021	23.723	−0.441	−0.293	−0.207	0.111	0.364
31	0.010	25.635	−0.335	−0.176	−0.076	0.254	0.520
32	0.035	26.835	−0.230	−0.059	0.053	0.396	0.675
33	0.054	28.588	−0.123	0.057	0.182	0.538	0.833
34	−0.011	29.747	−0.018	0.173	0.312	0.676	0.988
35	0.009	31.516	0.087	0.288	0.436	0.816	—
36	0.008	33.398	0.192	0.402	0.566	0.956	—
37	0.013	36.436	0.399	0.631	0.691	—	—
38	0.007	37.662	0.502	0.743	0.944	—	—
39	0.000	8.555	−1.616	−1.427	−1.418	−1.284	−1.243
40	0.000	12.947	−0.902	−0.754	−0.649	−0.492	−0.335
41	0.000	17.194	−0.711	−0.557	−0.423	−0.244	−0.058
42	−0.013	16.718	−0.455	−0.293	−0.137	0.058	0.268
43	0.000	18.031	−0.238	−0.065	0.107	0.319	0.549
44	0.000	17.927	−0.226	−0.055	0.120	0.333	0.564
45	0.000	19.577	0.102	0.291	0.493	0.735	0.995
46	0.000	22.440	0.411	0.627	0.858	—	—
47	—	—	—	—	—	—	—
48	—	—	—	—	—	—	—
49	—	—	—	—	—	—	—
50	−0.005	10.520	—	—	—	—	—
51	0.001	14.860	−1.128	−1.263	−1.034	−1.025	−0.794
52	−0.002	16.408	−1.079	−1.110	−0.921	−0.870	−0.653
53	0.001	21.054	−0.849	−0.745	−0.549	−0.431	−0.205
54	−0.002	22.950	−0.743	−0.626	−0.426	−0.286	−0.051
55	−0.002	25.477	−0.529	−0.391	−0.175	0.001	0.259
56	−0.003	27.209	−0.423	−0.274	−0.048	0.143	0.415
57	0.000	28.860	−0.317	−0.158	0.083	0.285	0.567
58	0.003	30.076	−0.211	−0.042	0.210	0.426	0.722
59	−0.001	32.320	−0.106	0.073	0.336	0.565	0.875
60	0.129	11.335	—	—	−1.258	−1.147	−0.914
61	0.060	13.259	—	−1.159	−1.100	−0.990	−0.780
62	0.079	16.177	−1.003	−0.980	−0.823	−0.694	−0.489
63	0.038	17.795	−0.929	−0.844	−0.691	−0.549	−0.338
64	0.078	19.506	−0.811	−0.723	−0.558	−0.401	−0.181
65	0.073	21.383	−0.705	−0.602	−0.425	−0.257	−0.027
66	0.069	22.582	−0.606	−0.483	−0.296	−0.114	0.127
67	0.051	24.440	−0.501	−0.365	−0.164	0.030	0.284
68	0.069	26.170	−0.397	−0.248	−0.032	0.170	0.440
69	0.069	27.396	−0.294	−0.133	0.098	0.314	0.595
70	0.077	29.117	−0.187	−0.024	0.226	0.458	0.753
71	0.067	30.461	−0.082	0.091	0.354	0.596	0.918

Table 1   (*Continued*)

No.	MIES	MIVW	280^a	260^a	240^a	220^a	200^a
72	0.060	32.072	0.023	0.208	0.481	0.736	1.075
73	0.047	33.677	0.125	0.330	0.609	0.875	—
74	0.052	35.107	0.228	0.444	0.737	1.014	—
75	0.060	37.044	0.317	0.663	0.862	—	—
76	0.062	38.249	0.434	0.775	0.991	—	—
77	0.063	39.896	0.537	0.887	1.117	—	—
78	0.059	41.760	0.639	—	—	—	—
79	—	no data	—	—	—	—	—
—	—	—	—	—	—	—	—
109	—	no data	—	—	—	—	—

	180^a	160^a	280^b	240^b	200^b	160^b
1	—	−0.988	—	—	—	—
2	−0.925	−0.767	—	—	−1.338	−1.046
3	−0.736	−0.557	—	—	−1.142	−0.839
4	−0.549	−0.353	—	−1.252	−0.981	−0.633
5	−0.368	−0.156	−1.335	−1.110	−0.818	−0.431
6	−0.188	0.044	−1.243	−0.983	−0.658	−0.234
7	−0.008	0.241	−1.130	−0.849	−0.499	−0.036
8	0.170	0.438	−1.018	−0.717	−0.339	0.160
9	0.346	0.633	−0.909	−0.583	−0.180	0.356
10	0.522	0.828	−0.801	−0.449	−0.022	0.550
11	0.697	1.021	−0.691	−0.326	0.135	0.743
12	0.870	—	−0.584	−0.197	0.292	0.934
13	1.021	—	−0.477	−0.068	0.447	—
14	—	—	−0.370	0.061	0.602	—
15	—	—	−0.265	0.189	0.757	—
16	—	—	−0.158	0.316	0.910	—
17	—	—	−0.055	0.442	—	—
18	—	—	0.050	0.568	—	—
19	—	—	0.153	0.694	—	—
20	—	—	0.257	—	—	—
21	−1.043	−0.893	—	—	—	−1.178
22	−0.849	−0.663	—	—	−1.225	−0.957
23	−0.653	−0.467	—	−1.326	−1.049	−0.745
24	−0.463	−0.256	−1.453	−1.180	−0.876	−0.531
25	—	—	—	—	—	—
26	−0.096	0.144	−1.196	−0.900	−0.563	−0.131
27	—	—	—	—	—	—
28	0.264	0.540	−0.960	−0.629	−0.240	0.265
29	0.443	0.736	−0.845	−0.496	−0.080	0.460
30	0.620	0.942	−0.732	−0.364	0.078	0.655
31	0.793	1.137	−0.623	−0.233	0.235	0.848
32	0.967	—	−0.513	−0.102	0.392	—
33	—	—	−0.404	0.027	0.548	—
34	—	—	−0.296	0.156	0.704	—
35	—	—	−0.190	0.285	0.858	—
36	—	—	−0.084	0.412	—	—
37	—	—	0.126	0.664	—	—
38	—	—	0.230	0.790	—	—
39	−1.143	−1.032	−1.668	−1.620	−1.465	−1.288

Table 1 (*Continued*)

	180^a	160^a	280^b	240^b	200^b	160^b
40	−0.164	0.036	−1.140	−0.889	−0.598	−0.224
41	0.145	0.378	−0.946	−0.665	−0.318	0.124
42	0.494	0.752	−0.691	−0.376	0.009	0.495
43	0.799	1.079	−0.478	−0.132	0.289	0.822
44	0.814	1.097	−0.463	−0.118	0.305	0.840
45	—	—	−0.140	0.260	0.734	—
46	—	—	0.172	0.626	—	—
47	—	—	—	—	—	—
48	—	—	—	—	—	—
49	—	—	—	—	—	—
50	−1.331	−1.013	—	—	—	−1.278
51	−0.834	−0.583	—	−1.276	−1.169	−0.852
52	−0.626	−0.374	—	−1.140	−1.001	−0.650
53	−0.054	0.223	−1.094	−0.864	−0.516	−0.053
54	0.127	0.419	−0.991	−0.733	−0.353	0.145
55	0.483	0.807	−0.785	−0.468	−0.036	0.534
56	0.658	1.001	−0.681	−0.338	0.122	0.729
57	0.832	—	−0.577	−0.209	0.278	0.920
58	1.015	—	−0.472	−0.079	0.434	—
59	—	—	−0.368	0.049	0.589	—
60	−0.939	−0.743	—	—	−1.294	−1.008
61	−0.748	−0.542	—	—	−1.129	−0.809
62	−0.363	−0.141	−1.298	−1.077	−0.800	−0.409
63	−0.187	0.058	−1.199	−0.953	−0.638	−0.210
64	−0.007	0.256	−1.095	−0.821	−0.477	−0.012
65	0.171	0.452	−0.991	−0.690	−0.318	0.183
66	0.348	0.648	−0.885	−0.561	−0.154	0.379
67	0.523	0.843	−0.788	−0.431	0.003	0.574
68	0.698	1.036	−0.673	−0.303	0.163	0.766
69	0.872	—	−0.567	−0.175	0.313	—
70	—	—	−0.462	−0.049	0.621	—
71	—	—	−0.357	0.079	—	—
72	—	—	−0.252	0.206	—	—
73	—	—	−0.148	0.330	—	—
74	—	—	−0.044	—	—	—
75	—	—	0.060	—	—	—
76	—	—	0.163	—	—	—
77	—	—	0.265	—	—	—
78	—	—	0.368	—	—	—
79	—	no data	—	—	—	—
—	—	—	—	—	—	—
109	—	no data	—	—	—	—

^{a}Log k values measured at temperatures in the range 280–160 °C on CPSil5, from ref. 1.
^{b}Log k values measured at temperatures in the range 280–160 °C on DB1, from ref. 1.

Table 2 Retention time (log k) and molecular interaction energy values for a polyethyleneglycol (Carbowax™) phase (CPWAX, DBWAX). Reproduced by permission of Institute for Chromatography, ref. 13.

No.	MIFS	MIHB	MIES	MIVW	240^a	200^a	240^b	200^b
1	—	—	—	—	—	—	—	—
2	—	—	—	—	—	—	—	—
3	—	—	—	—	—	—	—	—
4	—	—	—	—	—	—	—	—
5	14.7279	0.000	−1.150	14.976	—	−1.307	—	−1.312
6	16.4348	0.000	−1.150	16.737	—	−1.145	−1.478	−1.168
7	17.2929	0.000	−1.150	17.575	−1.300	−0.991	−1.358	−1.024
8	18.3794	0.000	−1.150	18.619	−1.170	−0.839	−1.221	−0.880
9	19.4720	0.000	−1.150	19.694	−1.043	−0.689	−1.093	−0.733
10	20.6851	0.000	−1.150	20.886	−0.916	−0.545	−0.968	−0.589
11	22.0013	0.000	−1.150	22.287	−0.793	−0.394	−0.844	−0.445
12	23.7104	0.000	−1.150	23.975	−0.669	−0.249	−0.723	−0.301
13	25.2255	0.000	−1.150	25.549	−0.550	−0.104	−0.605	−0.158
14	26.1911	0.000	−1.150	26.584	−0.426	0.041	−0.484	−0.017
15	27.5137	0.000	−1.150	27.944	−0.306	0.184	−0.366	0.124
16	28.6643	0.000	−1.150	29.024	−0.185	0.327	−0.249	0.265
17	30.1168	0.000	−1.150	30.570	−0.067	0.469	−0.133	0.404
18	31.7059	0.000	−1.150	32.402	0.051	0.610	−0.017	0.542
19	32.8353	0.000	−1.150	33.176	0.167	0.750	0.097	0.680
20	33.7937	0.000	−1.150	34.212	0.283	0.889	0.211	0.817
21	7.5361	0.048	−1.124	7.572	—	−1.121	—	−1.130
22	8.6249	0.056	−1.127	8.621	—	−0.955	—	−0.969
23	9.7138	0.054	−1.134	9.675	−1.164	−0.805	−1.134	−0.814
24	10.9677	0.056	−1.133	10.902	−1.021	−0.652	−1.025	−0.660
25	—	—	—	—	—	—	—	—
26	14.0514	0.058	−1.129	14.130	−0.763	−0.348	−0.761	−0.362
27	—	—	—	—	—	—	—	—
28	15.9110	0.059	−1.131	15.930	−0.507	−0.050	−0.397	0.080
29	16.9538	0.058	−1.134	16.947	−0.386	0.100	−0.278	0.254
30	18.1744	0.059	−1.131	18.203	−0.261	0.247	−0.159	0.428
31	19.4730	0.060	−1.129	19.658	−0.137	0.386	−0.041	0.602
32	21.1032	0.055	−1.138	21.106	−0.018	0.532	0.076	0.712
33	21.8017	0.055	−1.138	21.838	0.103	0.678	0.193	0.907
34	22.9217	0.055	−1.140	22.900	0.224	0.823	0.309	1.082
35	—	—	—	—	—	—	—	—
36	—	—	—	—	—	—	—	—
37	—	—	—	—	—	—	—	—
38	—	—	—	—	—	—	—	—
39	6.9931	0.000	−1.150	6.998	−1.651	−1.456	−1.758	−1.426
40	10.2538	0.000	−1.150	10.254	−0.603	0.024	−0.590	−0.236
41	11.9252	0.000	−1.150	10.654	−0.353	0.064	−0.346	0.061
42	14.0677	0.000	−1.134	14.094	0.047	0.508	0.058	0.510
43	13.7446	0.000	−1.150	13.854	0.426	0.946	0.462	0.951
44	13.7620	0.000	−1.150	13.767	0.433	0.952	0.449	0.957
45	15.8950	0.000	−1.150	15.926	0.949	—	0.970	—
46	—	—	—	—	—	—	—	—
47	—	—	—	—	—	—	—	—

Table 2 (*Continued*)

No.	MIFS	MIHB	MIES	MIVW	240[a]	200[a]	240[b]	200[b]
48	—	—	—	—	—	—	—	—
49	—	—	—	—	—	—	—	—
50	—	—	—	—	—	—	—	—
51	—	—	—	—	—	—	—	—
52	—	—	—	—	—	—	—	—
53	18.2859	0.000	−1.163	16.955	−1.205	−0.924	−1.286	−0.951
54	19.3193	0.000	−1.163	18.016	−1.076	−0.769	−1.152	−0.808
55	21.5834	0.000	−1.164	20.251	−0.835	−0.469	−0.900	−0.515
56	22.8993	0.000	−1.164	21.693	−0.716	−0.321	−0.778	−0.370
57	24.4546	0.000	−1.163	23.094	−0.595	−0.173	−0.656	−0.227
58	25.1489	0.000	−1.163	23.769	−0.476	−0.027	−0.536	−0.084
59	26.3361	0.000	−1.164	24.973	−0.357	0.117	−0.418	0.057
60	—	—	—	—	—	—	—	—
61	11.4103	0.000	−1.086	11.576	—	−1.151	—	−1.190
62	13.6755	0.000	−1.087	13.718	−1.172	−0.842	−1.174	−0.849
63	14.6557	0.000	−1.100	14.812	−1.056	−0.691	−1.048	−0.706
64	16.3692	0.000	−1.102	16.605	−0.920	−0.543	−0.923	−0.560
65	17.2585	0.000	−1.103	17.406	−0.794	−0.394	−0.803	−0.412
66	18.2980	0.000	−1.098	18.428	−0.666	−0.246	−0.682	−0.268
67	19.4004	0.000	−1.098	19.500	−0.543	−0.099	−0.560	−0.125
68	20.5876	0.000	−1.099	20.685	−0.420	0.048	−0.440	0.017
69	21.8825	0.000	−1.091	21.984	−0.298	0.192	−0.322	0.159
70	23.4636	0.000	−1.102	23.569	−0.179	0.338	−0.199	0.301
71	24.1648	0.000	−1.102	24.274	−0.059	0.482	−0.082	0.442
72	25.3375	0.000	−1.102	25.443	0.062	0.627	0.034	0.582
73	26.3548	0.000	−1.101	26.457	0.182	0.770	0.146	0.721
74	27.6285	0.000	−1.101	27.819	0.303	0.912	0.262	0.860
75	28.9478	0.000	−1.100	29.253	0.419	1.194	0.378	0.998
76	30.2597	0.000	−1.101	30.412	0.539	—	0.491	—
77	31.1002	0.000	−1.101	31.413	0.655	—	0.606	—
78	—	—	—	—	0.770	—	0.718	—
79	—	no data	—	—	—	—	—	—
—	—	—	—	—	—	—	—	—
109	—	no data	—	—	—	—	—	—

[a]Log *k* values measured at 240 and 200 °C on CPWAX, from ref. 1.
[b]Log *k* values measured at 240 and 200 °C on DBWAX, from ref. 1.

Table 3 Retention time (log *k*) and molecular interaction energy values for a 50% methylphenylsilicone (OV17) phase and a methylsilicone (OV1) phase.

No.	MIFS	MIHB	MIES	MIVW	140[a]	150[a]	160[a]	170[a]
1	10.9139	0.000	8.001	10.707	—	—	—	—
2	12.3809	0.000	8.001	13.212	−0.703	−0.842	—	—
3	13.9489	0.000	8.001	13.806	−0.478	−0.607	−0.684	−0.955
4	14.9783	0.000	8.002	15.074	−0.243	−0.368	−0.466	−0.686
5	16.1510	0.000	8.000	16.211	−0.018	−0.145	−0.266	−0.491
6	17.5759	0.000	8.004	17.686	0.206	0.067	−0.066	−0.284
7	19.1123	0.000	8.004	19.177	0.431	0.280	0.136	−0.006

Table 3   (*Continued*)

No.	MIFS	MIHB	MIES	MIVW	140^a	150^a	160^a	170^a
8	20.2633	0.000	8.002	19.842	0.655	0.493	0.338	0.184
9	21.7054	0.000	8.002	21.235	0.877	0.704	0.538	0.373
10	23.1244	0.000	8.003	22.899	1.098	0.913	0.736	0.560
11	24.2915	0.000	8.003	24.174	—	1.129	0.939	0.853
12	25.8377	0.000	8.003	25.763	—	—	1.116	0.921
13	27.3444	0.000	8.003	27.205	—	—	—	1.106
14	28.2207	0.000	8.002	28.758	—	—	—	—
21	10.3179	2.372	8.133	8.349	—	—	—	—
22	11.2727	2.335	8.130	9.190	−0.623	−0.788	—	—
23	12.5791	2.238	8.126	10.657	−0.380	−0.529	−0.618	—
24	14.1059	2.203	8.122	12.313	−0.138	−0.290	−0.391	−0.588
25	15.3246	2.284	8.130	13.736	0.092	−0.059	−0.173	−0.351
26	16.4174	2.256	8.127	14.738	0.324	0.165	0.040	−0.132
27	17.8321	2.250	8.126	16.110	0.554	0.386	0.250	0.079
28	19.2733	2.199	8.130	17.647	0.780	0.603	0.456	0.282
29	20.5299	2.248	8.127	18.829	0.990	0.803	0.639	0.458
30	21.8868	2.247	8.122	20.079	1.211	1.014	0.839	0.650
31	23.3495	2.248	8.124	21.573	—	1.223	1.038	0.841
32	24.5127	2.235	8.126	22.805	—	—	1.237	1.031
33	26.0210	2.244	8.120	24.284	—	—	—	1.216
34	27.5335	2.241	8.121	25.832	—	—	—	—
35	28.4934	2.226	8.130	27.301	—	—	—	—
36	29.5993	2.225	8.129	28.314	—	—	—	—
39	6.4676	0.000	8.001	5.887	—	—	—	—
40	10.3119	0.000	7.997	9.295	0.170	0.031	−0.068	−0.226
41	15.0069	0.000	7.999	14.406	0.562	0.406	0.281	0.123
42	12.5029	0.000	7.990	11.642	0.964	0.797	0.655	0.493
43	13.6374	0.000	7.995	12.474	1.325	1.144	0.988	0.817
44	13.3267	0.000	7.996	13.356	1.343	1.171	1.000	0.836
45	14.3167	0.000	7.996	15.134	—	—	1.518	1.332
46	16.2131	0.000	7.996	15.134	—	—	—	—
47	15.8115	0.000	7.974	14.884	—	—	—	—
48	—	—	—	—	—	—	—	—
49	—	—	—	—	—	—	—	—
50	no data	—	—	—	—	—	—	—
—	—	—	—	—	—	—	—	—
78	no data	—	—	—	—	—	—	—
79	7.6354	0.000	7.995	7.103	−0.767	−0.870	—	—
80	10.0797	0.000	8.014	9.344	−0.541	−0.678	—	—
81	10.7948	0.000	8.015	10.105	−0.336	−0.472	—	—
82	12.0400	0.000	8.010	10.957	−0.103	−0.242	−0.385	−0.533
83	14.4813	0.000	9.010	13.526	0.333	0.182	0.051	−0.104
84	16.2569	0.000	8.012	15.547	0.557	0.396	0.255	0.097
85	17.6930	0.000	8.011	16.935	0.778	0.608	0.456	0.292
86	20.6991	0.000	8.010	17.939	1.001	0.819	0.657	0.483
87	20.0920	0.000	8.011	19.298	—	—	—	—
88	12.4540	0.000	8.007	10.394	−0.455	−0.600	−0.699	−1.022
89	9.5052	0.000	8.020	8.918	−0.520	−0.680	−0.757	−1.022
90	9.2220	0.000	7.989	8.747	−0.520	−0.680	−0.757	−0.602
91	12.2549	0.000	7.984	9.729	−0.162	−0.298	−0.437	−0.656
92	10.1343	0.000	7.973	9.034	−0.228	−0.361	−0.460	−0.701
93	10.5664	0.000	8.021	9.878	−0.286	−0.412	−0.513	−0.287

Table 3 (*Continued*)

No.	MIFS	MIHB	MIES	MIVW	140^a	150^a	160^a	170^a
94	14.1929	0.000	7.985	11.245	0.125	−0.018	−0.137	−0.363
95	13.5876	0.000	7.985	10.999	0.052	−0.087	−0.199	−0.376
96	11.9359	0.000	7.975	10.593	0.023	−0.108	−0.218	−0.051
97	16.1700	0.000	7.976	13.007	0.373	0.226	0.101	—
98	8.4914	0.000	8.029	7.959	−0.551	−0.721	−0.783	—
99	9.2744	0.000	8.012	8.720	−0.133	−0.273	−0.361	−0.472
100	9.3614	0.000	8.088	8.700	−0.223	−0.359	−0.456	−0.545
101	8.8158	0.000	7.893	8.223	−0.223	−0.359	−0.456	−0.772
102	10.7597	0.000	7.951	10.096	0.221	0.075	−0.035	−0.155
103	10.0193	0.000	8.014	9.340	0.162	0.022	−0.098	−0.202
104	10.5352	0.000	8.138	9.784	0.082	0.008	−0.134	−0.292
105	11.7375	0.000	7.912	11.319	0.557	0.405	0.281	0.149
106	11.3879	0.000	7.896	10.688	0.469	0.318	0.194	0.074
107	12.3552	0.000	7.912	11.646	0.469	0.318	0.194	0.074
108	12.8618	0.000	7.870	12.113	0.853	0.689	0.552	0.407
109	13.6763	0.000	7.820	13.009	1.216	1.039	0.887	0.731

No.	180^a	190^a	200^a	210^a	220^a	230^a
1	—	—	—	—	—	—
2	—	—	—	—	—	—
3	—	—	—	—	—	—
4	−0.917	—	—	—	—	—
5	−0.614	−0.714	−0.644	—	—	—
6	−0.368	−0.503	−0.488	−0.622	—	—
7	−0.167	−0.307	−0.330	−0.461	—	−0.614
8	0.029	−0.111	−0.170	−0.292	−0.380	−0.478
9	0.215	0.069	−0.009	−0.136	−0.235	−0.340
10	0.399	0.246	0.151	0.019	−0.089	−0.203
11	0.583	0.421	0.315	0.177	0.057	−0.061
12	0.745	0.579	0.460	0.316	0.194	0.064
13	0.922	0.749	0.618	0.467	0.337	0.199
14	1.096	0.915	0.774	0.616	0.479	0.333
21	—	—	—	—	—	—
22	—	—	—	—	—	—
23	—	—	—	—	—	—
24	−0.738	—	—	—	—	—
25	−0.494	−0.559	−0.660	—	—	—
26	−0.268	−0.364	−0.480	—	—	—
27	−0.059	−0.175	−0.288	−0.383	−0.490	—
28	0.128	0.008	−0.112	−0.216	−0.270	−0.388
29	0.308	0.168	0.051	−0.061	−0.232	−0.256
30	0.490	0.430	0.218	0.096	0.007	−0.120
31	0.671	0.515	0.384	0.252	0.151	0.018
32	0.852	0.687	0.546	0.408	0.295	0.156
33	1.029	0.855	0.707	0.560	0.437	0.291
34	1.206	1.023	0.866	0.712	0.580	0.428
35	—	1.189	1.025	0.863	0.722	0.563
36	—	—	1.183	1.013	0.863	0.678
39	—	—	—	—	—	—
40	−0.376	−0.494	−0.526	−0.701	—	−0.697
41	−0.017	−0.143	−0.221	−0.375	−0.360	−0.458
42	0.049	0.218	0.120	−0.024	−0.050	−0.163

Table 3 (*Continued*)

No.	180[a]	190[a]	200[a]	210[a]	220[a]	230[a]
43	0.664	0.523	0.412	0.265	0.211	0.091
44	0.683	0.540	0.426	0.280	0.224	0.105
45	1.161	1.002	0.872	0.714	0.630	0.498
46	—	1.467	1.316	1.140	1.033	0.884
47	—	1.502	1.349	1.174	1.055	0.915
48	—	—	—	—	—	1.302
49	—	—	—	—	—	—
50	no data	—	—	—	—	—
—	—	—	—	—	—	—
78	no data	—	—	—	—	—
79	—	—	—	—	—	—
80	—	—	—	—	—	—
81	—	—	—	—	—	—
82	−0.686	−0.833	—	—	—	—
83	−0.240	−0.383	−0.437	−0.514	—	−0.654
84	−0.044	−0.178	−0.260	−0.352	−0.406	−0.511
85	0.145	0.008	−0.089	−0.184	−0.260	−0.360
86	0.333	0.189	0.080	−0.027	−0.113	−0.221
87	0.516	0.365	−0.246	0.129	0.033	−0.083
88	−1.168	−1.237	—	—	—	—
89	−1.168	−1.237	—	—	—	—
90	−1.168	−1.237	—	—	—	—
91	−0.747	−0.883	−0.903	−0.939	−0.939	—
92	−0.836	−0.883	−0.903	−0.939	−0.939	—
93	−0.836	−0.883	−0.903	−0.939	−0.939	—
94	−0.413	−0.575	−0.620	−0.705	−0.705	—
95	−0.500	−0.575	−0.620	−0.705	−0.705	—
96	−0.500	−0.575	−0.620	−0.705	−0.705	—
97	−0.180	−0.331	−0.394	−0.483	−0.483	—
98	—	−1.678	−1.222	—	—	—
99	−0.662	−0.860	−0.801	−0.857	—	—
100	−0.772	—	—	—	—	—
101	−0.772	—	—	—	—	—
102	−0.302	−0.449	−0.463	−0.527	−0.445	−0.714
103	−0.384	−0.542	−0.595	−0.607	—	−0.714
104	−0.384	−0.542	−0.595	−0.607	—	−0.714
105	0.000	−0.128	−0.194	−0.277	−0.325	−0.431
106	−0.090	−0.231	−0.306	−0.359	−0.415	−0.514
107	−0.090	−0.231	−0.306	−0.359	−0.415	−0.514
108	0.256	0.123	0.037	−0.059	−0.127	−0.256
109	0.571	0.431	0.326	0.216	0.131	0.003

No.	160[b]	170[b]	180[b]	190[b]	200[b]	210[b]	220[b]	230[b]
1	—	—	—	—	—	—	—	—
2	—	—	—	—	—	—	—	—
3	—	—	—	—	—	—	—	—
4	−0.796	−0.896	−1.027	—	—	—	—	—
5	−0.592	−0.699	−0.824	—	—	—	—	—
6	−0.385	−0.504	−0.627	−0.724	—	—	—	—
7	−0.180	−0.309	−0.438	−0.545	−0.658	−0.759	−0.896	—
8	0.022	−0.117	−0.237	−0.361	−0.489	−0.599	−0.728	−0.801

Table 3 (*Continued*)

No.	160[b]	170[b]	180[b]	190[b]	200[b]	210[b]	220[b]	230[b]
9	0.231	0.080	−0.064	−0.196	−0.317	−0.438	−0.567	−0.654
10	0.432	0.271	0.120	−0.020	−0.149	−0.281	−0.412	−0.504
11	0.627	0.458	0.300	0.151	0.012	−0.127	−0.257	−0.364
12	0.820	0.642	0.476	0.320	0.171	0.027	−0.110	−0.225
13	1.016	0.829	0.654	0.490	0.333	0.181	0.038	−0.080
14	1.203	1.010	0.827	0.655	0.487	0.331	0.182	0.046
21	—	—	—	—	—	—	—	—
22	−0.893	—	—	—	—	—	—	—
23	−0.678	−0.785	—	—	—	—	—	—
24	−0.465	−0.583	−0.676	−0.762	−0.893	—	—	—
25	−0.256	−0.385	−0.488	−0.585	−0.714	−0.801	—	—
26	−0.048	−0.188	−0.301	−0.410	−0.541	−0.638	−0.747	—
27	0.156	0.010	−0.107	−0.228	−0.365	−0.471	−0.588	−0.697
28	0.372	0.204	0.077	−0.052	−0.195	−0.308	−0.433	−0.548
29	0.567	0.393	0.254	0.116	−0.032	−0.152	−0.284	−0.401
30	0.760	0.581	0.430	0.282	0.129	−0.001	−0.136	−0.260
31	0.950	0.767	0.602	0.446	0.288	0.148	0.010	−0.119
32	1.154	0.958	0.787	0.621	0.455	0.308	0.160	0.024
33	1.344	1.142	0.959	0.784	0.612	0.456	0.304	0.162
34	—	1.333	1.147	0.964	0.783	0.621	0.458	0.308
35	—	—	1.311	1.119	0.932	0.760	0.594	0.438
36	—	—	—	1.289	1.094	0.914	0.739	0.577
37	—	—	—	—	—	—	—	—
38	—	—	—	—	—	—	—	—
39	−1.276	—	—	—	—	—	—	—
40	0.037	−0.086	−0.204	−0.296	−0.411	−0.495	−0.565	−0.703
41	—	—	—	—	—	—	—	—
42	0.859	0.710	0.568	0.436	0.308	0.184	0.075	−0.049
43	1.279	1.115	0.960	0.812	0.670	0.539	0.416	0.286
44	1.281	—	0.970	0.819	0.672	0.541	0.419	0.292
45	—	—	—	1.369	1.204	1.055	0.916	0.774
46	—	—	—	—	—	—	1.382	1.227
47	—	—	—	—	—	—	—	1.243
48	—	—	—	—	—	—	—	—
49	—	—	—	—	—	—	—	—
50	no data	—	—	—	—	—	—	—
—	—	—	—	—	—	—	—	—
78	no data	—	—	—	—	—	—	—
79	—	—	—	—	—	—	—	—
80	—	—	—	—	—	—	—	—
81	−0.620	−0.693	−0.839	—	—	—	—	—
82	−0.405	−0.499	−0.623	−0.721	−0.833	—	—	—
83	0.014	−0.116	−0.246	−0.366	−0.488	−0.592	−0.703	−0.801
84	0.215	0.074	−0.061	−0.186	−0.321	−0.435	−0.551	−0.656
85	0.416	0.264	0.122	−0.010	−0.150	−0.276	−0.398	−0.511
86	0.574	0.423	0.279	0.146	0.010	−0.115	−0.234	−0.343
87	0.779	0.613	0.457	0.314	0.171	0.039	−0.087	−0.202
88	—	—	—	—	—	—	—	—
89	—	—	—	—	—	—	—	—
90	—	—	—	—	—	—	—	—
91	−0.614	—	—	—	—	—	—	—

Table 3 (*Continued*)

No.	160^b	170^b	180^b	190^b	200^b	210^b	220^b	230^b
92	−0.440	—	—	—	—	—	—	—
93	−0.539	—	—	—	—	—	—	—
94	−0.161	−0.287	−0.400	−0.481	−0.616	—	—	—
95	−0.265	−0.388	−0.495	−0.573	−0.712	—	—	—
96	−0.265	−0.388	−0.495	−0.573	−0.712	—	—	—
97	−0.180	−0.331	−0.394	−0.483	−0.483	—	—	—
98	−1.252	—	—	—	—	—	—	—
99	−0.335	−0.440	−0.558	−0.654	−0.818	—	—	—
100	−0.796	−0.893	−1.027	—	—	—	—	—
101	−0.435	−0.529	−0.656	—	—	—	—	—
102	0.063	−0.055	−0.175	−0.265	−0.381	−0.472	−0.569	—
103	−0.045	−0.163	−0.282	−0.365	−0.483	−0.570	−0.662	—
104	−0.188	−0.299	−0.419	−0.489	−0.607	−0.883	−0.975	—
105	0.401	0.270	0.145	0.030	−0.096	−0.318	−0.421	−0.516
106	—	—	—	—	—	—	—	—
107	—	—	—	—	—	—	—	—
108	0.660	0.517	0.384	0.304	0.128	0.015	−0.098	−0.203
109	1.023	0.869	0.722	0.583	0.444	0.320	0.197	0.083

[a] Log k values measured at temperatures in the range 140–230 °C on a 50% methylphenylsilicone phase (OV17).
[b] Log k values measured at temperatures in the range 160–230 °C on a methylsilicone phase (OV1) as reference values.

Table 4 Atomic partial charge (apc) values calculated by the AM1 program, and reference and predicted dissociation constants (pK_a values) of some phenolic compounds.

Analyte	$pK_a{}^a$	$pK_a{}^b$	$pK_a{}^c$	apc (O)	apc (H)	$pK_a{}^d$	$pK_a{}^e$
Phenol	10.02	10.47	9.92	−0.2472	0.2184	10.3175	10.0205
2-Methylphenol	10.32	10.50	10.21	−0.2490	0.2198	10.0334	10.0947
3-Methylphenol	10.09	10.57	10.05	−0.2472	0.2183	10.3378	10.0408
4-Methylphenol	10.27	10.57	10.23	−0.2464	0.2180	10.3986	10.1016
2,3-Dimethylphenol	10.54	10.60	10.34	−0.2499	0.2196	10.0740	10.1353
2,4-Dimethylphenol	10.60	10.69	10.52	−0.2488	0.2194	10.1146	10.1799
2,5-Dimethylphenol	10.41	10.70	10.34	−0.2490	0.2198	10.0334	10.0947
2,6-Dimethylphenol	10.63	—	10.50	−0.2498	0.2187	10.2566	10.3179
3,4-Dimethylphenol	10.36	10.55	10.37	−0.2509	0.2170	10.6016	10.3046
3,5-Dimethylphenol	10.19	10.60	10.19	−0.2499	0.2173	10.5407	10.2437
2,3,4-Trimethylphenol	—	—	10.66	−0.2532	0.2164	10.7233	10.4263
2,3,5-Trimethylphenol	—	10.59	10.48	−0.2516	0.2170	10.6016	10.3046
2,3,6-Trimethylphenol	—	—	10.63	−0.2552	0.2187	10.2566	9.9596
2,4,6-Trimethylphenol	10.88	—	10.81	−0.2520	0.2179	10.4189	10.4802
2,3,4,5-Tetramethylphenol	—	—	—	−0.2521	0.2184	10.3175	10.0205
2,3,5,6-Tetramethylpheno	—	—	—	−0.2536	0.2189	10.2160	9.9190
2,3,4,5,6-Pentamethylphenol	—	—	—	−0.2560	0.2179	10.4189	10.1219
2-Ethylphenol	10.20	10.58	10.12	−0.2555	0.2205	9.8914	9.9527
3-Ethylphenol	9.90	10.59	10.08	−0.2534	0.2166	10.6827	10.3857
4-Ethylphenol	10.00	10.61	10.25	−0.2529	0.2167	10.6624	10.3654

Table 4 (*Continued*)

Analyte	$pK_a{}^a$	$pK_a{}^b$	$pK_a{}^c$	apc (O)	apc (H)	$pK_a{}^d$	$pK_a{}^e$
4-*tert*-Butylphenol	—	10.67	10.25	−0.2506	0.2172	10.5610	10.2640
2-Chlorophenol	8.48	9.44	8.40	−0.2416	0.2299	7.9839	8.0452
3-Chlorophenol	9.02	9.83	9.10	−0.2426	0.2222	9.5464	9.2494
4-Chlorophenol	9.38	10.08	9.39	−0.2438	0.2209	9.8102	9.5132
2,3-Dichlorophenol	7.45	8.04	7.58	−0.2385	0.2323	7.4969	7.5582
2,4-Dichlorophenol	7.89	8.33	7.87	−0.2379	0.2320	7.5578	7.6191
2,5-Dichlorophenol	7.50	7.53	7.58	−0.2370	0.2326	7.4360	7.4973
2,6-Dichlorophenol	6.79	7.32	6.89	−0.2278	0.2331	7.3346	7.0376
3,4-Dichlorophenol	8.39	9.00	8.56	−0.2385	0.2244	9.1000	8.8030
3,5-Dichlorophenol	8.18	8.43	8.27	−0.2377	0.2251	8.9579	8.6609
2,3,4-Trichlorophenol	7.59	7.27	7.04	−0.2376	0.2336	7.2331	7.2944
2,3,5-Trichlorophenol	7.23	6.82	6.75	−0.2334	0.2350	6.9490	7.0103
2,3,6-Trichlorophenol	6.12	6.29	6.06	−0.2252	0.2354	6.8679	6.5709
2,4,5-Trichlorophenol	7.33	7.08	7.04	−0.2242	0.2351	6.9287	6.9900
2,4,6-Trichlorophenol	6.42	6.42	6.35	−0.2244	0.2351	6.9287	6.6317
3,4,5-Trichlorophenol	7.74	—	7.74	−0.2341	0.2268	8.6130	8.3160
2,3,4,5-Tetrachlorophenol	6.96	—	6.22	−0.2307	0.2364	6.6650	6.7263
2,3,5,6-Tetrachlorophenol	5.44	—	5.24	−0.2221	0.2376	6.4215	6.1245
2,3,4,5,6-Pentachlorophenol	5.26	4.84	4.70	−0.2193	0.2390	6.1374	5.8404
2-Bromophenol	8.44	8.90	8.36	−0.2434	0.2309	7.7810	7.8423
3-Bromophenol	9.03	9.67	9.05	−0.2454	0.2217	9.6479	9.3509
4-Bromophenol	9.36	10.22	9.43	−0.2447	0.2217	9.6479	9.3509
2,4-Dibromophenol	7.80	8.20	8.36	−0.2402	0.2335	7.2534	7.3147
2,6-Dibromophenol	6.60	7.05	6.80	−0.2239	0.2363	6.6852	6.3882
1,2-Dihydroxybenzene	—	—	9.83	−0.2295	0.2133	11.3524	11.0554
1,3-Dihydroxybenzene	9.81	9.78	9.63	−0.2450	0.2219	9.6073	9.3103
1,4-Dihydroxybenzene	10.35	9.97	10.77	−0.2500	0.2163	10.7436	10.4466
1-Hydroxynaphthalene	—	10.54	—	−0.2525	0.2201	9.9725	9.6755
2-Hydroxynaphthalene	—	10.70	—	−0.2514	0.2186	10.2769	9.9799
2-Hydroxyacetophenone	—	11.04	—	−0.2487	0.2234	9.3029	9.0059
4-Hydroxybutylbenzoate	—	—	—	−0.2418	0.2239	9.2014	8.9044
4-Hydroxypropylbenzoate	—	—	—	−0.2418	0.2239	9.2014	8.9044
2-Nitrophenol	7.23	7.29	6.80	—	—	8.30	—
3-Nitrophenol	8.40	8.56	8.27	—	—	8.71	—
4-Nitrophenol	7.15	7.23	8.18	—	—	7.88	—
2,4-Dinitrophenol	4.09	—	5.06	—	—	−2.21	—
2,5-Dinitrophenol	5.22	5.39	5.15	—	—	−1.38	—
2,6-Dinitrophenol	3.71	4.93	3.68	—	—	−2.21	—
3,4-Dinitrophenol	5.43	—	6.53	—	—	6.65	—
1-Hydroxy-2,4-dinitronaphthalene	—	—	—	—	—	—	−3.03
2-Chloro-5-methylphenol	—	9.73	8.54	—	—	9.12	—
4-Chloro-2-methylphenol	—	10.53	9.68	—	—	9.53	—
4-Chloro-3,5-dimethylphenol	—	10.63	9.65	—	—	9.94	—
4-Chloro-3-methylphenol	—	9.57	9.52	—	—	9.74	—

[a]Values from ref. 2.
[b]Values from ref. 3.
[c]Values from Hammett's equation, ref. 4.
[d]Predicted pK_a from the apc of hydrogen.
[e]Predicted pK_a values from the apc of hydrogen with *ortho*-effect.

Table 5 Measured and predicted pK_a values of aromatic acids (1).

Substituent	Benzoic acid					Phenylacetic acid				
	pK_a^a	pK_a^b	pK_a^c	pK_a^d	pK_a^e	pK_a^a	pK_a^b	pK_a^c	pK_a^d	pK_a^e
H	4.200	4.49	4.20	4.293	4.197	4.334	4.29	4.30	4.291	4.291
2-Methyl-	—	—	3.91	4.389	4.305	—	—	4.20	4.569	4.601
3-Methyl-	—	—	4.26	4.348	4.259	4.067	—	4.32	4.583	4.617
4-Methyl-	—	—	4.34	4.361	4.274	—	—	—	4.638	4.632
2,4-Dimethyl-	—	—	4.05	4.471	4.398	—	—	—	4.638	4.679
2,5-Dimethyl-	—	—	3.97	4.443	4.367	—	—	—	4.624	4.663
2,6-Dimethyl-	—	—	3.62	4.826	4.799	—	—	—	4.610	4.648
3,4-Dimethyl-	—	—	4.40	4.416	4.336	—	—	—	4.569	4.601
3,5-Dimethyl-	—	—	4.32	4.402	4.320	—	—	—	4.638	4.679
2,4,6-Trimethyl-	—	—	3.76	4.895	4.876	—	—	—	4.610	4.648
2-Ethyl-	—	—	—	4.389	4.305	—	—	—	4.555	4.586
3-Ethyl-	—	—	—	4.348	4.259	—	—	—	4.555	4.586
4-Ethyl-	4.388	—	4.35	4.375	4.290	—	—	—	4.596	4.632
2-Chloro-	—	—	2.92	4.320	4.228	—	—	3.74	4.416	4.431
3-Chloro-	—	—	3.83	3.951	3.811	—	—	—	4.291	4.291
4-Chloro-	3.894	—	3.96	4.006	3.873	4.185	—	4.22	4.277	4.275
2,4-Dichloro-	—	—	2.68	4.061	3.935	—	—	—	4.166	4.151
2,5-Dichloro-	—	—	2.55	4.143	4.027	—	—	—	4.124	4.105
2,6-Dichloro-	—	—	1.64	4.238	4.135	—	—	—	4.416	4.431
3,4-Dichloro-		—	3.50	3.705	−3.534	—	—	—	4.180	4.167
3,5-Dichloro-	—	—	2.46	3.650	3.472	—	—	—	4.124	4.105
2-Bromo-	—	—	2.85	4.416	4.336	—	—	—	4.458	4.477
3-Bromo-	—	—	3.86	3.896	3.750	—	—	—	4.249	4.244
4-Bromo-	3.864	—	3.98	3.937	3.796	—	—	—	4.291	4.291
2-Hydroxy-	2.807	3.06	3.07	2.734	2.438	4.225	4.14	—	3.958	3.919
3-Hydroxy-	4.093	4.16	4.07	4.129	4.012	—	—	—	4.347	4.353
4-Hydroxy-	4.682	4.53	4.58	4.279	4.182	4.404	4.40	4.43	4.541	4.570
2,4-Dihydroxy-	—	—	—	4.525	4.459	—	—	—	4.707	4.756
2,5-Dihydroxy-	—	—	—	4.457	4.382	3.964	4.06	—	4.610	4.648
2,6-Dihydroxy-	—	—	—	3.568	4.291	—	—	—	4.291	4.291
3,4-Dihydroxy-	4.456	4.49	4.45	4.143	4.353	4.356	4.39	4.39	4.347	4.353
3,5-Dihydroxy-	4.068	4.12	3.94	3.978	4.322	—	—	—	4.319	4.322
3,4,5-Trihydroxy-	4.364	4.40	4.32	3.992	4.307	—	—	—	4.305	4.307
2-Methoxy-	—	4.09	4.09	4.635	4.818	—	—	—	4.763	4.818
3-Methoxy-	—	—	—	4.252	4.446	4.319	4.27	4.26	4.430	4.446
4-Methoxy-	—	—	—	4.375	4.663	4.399	4.35	4.40	4.624	4.663
4-Hydroxy-3-methoxy-	4.474	4.48	4.47	4.266	4.462	4.425	4.37	4.39	4.444	4.462

[a]Measured by liquid chromatography, from ref. 5.
[b]Measured by titration, from ref. 6.
[c]Determined from Hammett's equation, from refs. 4 and 5.
[d]Based on pK_a by liquid chromatography, from ref. 5.
[e]Based on pK_a determined from Hammett's equation, from ref. 4.

Table 6 Measured and predicted pK_a values of aromatic acids (2).

Substituent	Mandelic acid					Cinnamic acid				
	$pK_a^{\,a}$	$pK_a^{\,b}$	$pK_a^{\,c}$	$pK_a^{\,d}$	$pK_a^{\,e}$	$pK_a^{\,a}$	$pK_a^{\,b}$	$pK_a^{\,c}$	$pK_a^{\,d}$	$pK_a^{\,e}$
H	—	3.42	3.38	3.681	3.681	4.376	4.37	4.38	4.197	4.523
2-Methyl-	—	—	—	3.786	3.800	—	—	—	4.566	4.570
3-Methyl-	—	—	—	3.734	3.741	—	—	—	4.551	4.554
4-Methyl-	—	—	—	3.734	3.741	—	—	—	4.580	4.586
2,4-Dimethyl-	—	—	—	3.747	3.756	—	—	—	4.523	4.523
2,5-Dimethyl-	—	—	—	3.655	3.651	—	—	—	4.523	4.523
2,6-Dimethyl-	—	—	—	4.075	4.129	—	—	—	4.537	4.539
3,4-Dimethyl-	—	—	—	3.799	3.815	—	—	—	4.608	4.617
3,5-Dimethyl-	—	—	—	3.786	3.800	—	—	—	4.580	4.586
2,4,6-Trimethyl-	—	—	—	4.114	4.174	—	—	—	4.594	4.602
2-Ethyl-	—	—	—	3.760	3.771	—	—	—	4.566	4.570
3-Ethyl-	—	—	—	3.747	3.756	—	—	—	4.551	4.554
4-Ethyl-	—	—	—	3.747	3.756	—	—	—	4.580	4.586
2-Chloro-	—	—	—	3.707	3.711	—	—	—	4.410	4.397
3-Chloro-	—	—	—	3.392	3.353	—	—	—	4.367	4.350
4-Chloro-	3.894	—	—	3.458	3.427	—	—	—	4.353	4.334
2,4-Dichloro-	—	—	—	3.497	3.472	—	—	—	4.254	4.224
2,5-Dichloro-	—	—	—	3.510	3.487	—	—	—	4.282	4.225
2,6-Dichloro-	—	—	—	3.655	3.651	—	—	—	4.339	4.318
3,4-Dichloro-	—	—	—	3.222	3.159	—	—	—	4.211	4.177
3,5-Dichloro-	—	—	—	3.222	3.159	—	—	—	4.169	4.130
2-Bromo-	—	—	—	3.760	3.771	—	—	—	4.381	4.366
3-Bromo-	—	—	—	3.340	3.293	—	—	—	4.169	4.130
4-Bromo-	—	—	—	3.405	3.368	—	—	—	4.310	4.287
2-Hydroxy-	—	—	—	3.130	3.054	—	—	—	4.736	4.759
3-Hydroxy-	—	—	—	3.550	3.532	—	—	—	4.381	4.366
4-Hydroxy-	—	—	—	3.668	3.666	4.595	4.57	4.63	4.580	4.586
2,4-Dihydroxy-	—	—	—	3.169	3.099	—	—	—	4.806	4.838
2,5-Dihydroxy-	—	—	—	3.077	2.994	—	—	—	4.636	4.649
2,6-Dihydroxy-	—	—	—	2.749	2.621	—	—	—	4.480	4.476
3,4-Dihydroxy-	—	—	3.49	3.550	3.532	4.554	4.53	4.55	4.452	4.444
3,5-Dihydroxy-	—	—	—	3.432	3.397	—	—	—	4.395	4.381
3,4,5-Trihydroxy-	—	—	—	3.418	3.383	—	—	—	4.395	4.381
2-Methoxy-	—	—	—	3.944	3.979	—	—	—	4.778	4.806
3-Methoxy-	3.334	3.47	3.33	3.589	3.577	—	—	4.31	4.452	4.444
4-Methoxy-	—	—	—	3.786	3.800	4.399	4.35	4.40	4.622	4.633
4-Hydroxy-3-methoxy-	—	3.55	3.50	3.628	3.621	4.570	4.56	4.56	4.523	4.523

[a]Measured by liquid chromatography, from ref. 5.
[b]Measured by titration, from ref. 6.
[c]Determined from Hammett's equation, from refs. 4 and 5.
[d]Based on pK_a by liquid chromatography, from ref. 5.
[e]Based on pK_a determined from Hammett's equation, from ref. 4.

Table 7 Measured and predicted pK_a values of aromatic acids (3).

Analyte	pK_a^a	pK_a^b	pK_a^c
3-Phenylpropionic acid	4.691	—	4.59
4-Phenylbutyric acid	—	—	4.72
Hippuric acid	—	3.58	4.30
2-Hydroxy-hyppuric aicd	3.506	3.58	—
Indole-3-acetic acid	4.590	4.65	3.80
5-Hydroxy-indole-3-acetic acid	—	4.57	—
Indolepropionic acid	—	4.81	5.01
Indolebutyric acid	4.781	—	—
2-Naphthtoic acid	—	4.25	4.24
3-Hydroxy-2-naphthoic acid	—	2.89	—

[a]Measured by liquid chromatography, from ref. 5.
[b]Measured by titration, from ref. 6.
[c]Determined from Hammett's equation, from refs. 4 and 5.

Table 8 Measured and predicted pK_a values of some nitrogen-containing compounds. Reproduced by permission of Elsevier, ref. 3.

Analyte	pK_a^a	pK_a^b	pK_a^c	pK_a^d	pK_a^e
Benzylamine	9.33	8.993	9.390	9.017	2.936
Aniline	4.63	2.695	4.580	4.664	4.206
2-Methylaniline	4.44	3.109	4.292	4.664	4.540
3-Methylaniline	4.73	2.523	4.753	4.906	4.429
4-Methylaniline	5.08	3.224	4.983	4.906	4.651
2,4-Dimethylaniline	—	3.393	4.695	4.785	4.651
4-Methoxyaniline	5.34	4.024	5.386	5.389	5.430
2,5-Diethoxyaniline	—	2.817	4.234	2.850	6.319
2-Chloroaniline	2.65	1.805	2.650	2.608	3.873
3-Chloroaniline	3.46	2.375	3.154	3.938	4.095
4-Chloroaniline	4.15	2.972	3.889	4.059	3.984
2,5-Dichloroaniline	—	3.874	1.585	1.761	3.428
3,4-Dichloroaniline	—	2.537	2.823	3.334	3.65
4-Bromoaniline	3.58	2.584	3.946	3.575	3.539
2-Nitroaniline	—	—	1.038	0.264	−0.464
3-Nitroaniline	2.466	2.858	2.449	2.729	3.650
4-Nitroaniline	1	—	2.334	0.552	0.982
5-Aminoindane	—	4.132	—	4.606	4.429
1-Aminoindane	9.21	5.643	—	9.138	3.650
1-Aminonaphthalene	3.92	2.924	3.850	4.422	4.318
2-Aminonaphthalene	4.16	2.933	4.290	3.575	4.318
1-Aminoanthracene	—	2.594	—	4.422	4.540
1-Aminopyrene	—	2.423	—	3.575	3.873
Dibenzylamine	—	7.046	—	6.478	4.985
N-Methylaniline	—	3.619	—	3.696	8.654
N-Ethylaniline	5.12	4.340	—	0.068	1.538
N-Butylaniline	—	3.551	—	3.938	8.877
5-Aminoindol	—	3.976	—	5.933	16.438
N,N-Dimethylaniline	5.15	3.713	5.060	—	20.142
N,N-Diethylaniline	6.61	4.500	—	—	29.280
Pyridine	5.25	4.786	5.250	—	5.622
2-Aminopyridine	6.82	4.980	6.929	—	3.575

Table 8 (*Continued*)

Analyte	pK_a^a	pK_a^b	pK_a^c	pK_a^d	pK_a^e
3-Aminopyridine	—	5.244	5.250	—	4.180
4-Aminopyridine	9.114	7.510	8.639	—	2.850
2-Methylpyridine	5.97	3.298	6.017	—	6.206
3-Methylpyridine	5.68	3.820	5.604	—	5.388
4-Methylpyridine	6.02	3.699	6.076	—	6.089
4-*tert*-Butylpyridine	—	4.014	6.135	—	5.972
2,4-Dimethylpyridine	6.99	4.313	6.929	—	6.556
2,5-Dimethylpyridine	—	4.438	6.371	—	5.855
2,6-Dimethylpyridine	—	4.973	6.784	—	6.439
Pyrazine	0.65	2.326	—	—	0.638
2-Methylpyrazine	1.45	1.908	—	—	0.365
2,5-Dimethylpyrazine	—	1.439	—	—	1.710
2,6-Dimethylpyrazine	—	2.875	—	—	1.768
Quinoline	4.9	2.647	4.896	—	4.804
8-Hydroxyquinoline	5.017	3.338	4.236	—	−0.044
2-Methylquinoline	5.83	3.888	5.662	—	5.388
4-Methylquinoline	5.67	3.194	5.722	—	5.271
8-Methylquinoline	—	2.155	4.827	—	4.921

[a]Values from ref. 2.
[b]Measured by liquid chromatography, from ref. 3.
[c]Determined from Hamnett's equations, from ref. 4.
[d]CAChe pK_a from hydrogen apc.
[e]CAChe pK_a from nitrogen apc.

Table 9 Molecular properties of some phenolic compounds. fs, hb, es, and vw represent the energy of the final (optimized) structure, the hydrogen-bonding energy, the electrostatic energy, and the van der Waals energy of each analyte (kcal mol^{-1}), respectively. Log P represents NlogP. FS1/complex, HB1/complex, ES1/complex, and VW1/complex represent the energy values of complexes with the butyl-bonded phase. FS2/complex, HB2/complex, ES2/complex, and VW2/complex represent the energy values of complexes with the pentyl-bonded phase. Reproduced by permission of Elsevier, ref. 16.

No.	Analyte	log P[a]	pK$_a$[b]	fs	hb	es	vw
1	1,2-Dihydroxybenzene	0.712	11.5210	−12.3175	−3.339	0.195	2.979
2	1,3-Dihydroxybenzene	0.719	9.3306	−12.4093	−2.987	−0.007	2.905
3	1,4-Dihydroxybenzene	0.724	10.4665	−12.3849	−2.937	0.026	2.917
4	1-Hydroxy-2,4-dinitronaphthalene	2.760	—	−17.3047	−4.276	−0.545	1.178
5	1-Hydroxynaphthalene	2.720	9.6957	−19.4351	−1.648	0.000	6.132
6	Pentachlorophenol	4.658	6.3080	1.4528	−1.985	5.525	8.713
7	Pentamethylphenol	3.487	10.5879	−6.4902	−1.355	−0.873	8.204
8	2,3,4,5-Tetrachlorophenol	4.098	7.2814	−0.1855	−1.538	5.340	6.857
9	2,3,4,5-Tetramethylphenol	3.152	10.4865	−8.7706	−1.500	−0.317	5.759
10	2,3,4,6-Tetrachlorophenol	—	—	−4.5807	−1.968	2.356	6.101
11	2,3,4-Trichlorophenol	3.493	7.8493	−2.9751	−1.918	4.380	5.600
12	2,3,4-Trimethylphenol	2.738	10.8922	−9.6638	−1.501	−0.371	4.812
13	2,3,5,6-Tetrachlorophenol	4.164	6.5920	−1.5860	−1.987	5.466	5.891
14	2,3,5,6-Tetramethylphenol	3.173	10.3851	−9.8685	−1.369	−0.905	5.538
15	2,3,5-Trichlorophenol	3.574	7.5653	−6.1418	−1.923	2.404	4.615
16	2,3,5-Trimethylphenol	2.874	10.7705	−11.1410	−1.503	−0.469	3.867
17	2,3,6-Trichlorophenol	3.538	7.0382	−3.3925	−1.972	4.988	4.860
18	2,3,6-Trimethylphenol	2.814	10.4256	−10.5167	−1.400	−0.961	4.761
19	2,3-Dibromophenol	—	—	−4.0723	−1.889	3.636	5.077
20	2,3-Dichlorophenol	2.870	8.1130	−4.3435	−1.921	4.459	4.396
21	2,3-Dimethylphenol	2.360	10.2431	−9.5418	−1.336	−0.308	4.359
22	2,4,5-Trichlorophenol	3.580	7.5451	−6.1987	−1.909	2.446	4.519
23	2,4,6-Trichlorophenol	3.632	7.0991	−12.6817	−1.980	−3.211	4.108
24	2,4,6-Trimethylphenol	2.897	10.5879	−11.9591	−1.384	−1.374	4.001
25	2,4-Dibromophenol	3.339	7.8696	−9.7825	−1.889	3.636	3.857
26	2,4-Dichlorophenol	2.971	8.1739	−10.4377	−1.918	−0.556	3.653

27	2,4-Dimethylphenol	2.474	10.2837	-10.9887	-1.369	-0.572	3.661
28	2,4-Dinitrophenol	1.782	—	-13.3234	-4.119	-0.532	5.548
29	2,5-Dichlorophenol	2.955	8.0522	-7.2850	-1.911	2.583	3.629
30	2,5-Dimethylphenol	2.482	10.2025	-10.8232	-1.374	-0.416	3.638
31	2,5-Dinitrophenol	1.779	—	-13.2561	-4.139	-0.423	5.472
32	2,6-Dibromophenol	3.432	6.8556	-7.2816	-1.962	2.042	4.182
33	2,6-Dichlorophenol	2.921	7.5047	-7.7317	-1.962	1.985	3.876
34	2,6-Dimethylphenol	2.410	10.4256	-11.1215	-1.385	-1.141	4.018
35	2,6-Dinitrophenol	1.675	—	-12.8426	-4.456	-0.708	6.332
36	2-Bromophenol	2.491	8.3970	-8.0640	-1.902	1.870	3.578
37	2-Chroro-6-methylphenol	2.696	—	-6.7884	-1.424	2.573	4.152
38	2-Chlorophenol	2.225	8.5998	-8.0500	-1.906	2.084	3.432
39	2-Ethylphenol	2.470	10.0605	-9.4427	-1.360	-0.466	3.928
40	2-Methylphenol	1.994	10.2025	-10.3036	-1.372	-0.465	3.642
41	2-Nitrophenol	1.665	—	-10.5406	-1.807	-0.213	4.624
42	2-Hydroxyacetophenone	0.803	9.0263	-3.6999	-1.529	-0.587	5.155
43	2-Hydroxynaphthalene	2.749	9.9999	-20.8860	-1.505	0.000	5.702
44	3,4,5-Trichlorophenol	3.569	8.3366	-5.0902	-1.471	2.184	5.228
45	3,4-Dichlorophenol	2.953	8.8235	-6.4075	-1.471	2.294	4.037
46	3,4-Dimethylphenol	2.437	10.3245	-10.5498	-1.460	0.171	3.416
47	3,4-Dinitrophenol	1.754	—	24.8735	-1.481	0.140	4.245
48	3,5-Dichlorophenol	3.033	8.6815	-12.4676	-1.474	2.792	3.392
49	3,5-Dimethylphenol	2.516	10.2636	-11.3779	-1.464	-0.078	2.972
50	3-Bromophenol	2.590	9.3711	-9.9042	-1.474	-0.108	3.255
51	3-Chlorophenol	2.318	9.2697	-10.0510	-1.472	-0.115	3.172
52	3-Ethylphenol	2.536	10.4056	-8.9201	-1.463	0.024	3.985
53	3-Methylphenol	2.059	10.0608	-10.7121	-1.465	0.024	2.977
54	3-Nitrophenol	1.731	—	-10.6460	-1.481	-0.002	4.103
55	4-Bromophenol	2.597	9.3711	-9.8803	-1.465	-0.083	3.233
56	4-Chloro-2-methylphenol	2.732	—	-9.3482	-1.352	0.005	3.834
57	4-Chloro-3,5-dimethylphenol	3.083	—	-10.7196	-1.462	-1.321	4.522
58	4-Chloro-3-methylphenol	2.698	—	-10.3893	-1.463	-0.633	3.753
59	4-Chlorophenol	2.316	9.5334	-10.0243	-1.481	-0.087	3.155
60	4-Ethylphenol	2.523	10.3853	-8.9664	-1.457	0.017	3.992
61	4-Methylphenol	2.060	10.1216	-10.7544	-1.460	0.017	2.975

Table 9 *(Continued)*

No.	Analyte	log P^a	pK$_a$^{b}	fs	hb	es	vw
62	4-Nitrophenol	1.751	—	-10.7284	-1.469	-0.045	4.091
63	4-Hydroxybutylbenzoate	1.114	8.9249	-4.8077	-1.468	-2.469	8.376
64	4-Hydroxypropylbenzoate	0.574	8.9249	-5.4543	-1.467	-2.470	7.925
65	4-*tert*-Butylphenol	3.180	10.2839	-6.3765	-1.450	0.017	5.425
66	Phenol	1.574	10.0405	-10.2105	-1.477	0.000	2.960
	Butyl-bonded phase	—	—	3373.0355	0.000	0.000	419.960
	Pentyl-bonded phase	—	—	-612.9215	0.000	-347.501	-327.343
	Acetonitrile phase	—	—	44.718	0.000	47.055	-8.952

No.	FS1/complex	HB1/complex	ES1/complex	VW1/complex	FS2/complex	HB2/complex	ES2/complex	VW2/complex
1	3353.7896	-3.336	0.196	415.946	-639.3091	-3.422	-347.244	-338.244
2	3353.6866	-2.982	-0.007	415.878	-639.4068	-3.048	-347.460	-338.269
3	3353.5691	-2.933	0.026	415.748	-639.5571	-3.167	-347.439	-338.271
4	3346.5489	-4.297	-0.542	421.456	-649.6602	-4.275	-348.027	-335.642
5	3344.7302	-1.650	0.000	416.936	-649.3120	-1.648	-347.465	-338.254
6	3363.4253	-1.988	5.522	416.874	-633.3094	-1.986	-342.163	-340.415
7	3355.9886	-1.354	-0.872	417.354	-640.0792	-1.362	-348.302	-340.368
8	3362.8143	-1.538	5.340	416.106	-633.5105	-1.550	-342.297	-340.967
9	3354.6385	-1.502	-0.316	415.730	-640.7284	-1.505	-348.066	-341.751
10	3358.2550	-1.967	2.353	415.411	-638.1964	-2.019	-345.266	-341.884
11	3360.9748	-1.918	4.388	416.369	-634.4629	-1.937	-342.283	-340.093
12	3354.2541	-1.499	-0.371	415.433	-640.6401	-1.501	-347.800	-340.443
13	3360.7637	-1.988	5.461	415.021	-633.8383	-1.993	-347.145	-341.170
14	3352.6168	-1.385	-0.908	414.419	-643.5258	-1.375	-348.375	-342.912
15	3356.8993	-1.924	2.401	414.404	-638.8230	-1.927	-345.299	-342.382
16	3352.1748	-1.505	-0.470	413.854	-642.8024	-1.512	-347.476	-342.476
17	3360.3884	-1.979	4.990	415.447	-635.6925	-1.977	-347.708	-341.571
18	3352.7224	-1.405	-0.961	414.680	-642.1185	-1.419	-348.409	-340.934
19	3359.2060	-1.886	3.626	415.022	-635.7717	-1.891	-343.937	-340.638
20	3360.7247	-1.924	4.467	416.145	-634.4212	-1.925	-343.116	-339.799
21	3355.4153	-1.345	-0.308	416.065	-639.5455	-1.347	-347.798	-339.433
22	3357.3741	-1.901	2.449	414.841	-638.4696	-1.962	-345.443	-342.443

23	−342.930	−350.917	−1.976	−645.3557	414.148	−3.207	−1.959	3350.7130
24	−341.879	−348.813	−1.451	−643.7867	413.566	−1.375	−1.386	3350.9811
25	−342.524	−347.844	−1.916	−641.6017	415.069	3.625	−1.885	3359.2295
26	−340.708	−348.187	−1.923	−640.5907	415.106	−0.555	−1.904	3354.3640
27	−340.211	−348.048	−1.383	−640.6324	415.139	−0.572	−1.369	3353.6989
28	−337.322	−348.047	−4.134	−642.2304	416.572	−0.530	−4.136	3350.8033
29	−341.466	−345.055	−1.933	−638.1759	415.084	2.583	−1.917	3357.3709
30	−341.491	−347.875	−1.379	−641.6203	414.761	−0.416	−1.379	3353.4991
31	−339.251	−347.885	−4.168	−643.4423	416.662	−0.425	−4.122	3351.1095
32	−340.984	−345.537	−1.977	−638.3781	414.368	2.045	−1.968	3356.1364
33	−341.006	−345.642	−1.977	−638.1091	415.568	1.984	−1.962	3357.1179
34	−341.085	−348.575	−1.399	−641.5880	414.949	−1.141	−1.393	3353.0000
35	−337.998	−348.236	−4.470	−642.9613	418.011	−0.707	−4.462	3351.9908
36	−339.413	−345.647	−1.918	−636.5530	415.599	1.872	−1.908	3357.1555
37	−340.917	−344.998	−1.438	−637.0000	415.189	2.571	−1.429	3357.4163
38	−338.997	−345.424	−1.914	−636.2936	416.043	2.084	−1.905	3357.6962
39	−340.659	−347.918	−1.376	−639.7829	415.160	−0.467	−1.359	3355.1076
40	−339.075	−347.935	−1.386	−638.8017	415.952	−0.465	−1.373	3355.2249
41	−337.795	−348.091	−1.557	−638.6554	417.176	−0.214	−1.813	3355.1995
42	−336.846	−348.091	−1.557	−632.2194	416.093	−0.589	−1.523	3360.6036
43	−336.211	−347.486	−1.510	−648.7683	416.056	0.000	−1.510	3342.7851
44	−339.580	−345.436	−1.473	−635.5660	416.249	2.184	−1.470	3359.1339
45	−340.454	−345.304	−1.503	−636.7197	414.826	2.295	−1.472	3357.6559
46	−340.513	−347.278	−1.462	−639.8363	415.203	0.170	−1.461	3354.4723
47	−333.363	−347.453	−5.018	−636.6775	415.681	0.139	−1.480	3389.5742
48	−342.232	−350.453	−1.477	−643.7295	414.631	−2.791	−1.477	3352.0456
49	−341.724	−347.524	−1.475	−641.8059	414.748	−0.079	−1.467	3353.6386
50	−340.539	−347.637	−1.479	−639.7106	415.260	−0.108	−1.474	3355.2828
51	−339.957	−347.687	−1.477	−638.8110	415.332	−0.116	−1.471	3355.2945
52	−341.193	−347.413	−1.470	−639.6519	414.357	0.023	−1.469	3354.3640
53	−339.763	−347.421	−1.474	−639.2601	414.947	0.024	−1.466	3354.4499
54	−338.036	−347.452	−1.489	−638.4505	416.175	−0.002	−1.480	3354.6275
55	−340.814	−347.648	−1.494	−639.8387	414.505	−0.083	−1.464	3354.6512
56	−340.885	−347.614	−1.436	−640.2602	415.486	0.007	−1.376	3355.2375
57	−341.562	−348.763	−1.468	−642.3404	415.333	−1.322	−1.464	3353.1956

Table 9 (*Continued*)

No.	FS1/complex	HB1/complex	ES1/complex	VW1/complex	FS2/complex	HB2/complex	ES2/complex	VW2/complex
58	3353.6792	−1.465	−0.633	414.502	−539.9979	−1.469	−346.225	−340.110
59	3355.2130	−1.462	−0.087	415.194	−539.9198	−1.481	−347.674	−339.964
60	3353.4848	−1.455	0.017	413.618	−639.5023	−1.484	−347.457	−340.608
61	3354.5361	−1.485	0.017	415.084	−639.4583	−1.486	−347.469	−339.949
62	3354.8881	−1.469	−0.045	416.463	−638.9058	−1.485	−347.516	−338.709
63	3354.1263	−1.461	−2.470	414.192	−642.4034	−1.479	−349.967	−344.414
64	3355.4685	−1.463	−2.454	415.344	−641.5960	−1.479	−349.943	−343.001
65	3355.7475	−1.448	0.017	414.300	−638.7314	−1.615	−347.475	−339.952
66	3356.0700	−1.487	0.000	416.126	−637.1331	−1.470	−347.623	−335.947

[a]from ref. 7.
[b]from ref. 8.

Table 10 Measured and predicted capacity ratios of some phenolic compounds. Capacity ratios were predicted from NlogP. MI represents molecular interactions. Reproduced by permission of Elsevier, ref. 16.

No.	Analyte	*Capacity ratio*					
		pH 4.01_{mes}	$k_{max}/$ *log* P	$k_{max}/$ *MI*	*pH* 8.49_{mes}	*pH 8.49/* *log* P^a	*pH 8.49/* MI^a
1	1,2-Dihydroxybenzene	—	0.345	0.637	—	0.345	0.636
2	1,3-Dihydroxybenzene	0.377	0.347	0.637	0.360	0.303	0.556
3	1,4-Dihydroxybenzene	0.261	0.348	0.659	0.280	0.344	0.652
4	1-Hydroxy-2,4-dinitro-naphthalene	–	—	1.262	1.820	—	—
5	1-Hydroxynaphthalene	1.181	1.230	1.119	1.159	1.155	1.054
6	Pentachlorophenol	3.132	4.198	2.911	0.084	0.027	0.019
7	Pentamethylphenol	—	2.000	2.317	—	1.984	2.299
8	2,3,4,5-Tetrachlorophenol	—	2.944	2.280	—	0.172	0.133
9	2,3,4,5-Tetramethylphenol	—	—	2.917	—	1.602	2.888
10	2,3,4,6-Tetrachlorophenol	—	1.618	2.328	—	—	—
11	2,3,4-Trichlorophenol	1.788	2.009	1.535	0.140	0.374	0.286
12	2,3,4-Trimethylphenol	—	1.245	1.390	—	1.240	1.384
13	2,3,5,6-Tetrachlorophenol	—	3.076	1.782	—	0.038	0.022
14	2,3,5,6-Tetramethylphenol	—	1.641	2.350	—	1.620	2.320
15	2,3,5-Trichlorophenol	2.099	2.113	1.941	0.095	0.225	0.206
16	2,3,5-Trimethylphenol	1.494	1.358	1.589	1.510	1.351	1.580
17	2,3,6-Trichlorophenol	1.800	2.065	1.799	0.042	0.070	0.061
18	2,3,6-Trimethylphenol	1.636	1.306	1.570	1.648	1.291	1.552
19	2,3-Dibromophenol	—	—	1.600	—	—	—
20	2,3-Dichlorophenol	1.220	1.355	1.164	0.310	0.401	0.344
21	2,3-Dimethylphenol	1.138	0.979	1.148	1.151	0.962	1.128
22	2,4,5-Trichlorophenol	2.009	2.123	1.791	0.110	0.216	0.183
23	2,4,6-Trichlorophenol	2.131	2.193	1.936	0.051	0.086	0.076
24	2,4,6-Trimethylphenol	1.696	1.377	1.641	1.710	1.366	1.628
25	2,4-Dibromophenol	1.657	1.824	1.637	0.650	0.320	0.316
26	2,4-Dichlorophenol	1.282	1.442	1.183	0.468	0.470	0.385
27	2,4-Dimethylphenol	1.173	1.054	1.069	1.180	1.037	1.052
28	2,4-Dinitrophenol	—	0.679	0.927	—	—	—
29	2,5-Dichlorophenol	1.326	1.429	1.365	0.390	0.382	0.365
30	2,5-Dimethylphenol	1.167	1.059	1.340	1.169	1.039	1.314
31	2,5-Dinitrophenol	0.832	0.678	1.189	0.029	—	—
32	2,6-Dibromophenol	1.588	1.932	1.422	0.086	0.044	0.032
33	2,6-Dichlorophenol	1.282	1.400	1.236	0.091	0.131	0.116
34	2,6-Dimethylphenol	1.274	1.012	1.256	1.279	1.000	1.242
35	2,6-Dinitrophenol	0.583	0.635	1.175	0.031	—	—
36	2-Bromophenol	0.964	1.064	0.853	0.750	0.475	0.381
37	2-Chroro-6-methylphenol	1.173	1.213	1.274	1.009	—	—
38	2-Chlorophenol	0.890	0.899	0.813	0.700	0.506	0.458
39	2-Ethylphenol	1.231	1.052	1.227	1.239	1.025	1.195
40	2-Methylphenol	0.901	0.778	0.855	0.910	0.763	0.839
41	2-Nitrophenol	1.148	0.631	0.793	0.110	—	—

Table 10　(*Continued*)

No.	Analyte	pH 4.01$_{mes}$	k$_{max}$/ log P	k$_{max}$/ MI	pH 8.49$_{mes}$	pH 8.49/ log P[a]	pH 8.49/ MI[a]
42	2-Hydroxyacetophenone	—	0.366	0.859	—	0.283	0.665
43	2-Hydroxynaphthalene	1.043	1.253	0.759	1.030	1.215	0.736
44	3,4,5-Trichlorophenol	—	2.109	1.259	—	0.870	0.519
45	3,4-Dichlorophenol	1.335	1.426	1.219	0.920	0.974	0.833
46	3,4-Dimethylphenol	1.005	1.028	0.998	1.021	1.013	0.983
47	3,4-Dinitrophenol	—	0.668	0.865	—	—	—
48	3,5-Dichlorophenol	1.643	1.500	1.469	0.680	0.913	0.894
49	3,5-Dimethylphenol	1.065	1.081	1.247	1.074	1.063	1.227
50	3-Bromophenol	1.049	1.135	1.104	0.940	1.003	0.976
51	3-Chlorophenol	0.965	0.955	0.899	0.880	0.819	0.771
52	3-Ethylphenol	1.082	1.096	1.324	1.074	1.083	1.308
53	3-Methylphenol	0.823	0.809	0.863	0.830	0.788	0.840
54	3-Nitrophenol	0.695	0.658	0.746	0.360	—	—
55	4-Bromophenol	1.017	1.134	1.138	0.970	1.006	1.005
56	4-Chloro-2-methylphenol	1.324	1.242	1.371	1.300	—	—
57	4-Chloro-3,5-dimethylphenol	1.594	1.549	1.574	1.560	—	—
58	4-Chloro-3-methylphenol	1.205	1.213	1.059	1.169	—	—
59	4-Chlorophenol	0.925	0.953	0.925	0.890	0.874	0.848
60	4-Ethylphenol	1.091	1.086	1.274	1.099	1.072	1.257
61	4-Methylphenol	0.823	0.811	0.891	0.830	0.792	0.871
62	4-Nitrophenol	0.634	0.667	0.804	0.040	—	—
63	4-Hydroxybutylbenzoate	—	0.445	5.070	—	0.325	3.707
64	4-Hydroxypropylbenzoate	—	0.316	3.811	—	0.231	2.786
65	4-*tert*-Butylphenol	1.730	1.730	1.820	1.734	1.703	1.791
66	Phenol	0.649	0.596	0.622	0.661	0.580	0.605

[a]Values from ref. 9 and 10.

Table 11　Molecular properties of some benzoic acid derivatives on a model carbon phase. FS1, VW1, HB1, and ES1 represent the energy value of the final (optimized) structure, the van der Waals energy, the hydrogen-bonding energy, and the electrostatic energy (kcal mol^{-1}) of the complexes between a model carbon phase and a benzoic acid derivative. Log k_2 values from ref. 11, log k_2 represents the capacity ratios of the molecular form, log k_{2i} represent the ionized form, values from ref. 12. Reproduced by permission of Elsevier, ref. 15.

No.	Analyte	fs	hb	es	vw	FS1	HB1
1	2,4,6-Trimethylbenzoic acid	−8.6252	−3.442	−4.252	5.538	704.7818	−8.667
2	2,4-Dichlorobenzoic acid	−19.5852	−3.568	−15.694	7.065	694.7570	−9.574
3	2,4-Dimethylbenzoic acid	−11.2572	−3.518	−5.423	6.100	702.9060	−9.229
4	2,4-Dihydroxybenzoic acid	−19.4110	−9.308	−6.717	5.293	691.5560	−27.177
5	2,5-Dichlorobenzoic acid	−16.3200	−3.561	−12.447	6.979	697.9774	−9.938
6	2,5-Dimethylbenzoic acid	−11.1240	−3.521	−5.296	6.132	703.5891	−9.113
7	2,5-Dihydroxybenzoic acid	−19.5298	−9.344	−6.797	5.321	692.2385	−24.595
8	2,6-Dichlorobenzoic acid	−23.4971	−3.452	−22.126	5.983	692.0870	−8.884

Table 11 *(Continued)*

No.	Analyte	fs	hb	es	vw	FS1	HB1
9	2,6-Dimethylbenzoic acid	−7.7343	−3.439	−3.949	5.532	707.5040	−8.889
10	2,6-Dihydroxybenzoicacid	−22.7430	−13.031	−7.809	5.950	688.6374	−30.245
11	2-Bromobenzoic acid	−16.0131	−3.564	−12.649	6.760	699.4113	−9.966
12	2-Chlorobenzoic acid	−17.5343	−3.570	−13.381	6.789	698.3831	−9.817
13	2-Ethylbenzoic acid	−9.1914	−3.545	−5.443	6.198	706.6502	−9.515
14	2-Methylbenzoic acid	−10.4933	−3.518	−5.251	6.113	705.5074	−9.614
15	2-Methoxybenzoic acid	−11.3306	−3.587	−6.585	5.932	710.4371	−8.392
16	2-Hydroxybenzoic acid	−17.1254	−7.822	−6.681	5.336	693.8632	−25.636
17	3,4,5-Trihydroxybenzoic acid	−23.1168	−11.337	−6.767	4.883	686.5325	−28.352
18	3,4-Dichlorobenzoic acid	−9.3448	−3.447	−3.612	6.005	705.1582	−9.532
19	3,4-Dimethylbenzoic acid	−14.3591	−3.462	−6.660	5.364	699.9775	−9.290
20	3,4-Dihydroxybenzoicacid	−18.3220	−6.869	−6.532	4.803	693.6499	−20.933
21	3,5-Dichlorobenzoic acid	−15.2967	−3.449	−8.568	5.360	699.1785	−9.619
22	3,5-Dimethylbenzoic acid	−15.2490	−3.466	−6.932	4.958	698.6293	−9.421
23	3,5-Dihydroxybenzoicacid	−18.4059	−6.537	−6.743	4.760	693.9818	−21.033
24	3-Bromobenzoic acid	−13.2092	−3.456	−6.363	5.198	702.3478	−9.809
25	3-Chlorobenzoic acid	−13.3325	−3.453	−6.350	5.111	702.7270	−9.727
26	3-Ethylbenzoic acid	−13.7274	−3.462	−6.736	5.233	701.7864	−9.623
27	3-Methylbenzoic acid	−14.5031	−3.462	−6.737	4.926	701.2304	−9.596
28	3-Methoxybenzoic acid	−12.4574	−3.470	−6.743	5.165	702.8158	−9.641
29	3-Hydroxybenzoic acid	−16.0948	−4.985	−6.655	4.820	696.4237	−19.737
30	4-Bromobenzoic acid	−13.2872	−3.453	−6.439	5.159	702.2165	−9.553
31	4-Chlorobenzoic acid	−13.2874	−3.453	−6.440	5.158	702.6337	−9.465
32	4-Ethylbenzoic acid	−13.7157	−3.458	−6.719	5.208	701.7422	−9.249
33	4-Methylbenzoic acid	−14.5173	−3.458	−6.719	4.878	701.3074	−9.331
34	4-Methoxybenzoic acid	−10.1574	−3.456	−6.671	6.213	705.1619	−9.330
35	4-Hydroxy-3-methoxybenzoic acid	−14.4088	−5.174	−6.647	5.166	694.5556	−19.047
36	4-Hydroxybenzoic acid	−15.5198	−5.182	−6.658	4.787	698.8075	−13.986
37	Benzoic acid	−13.9181	−3.459	−6.671	4.876	703.7209	−9.449
	Carbon phase	731.8482	0.000	0.130	70.778	—	—

Analyte	ES1	VW1	log k$_1$[a]	log k$_2$	logk$_{2i}$
2,4,6-Trimethylbenzoic acid	−3.990	63.292	1.116	—	—
2,4-Dichlorobenzoic acid	−15.438	65.479	1.178	—	—
2,4-Dimethylbenzoic acid	−5.431	64.577	1.117	—	−0.856
2,4-Dihydroxybenzoic acid	−6.370	72.171	—	—	—
2,5-Dichlorobenzoic acid	−12.195	65.918	1.110	—	—
2,5-Dimethylbenzoic acid	(5.056)[a]	64.787	1.117	—	—
2,5-Dihydroxybenzoic acid	−6.458	69.027	—	—	—
2,6-Dichlorobenzoic acid	−21.545	66.569	0.789	—	—
2,6-Dimethylbenzoic acid	−3.706	65.365	0.815	—	—
2,6-Dihydroxybenzoic acid	−7.762	70.179	—	—	—
2-Bromobenzoic acid	−12.398	67.007	0.774	—	—
2-Chlorobenzoic acid	−13.141	67.100	0.710	—	—
2-Ethylbenzoic acid	−5.060	66.486	—	—	—
2-Methylbenzoic acid	−4.994	66.384	0.824	—	—
2-Methoxybenzoic acid	−6.628	65.162	—	—	—
2-Hydroxybenzoic acid	−6.353	72.260	—	(0.912)[a]	(− 0.645)[a]
3,4,5-Trihydroxybenzoic acid	−6.519	67.273	—	−0.495	−1.987
3,4-Dichlorobenzoic acid	−3.375	64.671	1.371	—	—
3,4-Dimethylbenzoic acid	−6.421	63.639	1.082	—	—
3,4-Dihydroxybenzoic acid	−6.267	68.525	—	−0.181	−1.510
3,5-Dichlorobenzoic acid	−8.331	64.120	1.442	—	—

Table 11 *(Continued)*

Analyte	ES1	VW1	log k_1 [a]	log k_2	logk_{2i}
3,5-Dimethylbenzoic acid	−6.691	62.901	1.160	—	—
3,5-Dihydroxybenzoic acid	−6.474	69.527	—	−0.242	−1.287
3-Bromobenzoic acid	−9.809	−6.121	1.073	—	—
3-Chlorobenzoic acid	−9.727	−6.110	0.993	—	—
3-Ethylbenzoic acid	−9.623	−6.490	—	—	—
3-Methylbenzoic acid	−6.491	65.010	0.867	—	—
3-Methoxybenzoic acid	−6.496	64.832	—	0.883	−0.369
3-Hydroxybenzoic acid	−6.371	70.077	—	0.246	−0.908
4-Bromobenzoic acid	−6.201	64.926	1.096	1.429	−0.374
4-Chlorobenzoic acid	−6.190	65.270	1.010	1.325	0.163
4-Ethylbenzoic acid	−6.478	64.525	1.157	1.515	0.212
4-Methylbenzoic acid	−6.477	64.677	0.847	1.128	−0.167
4-Methoxybenzoic acid	−6.435	65.564	—	—	−0.187
4-Hydroxy-3-methoxybenzoic acid	−6.451	66.066	—	0.148	−1.084
4-Hydroxybenzoic acid	−6.396	63.782	—	0.262	−1.142
Benzoic acid	−6.438	66.762	0.574	0.765	−0.517

[a]uncertain values because of the dissociation constant.

Table 12 Molecular interaction energy values of benzoic acid derivatives with two model phases. 2 represents a butyl-bonded silica phase, 3 represents a pentyl-bonded silica phase. Reproduced by permission of Elsevier, ref. 15.

Analyte	FS2	HB2	ES2	VW2
2,4,6-Trimethylbenzoic acid	3354.1217	−3.443	−4.260	415.033
2,4-Dichlorobenzoic acid	3344.0379	−3.572	−15.695	417.350
2,4-Dimethylbenzoic acid	3352.6070	−3.518	−5.601	416.364
2,4-Dihydroxybenzoic acid	3345.7726	−9.284	−6.708	417.203
2,5-Dichlorobenzoic acid	3346.6643	−3.557	−12.451	416.686
2,5-Dimethylbenzoic acid	3352.4404	−3.505	−5.280	416.053
2,5-Dihydroxybenzoic acid	3344.9820	−9.375	−6.794	416.467
2,6-Dichlorobenzoic acid	3340.8967	−3.488	−22.069	417.507
2,6-Dimethylbenzoic acid	3355.6645	−3.388	−3.897	415.447
2,6-Dihydroxybenzoic acid	3350.5446	−13.035	−7.806	425.889
2-Bromobenzoic acid	3348.0342	−3.567	−12.646	417.569
2-Chlorobenzoic acid	3346.9728	−3.514	−13.370	416.965
2-Ethylbenzoic acid	3355.8856	−3.534	−5.273	417.904
2-Methylbenzoic acid	3354.0768	−3.461	−5.201	416.234
2-Methoxybenzoic acid	3355.3736	−3.582	−6.687	416.043
2-Hydroxybenzoic acid	3348.6404	−7.920	−6.623	417.646
3,4,5-Trihydroxybenzoic acid	3341.5103	−11.358	−6.776	416.315
3,4-Dichlorobenzoic acid	3353.8374	−3.465	−3.623	416.074
3,4-Dimethylbenzoic acid	3349.6937	−3.464	−6.662	415.998
3,4-dihydroxybenzoic acid	3346.9732	−6.890	−6.545	416.808
3,5-Dichlorobenzoic acid	3348.1026	−3.450	−8.567	415.519
3,5-Dimethylbenzoic acid	3347.8388	−3.458	−6.928	414.778
3,5-Dihydroxybenzoic acid	3346.3498	−6.556	−6.747	416.298
3-Bromobenzoic acid	3350.2403	−3.462	−6.365	415.520
3-Chlorobenzoic acid	3350.6834	−3.452	−6.348	415.951
3-Ethylbenzoic acid	3350.2947	−3.481	−6.749	416.102
3-Methylbenzoic acid	3349.0077	−3.456	−6.732	415.273

Table 12 (*Continued*)

Analyte	FS2	HB2	ES2	VW2
3-Methoxybenzoic acid	3350.8786	−3.475	−6.747	415.275
3-Hydroxybenzoic acid	3348.7300	−5.003	−6.659	416.438
4-Bromobenzoic acid	3350.9572	−3.456	−6.443	416.072
4-Chlorobenzoic acid	3351.0843	−3.444	−6.422	416.414
4-Ethylbenzoic acid	3350.0377	−3.460	−6.721	415.770
4-Methylbenzoic acid	3349.6076	−3.451	−6.714	415.712
4-Methoxybenzoic acid	3352.7495	−3.444	−6.663	416.004
4-Hydroxy-3-methoxybenzoic acid	3349.7273	−5.173	−6.736	416.828
4-Hydroxybenzoic acid	3349.5118	−5.183	−6.657	415.514
Benzoic acid	3351.6603	−3.475	−6.682	417.184
Butyl phase	3373.0354	0.000	0.000	419.957

Analyte	FS3	HB3	ES3	VW3
2,4,6-Trimethylbenzoic acid	665.7833	−3.480	−355.832	635.528
2,4-Dichlorobenzoic acid	655.0733	−3.597	−367.596	636.366
2,4-Dimethylbenzoic acid	663.0578	−3.518	−356.985	634.533
2,4-Dihydroxybenzoic acid	658.4374	−9.681	−358.449	638.581
2,5-Dichlorobenzoic acid	659.0773	−3.547	−364.095	637.426
2,5-Dimethylbenzoic acid	663.6032	−3.677	−357.228	635.171
2,5-Dihydroxybenzoic acid	659.9836	−9.486	−358.474	640.442
2,6-Dichlorobenzoic acid	653.9944	−3.493	−373.662	639.233
2,6-Dimethylbenzoic acid	668.7392	−3.466	−355.530	636.880
2,6-Dihydroxybenzoic acid	657.4710	−13.060	−359.425	641.281
2-Bromobenzoic acid	660.8043	−3.615	−364.393	639.593
2-Chlorobenzoic acid	660.0688	−3.695	−365.364	638.950
2-Ethylbenzoic acid	668.1971	−3.565	−357.072	639.155
2-Methylbenzoic acid	667.1022	−3.534	−356.815	639.154
2-Methoxybenzoic acid	667.9472	−3.577	−358.200	640.170
2-Hydroxybenzoic acid	663.2607	−7.835	−358.264	641.126
3,4,5-Trihydroxybenzoic acid	653.4582	−12.967	−358.548	638.154
3,4-Dichlorobenzoic acid	665.3251	−4.367	−355.139	635.672
3,4-Dimethylbenzoic acid	660.7502	−3.707	−358.156	635.563
3,4-Dihydroxybenzoic acid	659.4534	−7.234	−358.200	638.187
3,5-Dichlorobenzoic acid	660.3372	−3.668	−360.664	636.568
3,5-Dimethylbenzoic acid	659.5881	−3.653	−358.876	634.681
3,5-Dihydroxybenzoic acid	659.0355	−8.377	−358.041	638.253
3-Bromobenzoic acid	661.9330	−4.514	−357.892	635.332
3-Chlorobenzoic acid	662.9764	−4.364	−357.884	636.237
3-Ethylbenzoic acid	663.0791	−3.502	−358.425	637.353
3-Methylbenzoic acid	661.7913	−3.986	−358.120	636.001
3-Methoxybenzoic acid	664.2649	−3.474	−358.431	636.575
3-Hydroxybenzoic acid	662.6316	−5.091	−358.524	638.846
4-Bromobenzoic acid	662.1248	−3.507	−358.278	635.897
4-Chlorobenzoic acid	662.1190	−3.507	−358.279	635.861
4-Ethylbenzoic acid	660.1631	−3.533	−358.514	634.500
4-Methylbenzoic acid	661.6807	−3.746	−358.239	636.523
4-Methoxybenzoic acid	666.4812	−3.482	−358.429	638.078
4-Hydroxy-3-methoxybenzoic acid	662.3273	−5.512	−358.338	637.464
4-Hydroxybenzoic acid	662.3578	−5.509	−358.471	638.391
Benzoic acid	665.0985	−3.951	−358.075	639.139
Pentyl phase	693.5732	0.000	−350.055	649.282

Table 13 Molecular properties of some ionized benzoic acid derivatives. fsi, hbi, esi, and vwi are final structure, hydrogen bonding, electrostatic, and van der Waals energy values of ionized benzoic acid derivatives. FSi, HBi, ESi, and VWi are energy values of complexes of ionized benzoic acid derivatives and a pentyl-bonded silica gel (kcal ml^{-1}). Reproduced by permission of Elsevier, ref. 15.

Analyte	fsi	hbi	esi	vwi	FS3i	HB3i	ES3i	VW3i
2,4,6-Trimethylbenzoic acid	5.9723	0.000	5.574	6.190	681.0784	0.000	−345.374	636.147
2,4-Dichlorobenzoic acid	−14.7810	0.000	−15.333	6.665	660.8674	0.000	−366.549	636.293
2,4-Dimethylbenzoic acid	1.2046	0.000	2.692	6.089	676.5859	0.000	−348.248	634.973
2,4-Dihydroxybenzoic acid	−7.1968	−2.813	−2.030	5.953	671.7559	−3.165	−352.926	639.703
2,5-Dichlorobenzoic acid	−11.4074	0.000	−11.965	6.695	665.6455	0.000	−363.459	638.731
2,5-Dimethylbenzoic acid	1.3154	0.000	2.799	6.130	677.4321	0.000	−348.719	636.905
2,5-Dihydroxybenzoic acid	−7.3230	−2.792	−2.156	5.979	672.9253	−2.935	−353.278	641.287
2,6-Dichlorobenzoic acid	−26.6594	0.000	−30.310	5.869	650.4113	0.000	−381.627	637.804
2,6-Dimethylbenzoic acid	6.8717	0.000	5.884	6.148	684.0866	0.000	−345.150	637.741
2,6-Dihydroxybenzoic acid	−6.4107	−2.651	−3.513	7.497	674.2711	−2.686	−354.571	643.041
2-Bromobenzoic acid	−10.6595	0.000	−11.721	6.714	666.8024	0.000	−362.886	639.259
2-Chlorobenzoic acid	−12.7746	0.000	−13.032	6.491	666.3541	0.000	−364.24	639.481
2-Ethylbenzoic acid	3.6903	0.000	2.980	6.572	681.8481	0.000	−348.02	639.299
2-Methylbenzoic acid	1.9783	0.000	2.871	6.105	680.3445	0.000	−348.181	639.300
2-Methoxybenzoic acid	−0.2372	0.000	−0.102	5.757	677.5379	0.000	−351.099	638.103
2-Hydroxybenzoic acid	−5.0058	−1.321	−2.075	6.045	674.6934	−1.334	−353.331	640.498
3,4,5-Trihydroxybenzoic acid	−11.7332	−7.754	−0.208	4.741	666.1980	−9.286	−351.258	638.525
3,4-Dichlorobenzoic acid	2.1915	0.000	3.213	5.874	677.9470	0.000	−348.583	636.130
3,4-Dimethylbenzoic acid	−3.0187	0.000	−0.015	5.231	673.1785	0.000	−351.224	636.233

3,4-dihydroxybenzoic acid	−7.0139	−3.339	0.000	4.668	672.0637	−3.686	−350.903	638.593
3,5-Dichlorobenzoic acid	−3.6542	0.000	−1.633	5.219	670.9216	0.000	−354.488	637.136
3,5-Dimethylbenzoic acid	−3.9390	0.000	−0.315	4.822	672.2625	0.000	−351.554	635.116
3,5-Dihydroxybenzoic acid	−7.1042	−2.963	−0.247	4.620	671.1521	−4.272	−351.840	638.867
3-Bromobenzoic acid	−1.7129	0.000	0.429	5.061	674.5229	0.000	−351.372	635.684
3-Chlorobenzoic acid	−1.8150	0.000	0.462	4.975	675.5756	0.000	−351.341	636.763
3-Ethylbenzoic acid	−2.3839	0.000	−0.095	5.105	675.6559	0.000	−351.065	637.721
3-Methylbenzoic acid	−3.1647	0.000	−0.095	4.794	673.9969	0.000	−351.548	636.300
3-Methoxybenzoic acid	−0.9413	0.000	0.022	5.027	676.8184	0.000	−350.937	636.793
3-Hydroxybenzoic acid	−4.8686	−1.466	−0.185	4.686	675.1146	−1.474	−351.345	639.481
4-Bromobenzoic acid	−1.9130	0.000	0.230	5.025	674.5226	0.000	−351.058	636.258
4-Chlorobenzoic acid	−1.9144	0.000	0.230	5.026	674.6309	0.000	−350.919	636.299
4-Ethylbenzoic acid	−2.3449	0.000	−0.051	5.078	672.9319	0.000	−351.077	634.953
4-Methylbenzoic acid	−3.1471	0.000	−0.051	5.078	674.3795	0.000	−351.180	637.044
4-Methoxybenzoic acid	1.2012	0.000	−0.026	6.087	678.5889	0.000	−351.075	637.998
4-Hydroxy-3-methoxy-benzoic acid	−3.0308	−1.690	0.003	5.042	674.9903	−2.036	−350.905	637.845
4-Hydroxybenzoic acid	−4.1908	−1.712	−0.046	4.657	675.1330	−1.984	−350.999	638.814
Benzoic acid	−2.5493	0.000	0.000	4.743	677.3723	0.000	−351.574	639.536
Pentyl phase	—	—	—	—	693.5732	0.000	−350.055	649.282

Table 14   Calculated energy values of acidic drugs and their complexes with a model phase. i represents the ionized form. Reproduced by permission of Oxford University Press, ref. 17.

No.	Acidic drug	FS1	HB1	ES1	VW1	FS2	HB2	ES2	VW2
1	p-Aminohippuric acid	3344.1074	-10.099	-8.912	416.750	6560.0353	-10.148	-404.144	-190.940
2	Amoxicillin	3394.5892	-8.567	0.365	407.435	6615.3485	-8.822	-394.697	-195.914
3	Barbituric acid	3302.5911	-8.104	-75.203	414.402	6522.6549	-8.368	-470.373	-189.924
4	Benzoic acid	3351.5351	-3.457	-6.672	417.054	6570.5402	-3.508	-401.866	-187.601
5	Furosemide	3368.2982	-5.526	-1.163	410.582	6583.5931	-5.955	-396.707	-197.270
6	p-Hydroxybenzoic acid	3349.1916	-4.928	-6.669	416.810	6567.7535	-5.048	-401.922	-188.219
7	Ibuprofen	3342.5102	-3.772	-5.058	412.100	6562.0379	-3.791	-400.425	-192.483
8	Indomethacin	3332.7620	-5.294	-12.543	409.513	6553.4435	-5.312	-407.812	-193.346
9	Iopanoic acid	3348.7604	-5.636	-4.668	410.919	6569.9775	-5.638	-400.012	-193.231
10	Mefenamic acid	3348.8393	-8.633	0.848	411.812	6574.6329	-4.919	-406.369	-187.177
11	Nalidixic acid	3320.6021	-4.052	-40.523	416.687	6539.7471	-4.064	-435.935	-187.407
12	Naproxen	3331.4418	-3.748	-5.020	412.699	6550.4294	-3.762	-400.201	-192.251
13	Nicotinic acid	3347.1021	-4.049	-10.520	416.077	6564.9971	-4.121	-405.849	-189.566
14	Phenylbutazone	3371.9024	0.000	-11.299	419.183	6591.4219	0.000	-406.505	-184.060
15	Probenecid	3354.0061	-3.442	-3.548	408.495	6574.2145	-3.461	-398.621	-193.997
16	Salicylic acid	3350.1086	-5.345	-6.434	417.712	6568.1128	-5.452	-401.762	-187.487
17	Sulfamethoxazole	3365.6830	-2.274	2.636	408.481	6582.1412	-2.276	-392.524	-198.790
18	Tolazamide	3352.7883	-3.174	-12.709	411.593	6569.4412	-2.814	-408.588	-197.404
19	Tolbutamide	3326.7263	-2.914	-25.673	408.060	6543.0870	-2.917	-421.246	-198.103
20	Warfarin	3342.8062	-2.841	-5.915	413.980	6562.1148	-2.870	-401.378	-189.830
	Butyl phase	3373.0369	0.000	0.000	419.941	—	—	—	—
	Pentyl phase	—	—	—	—	6594.9954	0.000	395.235	181.977

No.	Acidic drug	FS3	HB3	ES3	VW3	FS4	HB4	ES4	VW4
1	p-Aminohippuric acid	-651.9113	-10.260	-354.406	-339.926	-686.4470	-10.405	-412.426	-409.470
2	Amoxicillin	-604.6880	-9.256	-345.264	-349.126	-634.3374	-8.408	-403.125	-420.178
3	Barbituric acid	-685.8185	-9.209	-420.940	-331.687	-724.4791	-8.012	-478.554	-411.294
4	Benzoic acid	-638.4382	-3.731	-352.083	-331.782	-683.5664	-3.668	-410.117	-412.996
5	Furosemide	-625.4940	-5.487	-346.596	-344.768	-658.6313	-5.662	-403.923	-415.843
6	p-Hydroxybenzoic acid	-640.9142	-5.138	-352.038	-332.194	-681.1793	-5.301	-410.241	-411.627
7	Ibuprofen	-655.7320	-6.630	-350.743	-344.577	-693.1361	-3.750	-408.423	-419.931
8	Indomethacin	-663.7277	-9.048	-358.140	-342.018	-704.1475	-5.505	-415.957	-423.621
9	Iopanoic acid	-664.4756	-5.660	-350.167	-345.087	-686.9940	-5.643	-408.025	-420.321

(continuation of preceding table)

No.	Acidic drug								
10	Mefenamic acid	−638.0658	−4.960	−356.660	−335.679	−670.4021	−5.178	−414.604	−405.231
11	Nalidixic acid	−670.3637	−4.052	−386.092	−334.377	−710.5518	−4.159	−443.681	−411.480
12	Naproxen	−662.6026	−3.773	−350.314	−341.979	−701.8769	−3.770	−408.501	−417.874
13	Nicotinic acid	−641.5372	−4.049	−355.938	−331.860	−682.1029	−4.188	−413.822	−409.844
14	Phenylbutazone	−622.8831	0.000	−356.835	−336.159	−660.0197	0.000	−414.543	−410.521
15	Probenecid	−642.4373	−3.476	−349.401	−345.153	−672.2913	−3.450	−407.220	−415.511
16	Salicylic acid	−640.3111	−5.420	−351.835	−331.747	−684.4083	−5.563	−409.751	−411.829
17	Sulfamethoxazole	−625.8881	−2.205	−342.811	−344.607	−661.1186	−2.262	−400.829	−415.047
18	Tolazamide	−641.7513	−3.398	−358.438	−344.432	−684.0754	−3.437	−416.628	−420.787
19	Tolbutamide	−668.0298	−3.220	−371.252	−344.875	−704.1884	−3.450	−430.565	−418.878
20	Warfarin	−663.0003	−2.943	−351.341	−349.078	−690.4602	−2.924	−409.217	−414.657
	Monomethylpentyl phase	—	—	—	—	—	—	—	—
	Polydimethylpentyl phase	−608.4140	0.000	−345.403	−321.250	−648.6200	0.000	−403.451	−400.524

No.	Acidic drug	FS4i	HB4i	ES4i	VW4i	FSi	HBi	ESi	VWi
1	p-Aminohippuric acid	—	—	—	—	—	—	—	—
2	Amoxicillin	—	—	—	—	—	—	—	—
3	Barbituric acid	—	—	—	—	—	—	—	—
4	Benzoic acid	−671.2511	0.000	−403.332	−412.528	−2.5511	0.000	0.000	4.746
5	Furosemide	−657.0607	−2.056	−402.228	−416.554	13.8365	−2.594	−2.736	9.998
6	p-Hydroxybenzoic acid	−674.4414	−1.592	−402.960	−413.699	−4.9589	−0.050	−1.462	4.463
7	Ibuprofen	−679.3902	0.000	−399.952	−420.940	3.151	0.000	5.220	8.653
8	Indomethacin	−681.6643	−1.743	−407.909	−417.759	−7.2472	−4.273	0.000	6.009
9	Iopanoic acid	−672.4282	−2.175	−402.479	−417.147	2.9680	−2.158	0.411	7.674
10	Mefenamic acid	−660.0112	−1.262	−411.452	−403.430	20.2940	−8.420	−0.654	19.891
11	Nalidixic acid	−717.3566	0.000	−458.411	−411.646	−44.4161	0.000	−55.760	11.793
12	Naproxen	−683.0876	0.000	−399.852	−412.595	−13.5376	3.156	0.000	6.681
13	Nicotinic acid	−669.4499	0.000	−410.356	−410.764	−7.2772	−7.301	0.000	3.586
14	Phenylbutazone	−644.7365	0.000	−401.086	−410.792	33.3848	0.000	2.030	19.586
15	Probenecid	—	—	—	—	—	—	—	—
16	Salicylic acid	−670.3145	−1.806	−403.015	−409.905	−4.1495	−0.150	−1.487	5.234
17	Sulfamethoxazole	−669.4817	−2.061	−405.399	−419.458	1.4759	1.067	−2.230	3.175
18	Tolazamide	−664.8088	−0.117	−403.589	−418.244	13.5328	−0.089	−0.585	9.727
19	Tolbutamide	−676.9161	−0.125	−407.949	−416.260	−5.5479	−0.089	−4.856	5.974
20	Warfarin	−697.8105	−3.806	−408.928	−430.282	−17.6434	−2.951	−4.952	6.818

Table 15 Molecular properties of some basic drugs and their complexes with a model phase. Phase 1 is a butyl phase, Phase 2 an octyl phase, Phase 3 a dodecyl phase, Phase 4 is a dimethoxypentyl-bonded silica gel, and Phase 5 is a dimethoxyoctyl-bonded silica gel. m represents the molecular form. Units: kcal mol^{-1}. Reproduced by permission of Taylor & Francis, ref. 18.

Analyte	Phase 1		Phase 2		Phase 3	
	FS1	VW1	FS2	VW2	FS3	VW3
Ajmaline	3449.2940	419.012	3702.4403	928.626	3683.1781	802.938
Aniline	—	—	—	—	—	—
Atropine	3373.5110	416.630	3630.5949	932.968	3613.4250	799.359
Carbamazepine	3326.5340	425.659	3584.6978	936.349	3571.7635	820.432
Dextromethorphan	3388.3076	420.233	3642.3606	934.649	3626.1670	811.153
Homatropine	3398.1179	415.366	3646.4543	924.010	3632.1829	803.156
Imipramine	3361.9492	418.889	3612.4900	929.130	3603.4149	809.565
Isoproterenol	—	—	3611.9596	929.881	3589.9364	800.608
Lidocaine	3351.7982	419.903	3601.6674	930.212	3588.7415	804.765
Prazosin	3378.7577	412.230	3635.4741	919.998	3625.9502	800.423
Procaine	3368.0581	416.316	3620.2030	926.001	3608.9217	802.734
Pyridine	3370.5447	414.878	3629.5056	934.842	3627.8096	821.228
Quinine	3371.5502	417.277	3629.5188	930.133	3618.3498	811.841
Theobromine	3319.4073	412.363	3575.4821	328.801	3564.9648	807.361
Triamterene	3340.0077	415.322	3600.4204	931.428	3581.2900	804.556
Benzylamine	—	—	—	—	—	—
Phenethylamine	—	—	—	—	—	—
N,N-Dimethylaniline	—	—	—	—	—	—
Model phase	3375.0355	419.967	641.5884	47.116	636.3325	831.618

	Phase 4				Phase 5	
	FS4m	VW4m	FS4i	VW4i	FS5	VW5
Ajmaline	−586.2171	−412.489	−581.7239	−412.489	−444.492	−518.634
Aniline	−674.3700	−413.358	−670.4133	−414.091	−537.755	−525.313
Atropine	−663.7506	−416.743	−656.3950	−410.884	−526.630	−531.144
Carbamazepine	−711.1683	−405.831	−705.5730	−405.985	−574.687	−519.519
Dextromethorphan	−653.2251	−416.983	−652.8863	−415.706	−512.783	−528.443
Homatropine	−638.1184	−415.704	−638.0416	−416.332	−498.251	−528.148
Imipramine	−675.7248	−417.048	−676.9917	−418.246	−530.219	−519.314
Isoproterenol	−678.6589	−412.981	−687.9121	−421.162	−550.222	−532.496
Lidocaine	−689.7850	−417.076	−696.1161	−418.638	−551.447	−529.421
Prazosin	−650.7574	−415.567	−649.4318	−415.973	−518.306	−532.441
Procaine	−668.3104	−414.595	−661.7555	−414.490	−527.348	−522.378
Pyridine	−661.1491	−414.360	−662.5748	−413.368	−526.203	−526.468
Quinine	−668.7683	−417.033	−666.9434	−414.794	−521.414	−524.619
Theobromine	−713.9513	−417.599	−713.9513	−417.599	−581.891	−532.259
Triamterene	−691.7278	−411.881	−692.9782	−414.864	−561.068	−531.486
Triamterene	−681.0857	−416.515	−678.4571	−416.494	—	—
Benzylamine	−679.6221	−417.988	−679.5879	−418.284	—	—
Phenethylamine	−667.2389	−412.871	−671.2089	−415.362	—	—
N,N-Dimethylaniline	−648.6239	−400.533	−648.6239	−400.533	—	—
Model phase	−648.6200	−400.524	—	—	−518.739	−515.856

Table 16 Molecular properties of analytes and their complexes with a model phenylcarbamoylated cyclodextrin phase. Numbers 21*R*–29*S* were calculated as a complex with TFA. Reproduced by permission of Taylor & Francis, ref. 19.

No.	fs	hb	es	vw	FS	HB	ES	VW
1a	−42.0813	−12.524	−0.073	13.014	−67.8839	−171.808	−432.465	234.555
1b	−43.8243	−14.536	−0.301	13.571	−66.2554	−175.915	−434.109	236.240
2a	−24.3721	0.000	−0.025	10.778	−49.7791	−149.827	−429.851	225.261
2b	−24.3721	0.000	−0.025	10.778	−50.3818	−151.979	−431.146	224.300
3*R*2	−21.1775	−3.012	−0.987	6.693	−52.6655	−166.629	−434.383	223.241
3*S*2	−21.7950	−2.999	−0.997	6.205	−51.8824	−166.013	−434.256	222.861
4*RR*	95.0669	0.000	0.049	4.659	49.3740	−157.455	−430.741	216.306
4*SS*	95.0671	0.000	0.048	4.659	51.2440	−157.095	−430.799	217.666
5*RR*	94.8670	0.000	0.100	4.478	50.9872	−160.609	−432.813	217.433
5*SS*	94.8670	0.000	0.100	4.478	50.5432	−161.174	−432.757	217.441
6*R*	−9.5993	−5.511	−5.765	11.165	−37.7840	−168.238	−436.183	227.332
6*S*	−9.5969	−5.575	−5.763	11.120	−38.2176	−165.290	−439.355	225.183
7*R*	−9.7997	−4.837	−4.873	11.580	−34.1269	−160.034	−439.738	227.202
7*S*	−8.1314	−3.638	−5.712	12.017	−33.2391	−157.968	−437.211	227.234
8*R*	−21.0007	−5.159	−5.998	12.433	−53.2718	−164.958	−436.086	229.831
8*S*	−23.4939	−7.658	−4.948	13.203	−53.2752	−166.940	−436.696	221.244
9*R*	−10.4416	−5.357	−6.379	8.681	−38.3941	−163.862	−439.602	228.029
9*S*	−12.1933	−4.055	−6.236	8.948	−43.1784	−167.485	−443.336	220.553
10*R*	−4.0153	−4.845	21.819	11.971	−19.3381	−160.040	−410.851	235.402
10*S*	−4.0153	−4.845	21.819	11.971	−19.3381	−160.040	−410.851	235.402
11*R*	−32.9582	−5.540	−18.128	11.968	−58.3975	−163.424	−450.274	226.802
11*S*	−32.9061	−5.549	−18.109	11.878	−59.1601	−166.066	−452.227	225.866
12*R*	−43.6639	−5.563	−18.124	14.800	−82.2147	−166.448	−450.842	219.539
12*S*	−43.6674	−5.565	−18.126	14.729	−81.9278	−166.738	−450.738	219.525
13*R*	−41.3145	−6.928	−18.112	15.847	−64.1429	−158.267	−447.847	234.505
13*S*	−40.0954	−6.008	−18.151	15.579	−64.4172	−164.865	−448.736	234.827
14*RR*	18.3445	0.000	−23.696	15.446	−12.3379	−153.248	−455.193	224.052
14*SS*	18.3439	0.000	−23.696	15.445	−15.0529	−158.954	−456.903	221.900
15*R*	−8.4325	0.000	−6.532	8.768	−48.1309	−154.634	−438.562	213.951
15*S*	−8.4299	0.000	−6.533	8.770	−48.0557	−150.586	−437.523	209.890
16*R*	−1.6885	0.000	−1.960	9.998	−46.1194	−159.996	−437.252	210.920
16*S*	−1.6873	0.000	−1.960	9.998	−47.1236	−158.063	−435.012	211.393
17*R*	−0.4602	0.000	3.639	9.782	−40.9189	−149.936	−426.481	212.747
17*S*	−0.4599	0.000	3.638	9.784	−40.6286	−158.640	−431.986	211.664
18*R*	−13.7266	0.000	−2.134	7.301	−52.2646	−154.807	−433.845	215.776
18*S*	−13.7258	0.000	−2.135	7.302	−52.2357	−154.793	−433.834	215.040
19*R*	0.3921	−6.153	4.408	9.868	−26.7313	−165.027	−429.185	219.503
19*S*	0.3926	−6.152	4.407	9.880	−27.9734	−162.104	−427.098	220.028
20*R*	−8.4366	−1.874	2.774	10.201	−38.1711	−155.185	−429.171	221.843
20*S*	−9.2404	−2.113	2.157	10.265	−38.3116	−155.185	−429.345	221.682
21*R*	1.8345	−18.016	21.985	4.374	−202.9211	−179.134	−408.845	193.470
21*S*	1.8346	−18.015	21.985	4.374	−204.2144	−181.694	−406.656	191.996
22*R*	3.4301	−18.006	23.328	4.587	−202.0810	−183.613	−411.762	188.429
22*S*	3.4299	−18.019	23.336	4.592	−204.2233	−186.786	−410.690	190.346
23*R*	2.1194	−18.005	22.015	4.547	−204.2544	−193.177	−411.248	195.130
23*S*	2.1192	−18.015	22.020	4.550	−206.0514	−188.873	−412.308	191.496
24*R*	2.1132	−18.042	22.030	4.551	−205.6016	−181.340	−408.986	191.908
24*S*	2.1174	−17.996	22.001	4.533	−207.9372	−185.237	−411.609	192.950
25*R*	4.0319	−18.027	20.243	5.280	−206.7394	−189.863	−414.006	192.942
25*S*	3.9899	−18.066	20.225	5.281	−210.6729	−189.399	−412.033	189.134
26*R*	8.3509	−18.025	23.373	6.126	−202.7804	−189.943	−410.591	193.354

Table 16 (*Continued*)

No.	*fs*	*hb*	*es*	*vw*	*FS*	*HB*	*ES*	*VW*
26S	8.3583	−17.982	23.357	6.122	−202.3356	−189.478	−409.366	192.996
27R	4.1355	−18.017	22.365	3.791	−205.1706	−187.801	−408.633	187.063
27S	4.1356	−18.009	22.359	3.790	−207.5152	−188.110	−412.367	188.936
28R	−1.1987	−18.390	20.692	5.154	−209.6206	−187.837	−412.094	190.867
28S	−1.1311	−18.301	20.685	5.123	−208.9940	−189.729	−413.671	193.723
29R	3.4975	−21.714	30.126	4.345	−201.0337	−190.717	−404.385	194.710
29S	3.4970	−21.713	30.125	4.345	−198.8355	−189.106	−404.371	194.564

Table 17 Chromatographic data and molecular interaction energy values of β-cyclodextrin enantiomer complexes. α represents the selectivity. Numbers 21R–29S were calculated as a complex with TFA. Reproduced by permission of Taylor & Francis, ref. 19.

No.	*MIFS*	*MIHB*	*MIES*	*MIVW*	*k*	α	*ΔMIFS*	*ΔMIFS/ MIFS*
1	30.3745	−4.708	−6.427	18.679	3.38	1.19	3.3715	0.1250
11	27.0030	−2.613	−5.011	17.551	4.01	—	—	—
2	29.9789	−14.165	−8.993	25.737	0.53	1.35	0.6027	0.0200
22	30.5816	−12.013	−7.698	26.698	0.72	—	—	—
3R	36.0599	−0.375	−5.423	23.672	2.08	1.04	1.4006	0.0400
3S	34.6593	−0.978	−5.560	23.564	2.16	—	—	—
4RR	50.2648	−6.537	−8.029	28.573	0.19	1.62	1.8698	0.0386
4SS	48.3950	−6.897	−7.972	27.213	0.31	—	—	—
5RR	48.4517	−3.383	−5.906	27.265	0.17	1.00	0.4440	0.0091
5SS	48.8957	−2.818	−5.962	27.257	0.17	—	—	—
6R	32.7566	−1.265	−8.401	24.053	1.89	1.09	0.4360	0.0133
6S	33.1926	−4.277	−5.227	26.157	2.07	—	—	—
7R	28.8991	−8.795	−3.954	24.598	1.40	1.11	0.7805	0.0270
7S	29.6796	−9.662	−7.320	25.003	1.56	—	—	—
8R	36.8430	−4.193	−8.731	22.822	2.02	1.35	2.4898	0.0725
8S	34.3532	−4.710	−7.071	32.179	2.74	—	—	—
9R	32.5244	−5.487	−5.596	20.872	5.37	1.07	3.0326	0.0932
9S	35.5570	−0.562	−1.719	28.615	5.73	—	—	—
10R	19.8947	−8.797	−6.149	16.789	1.54	1.13	0.3207	0.0161
10S	19.5740	−6.786	−4.497	16.991	1.36	—	—	—
11R	30.0112	−6.108	−6.673	25.386	19.74	1.26	0.8147	0.0271
11S	30.8259	−3.475	−4.701	26.232	24.90	—	—	—
12R	43.1227	−3.107	−6.101	35.481	1.43	1.09	0.2904	0.0067
12S	42.8323	−2.819	−6.207	35.424	1.56	—	—	—
13R	27.4003	−12.653	−9.084	21.562	1.24	1.21	1.4934	0.0545
13S	28.8937	−5.135	−8.234	20.972	1.50	—	—	—
14RR	35.2543	−10.744	−7.322	31.614	0.94	1.16	0.9285	0.0263
14SS	37.9687	−5.038	−5.612	33.765	1.09	—	—	—
15R	44.2703	−9.358	−6.789	35.037	0.55	1.00	0.0726	0.0016
15S	44.1977	−13.406	−7.829	39.100	0.55	—	—	—
16R	49.0028	−3.996	−3.527	39.298	1.18	1.17	1.0054	0.0205
16S	50.0082	−5.929	−5.767	38.825	1.30	—	—	—
17R	45.0306	−14.056	−8.699	37.255	0.67	1.00	0.2900	0.0065
17S	44.7406	−5.352	−3.195	38.340	0.67	—	—	—

Table 17 *(Continued)*

No.	MIFS	MIHB	MIES	MIVW	k	α	ΔMIFS	ΔMIFS/ MIFS
18R	43.1099	−9.185	−7.108	31.745	0.62	1.00	0.0281	0.0006
18S	43.0818	−9.199	−7.120	32.482	0.62	—	—	—
19R	31.6953	−5.118	−5.226	30.585	2.69	1.12	1.2426	0.0392
19S	32.9379	−8.040	−7.314	30.072	3.03	—	—	—
20R	34.3064	−10.681	−6.874	28.578	1.73	1.05	0.6633	0.0197
20S	33.6431	−10.920	−7.317	28.803	1.82	—	—	—
21R	209.3275	−2.874	−7.989	51.124	0.97	1.18	1.2934	0.0062
21S	210.6209	−0.313	−10.178	52.598	0.82	—	—	—
22R	210.0830	1.615	−3.729	56.378	0.95	1.18	2.1421	0.0102
22S	212.2251	4.775	−4.793	54.466	0.80	—	—	—
23R	210.9457	11.180	−5.556	49.637	0.88	1.09	1.7968	0.0085
23S	212.7425	6.866	−4.491	53.274	0.81	—	—	—
24R	212.2867	−0.694	−7.803	52.863	1.00	1.06	2.3398	0.0110
24S	214.6265	3.249	−5.209	51.803	0.94	—	—	—
25R	215.3432	7.844	−4.570	52.558	0.91	1.10	3.8915	0.0181
25S	219.2347	7.341	−6.561	56.367	0.83	—	—	—
26R	215.7032	7.926	−4.855	52.992	0.79	1.00	0.4374	0.0020
26S	215.2658	7.504	−6.096	53.346	0.79	—	—	—
27R	213.8780	5.792	−7.821	56.948	0.82	1.00	2.3447	0.0109
27S	216.2227	6.109	−4.093	55.074	0.82	—	—	—
28R	212.9938	5.455	−6.033	54.507	0.67	1.00	0.5590	0.0026
28S	212.4348	7.436	−4.463	51.620	0.67	—	—	—
29R	209.1031	5.011	−4.308	49.855	6.44	1.08	2.1987	0.0106
29S	206.9044	3.401	−4.323	50.001	6.95	—	—	—

Table 18 Human serum albumin–drug binding affinity and drug properties. nK_1 represents log k measured using an immobilized-HSA phase. nK_2 represents predicted log K_{HSA}. k_A represents log k of acidic compounds at pH 7.4. k_B represents log k of basic compounds at pH 7.4. MIF_A and MIF_B represents the molecular interaction energy values of acidic and basic compounds, resepectively. nK_3 represents log nK measured using a modified Hummel–Dreyer method. nK_4, nK_5, nK_6 and nK_7 represent values. PB_1 and PB_2 represent the binding %. Log P_C are predicted log P values, and log P_M are measured values. $_{7.4}$ represents pH 7.4. Reproduced by permission of Bentham Science, ref. 20.

Drug	nK_1	nK_2	k_A	k_B	MIF_A	MIF_B	nK_3
Acebutolol	−0.21	−0.04	—	—	—	—	—
Acenocoumarol	—	—	—	—	—	—	—
Acetaminophen	−0.81	−0.57	—	—	—	—	—
Acetylsalicylate	−1.39	−0.64	—	—	—	—	—
Acrivastine	−0.02	0.20	—	—	—	—	—
Ajmaline	—	—	—	1.170	—	—	—

Table 18　(*Continued*)

Drug	nK_1	nK_2	k_A	k_B	MIF_A	MIF_B	nK_3
Alloprinol	—	—	—	−0.657	—	—	—
Alprenolol	0.04	−0.10	—	—	—	—	—
p-Aminohippurate	—	—	—	—	—	—	—
Amoxicillin	−1.21	−1.08	—	−1.083	—	—	—
Aniline	—	—	—	−0.173	—	14.271	—
Atenolol	−0.48	−0.22	—	—	—	—	—
Atropine	—	—	—	0.750	—	22.99	—
Azapropazone	—	—	—	—	—	—	—
Barbital	—	—	—	—	—	—	—
Benzoic acid	—	—	—	—	—	—	—
Benzylamine	—	—	—	0.471	—	21.37	—
Benzylthiouracil	—	—	—	—	—	—	—
Bilirubin	—	—	—	—	—	—	—
Bumetanide	−0.03	−0.09	—	−0.289	—	—	—
Bupropion	−0.05	0.08	—	—	—	—	—
Caffeine	−0.92	−0.88	—	—	—	—	—
Camptothecin	−0.08	−0.18	—	—	—	—	—
Canrenoate	—	—	—	—	—	—	—
Captopril	−2.69	—	—	—	—	—	—
Carbamazepine	−0.10	−0.10	—	0.360	—	—	—
(*S*)-Carbenicillin	—	—	—	—	—	—	—
3-Carboxy-4-methyl-5-propyl-2-furanpropanoate	—	—	—	—	—	—	—
Carprofen	—	—	—	—	—	—	—
Cefuroxime	−1.33	−0.90	—	—	—	—	—
Cefuroxime axetil	−0.56	−0.90	—	—	—	—	—
Cephalexin	−1.11	−1.08	—	—	—	—	—
Chlofibrate	—	—	—	—	—	—	—
Chloramphenicol	−0.46	−0.70	—	0.055	—	—	—
Chlorideion	—	—	—	—	—	—	—
Chlorothiazole	—	—	—	—	—	—	—
Chlorpheniramine	—	—	—	—	—	—	—
Chlorpromazine	1.10	0.83	—	—	—	—	—
Chlorpropamide	−0.44	−1.42	—	—	—	—	—
Cimetidine	−0.44	−0.59	—	—	—	—	—
Ciprofloxacin	0.14	0.10	—	—	—	—	—
Clofibrate	0.27	−0.12	—	—	—	—	—
Clonidine	−0.13	−0.47	—	—	—	—	—
Clotrimazole	1.34	1.05	—	—	—	—	—
Cromolyn	−1.07	−0.45	—	—	—	—	—
Dansylglycine	−0.26	0.12	—	—	—	—	—
Desipramine	0.61	0.72	—	—	—	—	—
Dextromethorphan	—	—	—	1.491	—	35.30	4.00
Diazepam	—	—	—	0.895	—	—	—
Diclofenac	—	—	—	—	—	—	—
Digitoxin	0.13	0.49	—	—	—	—	—
Dimethylaniline	—	—	—	0.394	—	21.00	—
Doxycycline	0.01	−0.38	—	—	—	—	—
Droperidol	0.43	0.47	—	—	—	—	—
Ebselen	−1.04	−0.35	—	—	—	—	—
Estradiol	0.68	0.36	—	—	—	—	—

Table 18 (*Continued*)

Drug	nK_1	nK_2	k_A	k_B	MIF_A	MIF_B	nK_3
Ethacrynate	—	—	—	—	—	—	—
Ethambutol	—	—	—	—	—	—	—
(*S*)-Etodolac	—	—	—	—	—	—	—
Etoposide	−0.49	−0.64	—	—	—	—	—
5-Fluorocytosine	−1.11	−0.79	—	—	—	—	—
Furosemide	−0.13	−0.64	−0.202	—	31.05	—	5.54
Fusidate	—	0.33	0.72	—	—	—	—
Glibenclamide	0.68	0.58	—	—	—	—	—
Hippurate	—	—	—	—	—	—	—
Homatropine	—	—	—	0.617	—	23.89	—
Hydrochloro-thiazide	−0.42	−0.76	—	—	—	—	—
Hydrocortisone	−0.40	−0.23	—	—	—	—	—
p-Hydroxybenzoate	—	—	—	—	20.72	—	—
Ibuprofen	—	—	0.197	—	32.56	—	—
Imipramine	0.75	0.91	—	1.696	—	39.25	4.26
Indole-3-acetate	—	—	—	—	—	—	—
Indomethacin	0.47	0.16	0.619	—	47.10	—	—
Indoxylsulfate	—	—	—	—	—	—	—
Iodipamide	—	—	—	—	—	—	—
Iopanoate	—	—	0.704	—	39.65	—	—
Iophenoxate	—	—	—	—	—	—	—
Isoproterenol	—	—	—	0.237	—	25.94	—
Itraconazole	1.04	1.5	—	—	—	—	—
Ketoconazole	0.84	0.76	—	—	—	—	—
Ketoprofen	0.03	−0.01	—	—	—	—	—
Labetalol	0.14	0.24	—	—	—	—	—
Lamotrigine	−0.13	−0.40	—	—	—	—	—
Levofloxacin	0.14	—	—	—	—	—	—
Lidocaine	−0.23	0.15	—	0.778	—	26.72	3.74
Mefenamate	—	—	0.602	—	41.96	—	—
Methotrexate	−0.77	−0.35	—	—	—	—	—
Methylprednisolone	−0.22	−0.30	—	—	—	—	—
Metoprolol	−0.29	−0.10	—	—	—	—	—
Mexiletine	—	—	—	—	—	—	—
Minocycline	0.21	−0.01	—	—	—	—	—
Monooleoylglycerol	—	—	—	—	—	—	—
(*S*)-Nadolol	−0.40	0.08	—	—	—	—	—
Nalidixate	—	—	−0.305	—	27.92	—	—
Naproxen	0.25	−0.01	−0.016	—	31.87	—	5.81
Naproxen	—	—	—	—	—	—	—
N-Ethylphenylamine	—	—	—	0.514	—	23.59	—
Nicotinate	—	—	−1.256	—	24.79	—	—
Norfloxacin	0.14	0.12	—	—	—	—	—
Novobiocin	0.35	0.13	—	—	—	—	—
Octanoate	—	—	—	—	—	—	—
Ofloxacin	0.14	0.21	—	—	—	—	—
Ondansetron	0.37	0.18	—	—	—	—	—
Oxprenolol	−0.15	−0.20	—	—	—	—	—
Oxyphenbutazone	−0.02	0.09	—	—	—	—	—
Phenobarbital	—	—	—	—	—	—	—

Table 18 (*Continued*)

Drug	nK_1	nK_2	k_A	k_B	MIF_A	MIF_B	nK_3
Phenoxymethyl-penicillinic acid	−0.69	−0.71	—	—	—	—	—
Phenylbutazone	0.19	0.20	0.255	—	41.98	—	5.95
Phenytoin	0.00	−0.12	—	—	—	—	—
Pindolol	−0.13	−0.15	—	—	—	—	—
Piretanide	—	—	—	—	—	—	—
Pirprofen	—	—	—	—	—	—	—
Prazosin	0.06	−0.06	—	0.547	—	26.69	—
Prednisolone	−0.40	−0.40	—	0.341	—	—	—
Procaine	−0.19	−0.15	—	−0.014	—	—	—
Progesterone	0.59	0.30	—	—	—	—	—
Promazine	0.92	0.77	—	—	—	—	—
Probenecid	—	—	−0.017	—	37.32	—	—
Propranolol	0.28	0.26	—	—	—	—	—
Propafenone	—	—	—	—	—	—	—
Propylthiouracil	−0.75	−0.83	—	—	—	—	—
Pyridine	—	—	—	−0.182	—	14.67	—
Quercetin	—	—	—	—	—	—	—
Quinidine	0.44	0.57	—	—	—	—	—
Quinine	0.49	0.57	—	1.383	—	34.11	3.78
Ranitidine	−0.1	−0.30	—	—	—	—	—
Rifampicin	—	—	—	0.993	—	—	—
Salbutaminol	—	—	—	—	—	—	—
Salicylate	−0.66	−0.69	−0.551	—	30.65	—	4.81
Sancicline	0.21	−0.24	—	—	—	—	—
Scopolamine	−0.34	−0.17	—	0.197	—	23.07	3.18
Sotalol	−0.44	−0.22	—	—	—	—	—
Spironolactone	—	—	—	—	—	—	—
Sulbenicillin	—	—	—	—	—	—	—
Sulfadimethoxine	—	—	—	—	—	—	4.02
Sulfamethoxazole	—	—	—	—	—	—	—
Sulfaphenazole	−0.21	−0.13	—	—	—	—	—
Sulfasalazine	0.56	−0.04	—	—	—	—	—
Sulfathiazole	—	—	—	—	—	—	—
Sumatriptan	−0.05	0.19	—	—	—	—	—
Tenoxican	—	—	—	—	—	—	—
Terazosin	−0.16	−0.26	—	—	—	—	—
Terbinafine	1.17	0.71	—	—	—	—	—
Terbutaline	—	—	—	0.243	—	25.09	—
Testosterone	0.74	0.20	—	—	—	—	—
Tetracaine	0.32	0.16	—	—	—	—	—
Tetracycline	−0.08	−0.24	—	—	—	—	—
Theobromine	—	—	—	0.248	—	—	—
Theophylline	—	—	—	−0.433	—	—	—
(S)-Thiamylal	—	—	—	—	—	—	—
L-Thyroxine	—	—	—	—	—	—	—
Timolol	−0.33	−0.38	—	—	—	—	—
Tolazamide	−0.42	−0.14	−0.028	—	34.14	—	—
Tolbutamide	−0.22	−0.27	−0.150	—	33.88	—	5.26
Triamterene	—	—	—	0.224	—	22.06	—
Triflupromazine	1.06	1.42	—	—	—	—	—

Table 18 (*Continued*)

Drug	nK_1	nK_2	k_A	k_B	MIF_A	MIF_B	nK_3
Trimethoprim	−0.26	−0.22	—	—	—	—	—
L-Tryptophan	−0.78	−0.56	—	—	—	—	—
Varproate	—	—	—	—	—	—	—
Verapamil	0.52	1.16	—	—	—	—	—
Warfarin	−0.04	0.05	0.117	—	31.95	—	5.63
Zidovudine	−1.02	−1.02	—	—	—	—	—

Drug	nK_4	nK_5	nK_6	nK_7	PB_1	PB_2	pK_a^d	$log\ P_C$	$log\ P_M$
Acebutolol	—	—	—	—	—	—	9.20	1.61	1.77
Acenocoumarol	—	—	5.34	—	—	—	4.7	1.31	—
Acetaminophen	—	—	—	—	<5	23	9.51^e	0.49	0.51
	—	—	—	—	—	—	9.71	—	—
Acetylsalicylate	—	—	—	—	72	—	3.5	1.10	1.19
Acrivastine	—	—	—	—	—	—	—	1.12^h	$0.10_{7.4}$
Ajmaline	—	3.13	—	—	—	—	8.2	1.26	—
Alloprinol	—	2.37	—	—	—	—	9.4	−0.92	−0.55
Alprenolol	—	—	—	—	—	—	9.65	2.59	3.10
p-Aminohippurate	3.12	—	—	—	—	—	3.85	−0.25	—
	—	—	—	—	—	—	5.90^f	0.23^g	—
Amoxicillin	3.12	—	—	—	17	—	1.45	0.33	0.38
	—	—	—	—	—	—	2.4	-2.50^v	—
	—	—	—	—	—	—	8.53^f	—	—
	—	—	—	—	—	—	9.6	—	—
Aniline	—	3.05	—	—	—	33	4.63	—	—
Atenolol	—	—	—	—	—	—	9.6	−0.11	0.16
Atropine	—	3.51	—	—	—	—	9.8	—	—
Azapropazone	—	—	5.45	—	—	—	—	—	—
Barbital	3.57	—	—	—	—	—	7.97	0.65	0.65
Benzoic acid	4.09	—	—	—	—	—	4.20	1.88	1.87
	—	—	—	—	—	—	5.73^f	1.49^g	—
Benzylamine	—	3.44	—	—	—	—	9.33	—	—
Benzylthiouracil	—	—	4.61	—	—	—	—	—	—
Bilirubin	—	—	7.98	—	—	—	—	—	—
Bumetanide	—	—	—	5.04^b	—	—	5.2	4.06	—
	—	—	—	—	—	—	10.0	—	—
Bupropion	—	—	—	—	—	—	—	3.21	—
Caffeine	—	2.75	—	—	—	—	0.6	0.07	−0.07
	—	—	—	—	—	—	14.0	—	—
Camptothecin	—	—	—	6.08^b	—	—	10.83	1.31	1.74
	—	—	—	6.90	—	—	—	—	—
Canrenoate	—	—	5.30	—	—	—	—	3.3	—
Captopril	—	—	—	—	—	—	3.7	1.02	—
	—	—	—	—	—	—	9.8	—	—
Carbamazepine	—	3.36	—	—	72.1	—	—	1.98	2.45
(S)-Carbenicillin	—	—	3.38	—	—	—	2.7	1.57	1.13
3-Carboxy-4-methyl 5-propyl-2-furanpropanoate	—	—	7.11	—	—	—	—	—	—
Carprofen	—	—	>6	—	—	—	—	3.93	—
Cefuroxime	—	—	—	—	—	—	—	−0.31	−0.16
Cefuroxime axetil	—	—	—	—	—	—	—	0.14	0.89
Cephalexin	—	—	—	—	15	—	3.20	0.23	0.65

Table 18   (*Continued*)

Drug	nK_4	nK_5	nK_6	nK_7	PB_1	PB_2	$pK_a{}^d$	$\log P_C$	$\log P_M$
Chlofibrate	—	—	5.88	—	—	—	—	—	—
Chloramphenicol	5.24	—	—	—	60	—	5.5^e	—	1.14
Chlorideion	—	—	2.86	—	—	—	—	—	—
Chlorothiazole	—	—	4.74	—	—	—	—	—	—
Chlorpheniramine	—	—	—	3.11^a	—	—	9.16	2.73	3.39
Chlorpromazine	—	—	—	5.28	95.7	—	9.30	5.20	5.35
Chlorpropamide	—	—	5.52	4.65	72.4	—	4.92	2.23	2.27
	—	—	—	—	—	—	—	6.6^e	—
Cimetidine	—	—	—	—	—	—	6.80	0.21	0.40
Ciprofloxacin	—	—	—	—	—	—	—	0.8	—
Clofibrate	—	—	—	4.40	—	—	3.46	2.86	2.57
	—	—	—	—	—	—	3.65^f	—	—
Clonidine	—	—	—	—	—	—	8.05	—	1.59
Clotrimazole	—	—	—	—	—	—	—	5.2	—
Cromolyn	—	—	—	—	—	—	1.10	1.95	1.92
	—	—	—	—	—	—	1.90	—	—
Dansylglycine	—	—	—	5.66	—	—	—	—	—
Desipramine	—	—	—	4.85	82.7	—	10.2^e	4.09	4.9
	—	—	—	—	—	—	10.44	—	—
Dextromethorphan	—	4.05	—	—	—	—	8.3	3.99	—
Diazepam	7.02	—	5.58	5.82^c	96.8	—	3.3	3.18	2.80
	—	—	—	5.69^c	—	—	—	—	3.4^e
Diclofenac	—	—	6.52	—	—	—	4.5	4.77	4.40
Digitoxin	—	—	—	4.63	93.4	—	—	1.39	1.76
Dimethylaniline	—	3.35	—	—	—	—	5.15	—	—
Doxycycline	—	—	—	—	82	—	3.4^e	—	$-0.22_{7.4}$
	—	—	—	—	—	—	3.5	—	—
	—	—	—	—	—	—	7.7	—	—
	—	—	—	—	—	—	9.5	—	—
	—	—	—	—	—	—	9.7^e	—	—
Droperidol	—	—	—	—	—	—	7.6	3.50	3.50
Ebselen	—	—	—	—	—	—	—	—	—
Estradiol	—	—	—	5.00^b	—	—	—	3.78	4.01
Ethacrynate	—	—	6.23	—	—	—	3.5	3.19	—
Ethambutol	—	2.59	—	—	39	—	6.3	0.12	—
	—	—	—	—	—	—	9.5	—	—
	—	—	—	—	—	—	10.0^e	—	—
(S)-Etodolac	—	—	5.30	—	—	—	—	3.62	—
Etoposide	—	—	—	—	—	—	—	-1.12	—
5-Fluorocytosine	—	—	—	—	—	—	—	—	—
Furosemide	5.70	—	5.28	4.18^c	—	—	3.9	2.29	—
	—	—	—	4.86^c	—	—	5.83^f	1.90^g	—
	—	—	—	5.02^c	—	—	—	—	—
	—	—	—	5.23	—	—	—	—	—
	—	—	—	5.32	—	—	—	—	—
	—	—	—	5.41^c	—	—	—	—	—
	—	—	—	5.43^c	—	—	—	—	—
Fusidate	—	—	—	4.89	—	—	5.4	5.6	—
Glibenclamid	—	—	—	5.89	—	—	—	—	—
Hippurate	—	—	4.00	—	—	—	—	—	—
Homatropine	—	3.42	—	—	—	—	9.9	1.45	—

Table 18 (*Continued*)

Drug	nK$_4$	nK$_5$	nK$_6$	nK$_7$	PB$_1$	PB$_2$	pK$_a{}^d$	log P$_C$	log P$_M$
Hydrochloro-	—	—	—	—	—	—	7.0	−0.15	−0.07
thiazide	—	—	—	—	—	—	9.2	—	—
Hydrocortisone	—	—	—	—	—	—	—	1.86	1.61
p-Hydroxybenzoate	4.06	—	—	—	—	—	8.69^f	1.00^g	—
	—	—	—	—	—	—	9.46^f	—	—
Ibuprofen	6.61	—	6.43	4.38^c	—	—	4.4	3.68	3.50
	—	—	—	5.20^c	—	—	5.2	3.55^g	—
	—	—	—	—	—	—	6.15	—	—
Imipramine	—	4.20	—	4.38	9.58	—	8.0^e	4.41	4.80
	—	—	—	4.50^c	—	—	9.5	—	—
Indole-3-acetate	—	—	5.32	—	—	—	—	—	—
Indomethacin	7.32	—	6.15	6.00	94	—	4.5	4.23	4.27
	—	—	—	5.48	—	—	—	—	—
	—	—	—	6.00^b	—	—	—	—	—
Indoxylsulfate	—	—	6.20	—	—	—	—	—	—
Iodipamide	—	—	5.00	—	—	—	—	—	—
Iopanoate	7.84	—	6.83	—	—	—	—	5.22	—
Iophenoxate	—	—	7.89	—	—	—	—	5.21	—
Isoproterenol	3.12	3.25	—	—	—	—	8.6	0.08	—
	—	—	—	—	—	—	10.1	—	—
	—	—	—	—	—	—	12.0	—	—
Itraconazole	—	—	—	—	—	—	—	—	—
Ketoconazole	—	—	—	—	—	—	2.9	4.78	4.34
	—	—	—	—	—	—	6.51	—	—
Ketoprofen	—	—	6.40	^{R}5.47^b	—	—	4.60	2.79	3.12
	—	—	—	^{S}5.45^b	—	—	—	—	—
Labetalol	—	—	—	—	—	—	7.4	2.18	3.09
	—	—	—	—	—	—	8.7	—	—
Lamotrigine	—	—	—	—	—	—	—	2.08	—
Levofloxacin	—	—	—	—	—	—	—	—	—
Lidocaine	—	3.68	—	2.52^a	66	—	7.9	1.98	2.26
	—	—	—	5.11	—	—	—	—	1.95^a
	—	—	—	5.41^c	—	—	—	—	—
Mefenamate	7.84	—	—	—	—	—	4.2	5.34	5.12
	—	—	—	—	—	—	6.43^f	4.97^g	—
Methotrexate	—	—	—	—	25	—	3.8	−0.46	—
	—	—	—	—	—	—	4.3^e	—	—
	—	—	—	—	—	—	4.8	—	—
	—	—	—	—	—	—	5.5^e	—	—
	—	—	—	—	—	—	5.6	—	—
Methylprednisolone	—	—	—	—	—	—	—	2.7	—
Metoprolol	—	—	—	—	11	—	9.7	1.20	1.88
Mexiletine	—	—	—	2.64^a	—	—	9.0	2.57	2.57^a
Minocycline	—	—	—	—	76	—	2.8	1.37	—
	—	—	—	—	—	—	5.0	—	—
	—	—	—	—	—	—	7.8	—	—
	—	—	—	—	—	—	9.5	—	—
Monooleoylglycerol	—	—	5.60	—	—	—	—	—	—
Nadolol	—	—	—	—	—	—	9.37	0.23	0.71
Nalidixate	4.18	—	—	—	93	—	6.0	0.97^g	1.41
	—	—	—	—	—	—	6.99^f	—	—

Table 18 (*Continued*)

Drug	nK_4	nK_5	nK_6	nK_7	PB_1	PB_2	$pK_a{}^d$	$\log P_C$	$\log P_M$
Naproxen	5.75	—	—	4.58^c	—	—	4.2	2.82	3.18
	—	—	—	4.95^c	—	—	6.13^f	—	—
	—	—	—	6.35^c	—	—	—	—	—
(S)-Naproxen	—	—	6.58	—	—	—	—	—	—
N-Ethylphenylamine	—	3.47	—	—	—	—	5.12	—	—
Nicotinate	3.20	—	—	—	—	—	4.95^f	0.48^g	—
	—	—	—	—	—	—	6.27^f	—	—
Norfloxacin	—	—	—	—	—	—	—	0.42	—
Novobiocin	—	—	—	5.74	—	—	4.3	2.42	—
	—	—	—	—	—	—	9.1	—	—
Octanoate	—	—	6.20	—	—	—	4.88	2.94	3.05
Ofloxacin	—	—	—	—	—	—	—	0.92	—
Ondansetron	—	—	—	—	—	—	—	3.17	—
Oxprenolol	—	—	—	—	—	—	—	—	—
Oxyphenbutazone	—	—	5.54	5.55^b	—	—	4.7	2.50	2.72
Phenobarbital	3.12	—	—	—	50.7	—	7.4	1.36	1.47
	—	—	—	—	—	—	7.52^e	—	—
Phenoxymethyl- penicillinic acid	—	—	—	—	—	—	—	—	—
Phenylbutazone	6.05	—	6.18	5.36	—	—	4.4	3.17	3.16
	—	—	—	5.38^b	—	—	6.44^f	3.25^g	—
	—	—	—	5.77^b	—	—	—	—	—
Phenytoin	—	—	—	4.23^b	89	82	8.3	2.09	2.47
Pindolol	—	—	—	—	—	—	8.8	1.65	1.75
	—	—	—	—	—	—	9.7	—	—
Piretanide	—	—	4.98	—	—	—	—	3.93	—
Pirprofen	—	—	5.59	—	—	—	—	3.05	—
Prazosin	—	3.20	—	—	—	—	6.5	2.16	—
Prednisolone	—	3.39	—	—	—	—	—	1.6	1.62
Procaine	—	3.37	—	3.49	—	—	8.11	2.24	1.87
	—	—	—	3.79^c	—	—	8.80	—	—
Progesterone	—	—	—	5.57^b	—	—	—	3.85	3.87
Promazine	—	—	—	4.93	—	—	9.4	4.28	4.55
Probenecid	5.29	—	—	—	—	—	3.4	3.37	3.21
Propranolol	—	—	—	3.00^a	93.2	—	9.45	2.75	3.56
	—	—	—	—	—	—	—	—	2.75^a
Propafenone	—	—	—	3.16^a	—	—	—	3.21	3.64^a
Propylthiouracil	—	—	—	—	—	—	7.8	—	—
	—	—	—	—	—	—	8^e	—	—
	—	—	—	—	—	—	8.3	—	—
Pyridine	—	2.99	—	—	—	—	5.19	—	—
Quercetin	—	—	5.43	—	—	—	—	—	—
Quinidine	—	—	—	3.15	82	78	4.2	3.2	3.44
	—	—	—	—	—	—	5.4^e	—	—
	—	—	—	—	—	—	7.9	—	—
	—	—	—	—	—	—	10.0^e	—	—
Quinine	—	3.96	—	3.88	70	82	4.1	3.2	3.44
	—	—	—	3.88^c	—	—	8.5	—	—
	—	—	—	—	—	—	8.7^e	—	—
Ranitidine	—	—	—	—	—	—	2.3	0.27	0.27
	—	—	—	—	—	—	8.2	—	—

Table 18 (*Continued*)

Drug	nK_4	nK_5	nK_6	nK_7	PB_1	PB_2	$pK_a{}^d$	$\log P_C$	$\log P_M$
Rifampicin	—	3.86	—	—	87	—	—	—	—
Salbutaminol	—	—	—	2.43^a	—	—	—	—	0.06^a
Salicylate	—	—	5.28	4.85	—	87	2.97	2.19	2.26
	—	—	—	5.01	—	—	13.4	1.06^g	—
	—	—	—	5.11	—	—	—	—	—
	—	—	—	5.34	—	—	—	—	—
	—	—	—	5.93	—	—	—	—	—
	—	—	—	4.11^c	—	—	—	—	—
	—	—	—	4.96^c	—	—	—	—	—
	—	—	—	5.28^c	—	—	—	—	—
	—	—	—	5.38^c	—	—	—	—	—
	—	—	—	5.41^c	—	—	—	—	—
Sancicline	—	—	—	—	—	—	—	−1.72	—
Scopolamine	—	3.18	—	—	—	—	7.75	−0.20	1.2
Sotalol	—	—	—	—	—	—	8.15	0.23	0.24
	—	—	—	—	—	—	9.05	—	—
Spironolactone	—	—	3.48	—	—	—	—	2.8	2.26
Sulbenicillin	—	—	3.72	—	—	—	0.63	—	
	—	—	—	—	—	—	—	−3.5	—
Sulfadimethoxine	—	—	4.95	—	99	—	2.02	1.29	1.63
	—	—	—	—	—	—	5.6^e	—	—
	—	—	—	—	—	—	6.70	—	—
Sulfamethoxazole	4.15	—	—	2.78^a	68	—	5.81	0.86	0.89
	—	—	—	3.80^c	—	—	6.0^e	0.79^g	-0.53^a
Sulfaphenazole	—	—	—	4.96	—	—	1.9	2.10	1.52
	—	—	—	—	—	—	6.5	—	—
Sulfasalazine	—	—	—	—	—	—	2.4	4.25	—
	—	—	—	—	—	—	9.7	—	—
	—	—	—	—	—	—	11.8	—	—
Sulfathiazole	—	—	4.40	—	—	—	2.36	0.64	0.05
	—	—	—	—	—	—	7.23	—	—
Sumatriptan	—	—	—	—	—	—	—	0.79	—
Tenoxican	—	—	5.57	—	—	—	—	—	—
Terazosin	—	—	—	—	—	—	—	2.29	—
Terbinafine	—	—	—	—	—	—	—	5.42	—
Terbutaline	3.52	3.26	—	2.40^a	—	—	8.8	0.48	0.48^a
	—	—	—	—	—	—	10.1	—	—
	—	—	—	—	—	—	11.2	—	—
Testosterone	—	—	—	—	—	—	—	3.35	3.32
Tetracaine	—	—	—	—	—	—	8.39	3.65	3.73
Tetracycline	3.12	—	—	3.15^c	55	—	3.3	−2.56	—
	—	—	—	—	—	—	7.7	—	—
	—	—	—	—	—	—	8.3^e	—	—
	—	—	—	—	—	—	9.7	—	—
Theobromine	—	2.52	—	—	—	—	0.12	−1.01	−0.78
	—	—	—	—	—	—	10.05	—	—
Theophylline	—	2.68	—	—	59	—	3.5	−0.25	−0.02
	—	—	—	—	—	—	8.6	—	—
	—	—	—	—	—	—	8.75^e	—	—
(S)-Thiamylal	—	—	4.94	—	—	—	7.48	2.97	3.23
L-Thyroxine	—	—	5.48	—	—	—	—	—	—

Table 18 (*Continued*)

Drug	nK_1	nK_5	nK_6	nK_7	PB_1	PB_2	pK_a[d]	$log\ P_C$	$log\ P_M$
Timolol	—	—	—	—	—	—	9.21	1.63	1.91
Tolazamide	5.16	—	—	4.94	—	—	3.1	1.79	—
	—	—	—	4.94[c]	—	—	5.7	1.45[g]	—
	—	—	—	—	—	—	7.29[f]	—	—
Tolbutamide	5.29	—	4.60	5.34	86.6	—	5.27	2.30	2.34
	—	—	—	5.15[c]	—	—	5.5[e]	2.27[g]	—
	—	—	—	5.64[c]	—	—	6.62[f]	—	—
Triamterene	—	3.24	—	—	—	70	6.2	1.99	1.11
Triflupromazine	—	—	—	4.74	—	—	9.2	5.52	5.19
Trimethoprim	—	—	—	—	70	—	7.13	0.73	0.91
	—	—	—	—	—	—	7.2[e]	—	—
L-Tryptophan	—	—	4.64	4.20[b]	—	—	2.38	−1.58	−1.06
	—	—	—	4.80[b]	—	—	9.39	—	—
Varproate	—	—	5.45	—	—	—	—	—	—
Verapamil	—	—	—	—	—	—	8.92	3.53	3.79
Warfarin	5.38	—	5.53	^{R}5.59[b]	97	—	5.1	1.57	2.52
	—	—	—	^{S}5.86[b]	—	—	5.05[e]	2.87[g]	—
	—	—	—	5.17[c]	—	—	6.76[f]	—	—
	—	—	—	5.23[c]	—	—	—	—	—
	—	—	—	5.25[c]	—	—	—	—	—
	—	—	—	5.40[c]	—	—	—	—	—
	—	—	—	5.46[c]	—	—	—	—	—
	—	—	—	5.52[c]	—	—	—	—	—
	—	—	—	5.53[c]	—	—	—	—	—
	—	—	—	6.15[c]	—	—	—	—	—
Zidovudine	—	—	—	—	—	—	—	—	—

[a]Values from ref. 21.
[b]Values from ref. 22.
[c]Values from ref. 23.
[d]Values from ref. 24.
[e]Values from ref. 25.
[f]Measured in liquid chromatography, values from ref. 26.
[g]Calculated using the VlogP program.
[h]Zwitter ion.

References

1. T. Hanai and C. Hong, Structure-retention correlation in CGC, *J. High Res. Chromatogr.*, 1989, **12**, 327–332.
2. L. Lepri, P. G. Desideri and D. Heimler, Reversed-phase and soap thin-layer chromatography of phenols, *J. Chromatogr.*, 1980, **195**, 339–348.
3. T. Hanai, K. Koizumi, T. Kinoshita, R. Arora and F. Ahmed, Prediction of pK_a values of phenolic and nitrogen-containing compounds by computational chemical analysis compared to those measured by liquid chromatography, *J. Chromatogr., A*, 1997, **762**, 55–61.
4. D. D. Perrin, B. Dempsey, E. P. Serjeant, *pKa Prediction for Organic Acids and Bases*, Chapman and Hall, London, 1981.

5. T. Hanai and J. Hubert, Optimization of retention time of aromatic acids in liquid chromatography from log P and predicted pK_a values, *J. High Res. Chromatogr. Chromatogr. Commun.*, 1984, 7, 524–528.

6. T. Hanai, K. C. Tran and J. Hubert, Prediction of retention times for aromatic acids in liquid chromatography, *J. Chromatogr.*, 1982, **239**, 385–395.

7. N. Boder, Z. Gabanyi and C.-K. Wong, A new method for the estimation of partition-coefficient, *J. Am. Chem. Soc.*, 1989, **111**, 3783–3786.

8. CAChe[a] Manuals, Fujitsu, Tokyo, 1994 and 2002.

9. T. Hanai, K. Koizumi and T. Kinoshita, Prediction of retention factors of phenolic and nitrogen-containing compounds in reversed-phase liquid chromatography based on log P and pK_a obtained by computational chemical calculation, *J. Liq. Chromatogr. Relat. Technol.*, 2000, **23**, 363–385.

10. T. Hanai, Quantitative structure-retention relationships of phenolic compounds without Hammett's equations, *J. Chromatogr., A*, 2003, **985**, 343–349.

11. Y. Arai, J. Yamaguchi and T. Hanai, Enthalpy effect in the retention of aromatic acids on an octadecyl-bonded silica gel, *J. Chromatogr.*, 1987, **400**, 21–26.

12. T. Hanai, Chi. Mizutani and H. Homma, Computational chemical simulation of chromatographic retention of phenolic compounds, *J. Liq. Chromatogr. Relat. Technol.*, 2003, **26**, 2031–2039.

13. T. Hanai, Quantitative in silico analysis of retention time on methylsilicone and polyethyleneglycol phases in capillary gas chromatography. http://www.internet-chromatography.com/html/toshihikbeitrage.html.

14. T. Hanai and H. Homma, Computational chemical prediction of the retention factor of aromatic acids, *J. Liq. Chromatogr. Relat. Technol.*, 2002, **25**, 1661–1676.

15. T. Hanai, Simulation of chromatography of phenolic compounds with a computational chemical method, *J. Chromatogr., A*, 2005, **1087**, 45–51.

16. T. Hanai, Analysis of the mechanism of retention on graphitic carbon by a computational chemical method, *J. Chromatogr., A*, 2004, **1027**, 279–287.

17. T. Hanai, R. Miyazaki, A. Koseki and T. Kinoshita, Computational chemical analysis of the retention of acidic drugs on a pentyl-bonded silica gel in reversed-phase liquid chromatography, *J. Chromatogr. Sci.*, 2004, **42**, 354–360.

18. T. Hanai, Chromatography In Silico for Basic Drugs, *J. Liq. Chromatogr. Rel. Technol.*, 2005, **28**, 2163–2177.

19. F. Tazerouti, A. Y. Badjah-Hadj-Ahmed and T. Hanai, Analysis of the mechanism of retention on a modified b-cyclodextrin/silica chiral stationary phase using a computational chemical method, *J. Liq. Chromatogr. Relat. Technol.*, 2007, **30**, 3043–3057.

20. T. Hanai, Evaluation of Measuring Methods of Human Serum Albumin-Drug Binding Affinity, *Curr. Pharm. Anal.*, 2007, **3**, 205–212.

21. H. Xu, X.-D. Yu, X.-D. Li and H. Y. Chen, Determination of binding constants for basic drugs with serum albumin by affinity capillary electrophoresis with the partial filling technique, *Chromatographia*, 2005, **61**, 419–422.
22. Personal Collection of Y. Matsushita, Kitasato University, School of Pharmaceutical Sciences, Tokyo.
23. K. Koizumi, C. Ikeda, M. Ito, J. Suzuki, T. Kinoshita, K. Yasukawa and T. Hanai, Influence of glycosylation on the drug binding of human serum albumin, *Biomed. Chromatogr.*, 1998, **12**, 203–210.
24. T. Hanai, A. Koseki, R. Yoshikawa, M. Ueno, T. Kinoshita and H. Homma, Prediction of human serum albumin–drug binding affinity without albumin, *Anal. Chim. Acta*, 2002, **454**, 101–108.
25. W. A. Ritschel, *Handbook of Basic Pharmacokinetics*, Drug Intelligence Publications, Hamilton Press, Femandina Beach, FL, 1980.
26. T. Hanai, R. Miyazaki, E. Kamijima, H. Homma and T. Kinoshita, Computational prediction of drug–albumin binding affinity by modeling liquid chromatographic interaction, *Internet Electron. J. Mol. Des.*, 2003, **2**, 702–711.

Subject Index